Springer
Berlin
Heidelberg
New York
Barcelona
Budapest
Hong Kong
London
Milan
Paris
Santa Clara
Singapore
Tokyo

J. M. Rüeger

Electronic Distance Measurement

An Introduction

Fourth Edition

With 56 Figures and 18 Tables

Dr. J. M. Rüeger
School of Geomatic Engineering
University of New South Wales
Sydney NSW 2052
Australia

Library of Congress Cataloging-in-Publication Data

Rüeger, J. M. (Jean M.), 1944-
    Electronic distance measurement : an introduction / J.M. Rüeger. -
- 4th ed.
        p.   cm.
    Includes bibliographical references and index.

    1. Geodesy--Instruments.   2. Distances--Measurement.   I. Title.
QB328.A1R84   1996
526'.028--dc20   96-16952
                                                               CIP

ISBN-13: 978-3-540-61159-2   e-ISBN-13: 978-3-642-80233-1
DOI: 10.1007/978-3-642-80233-1

The use of general descriptive names, registered names, trademarks, etc. in this publication does not imply, even in the absence of a specific statement, that such names are ex-
empt from the relevant protective laws and regulations and therefore free for general
use.

Typesetting: K + V Fotosatz GmbH, Beerfelden

SPIN 10521163   2132/3136 – 543210 – Printed on acid-free paper

Preface to the 1996 Edition

As the basic principles of EDM instruments have changed little since the third edition of 1990, there was no need for significant changes. This edition differs from its predecessor in that it contains corrections of a number of errors and misprints, totally revised tables in Appendices D, E and F and a new note in Section 2.4.3 on the introduction of the new temperature scale in 1990. The author is indebted to the many readers who reported the many small errors and misprints. T. Black, H. Buchanan, R. Da-Col, R. Köchle, P. H. Lam, J. Nolton, J. R. Pollard and A. Quade were particularly helpful. All known errors have been corrected. The assistance provided by most manufacturers (or their agents) with the updating of the tables with the instrument data was greatly appreciated.

Sydney, February 1996   J. M. RÜEGER

Preface

The book has evolved from the author's continuing teaching of the subject and from two editions of a text of the same title. The first edition was published in 1978 by the School of Surveying, University of New South Wales, Sydney, Australia. Like its predecessors, this totally revised third edition is designed to make the subject matter more readily available to students proceeding to degrees in Surveying and related fields. At the same time, it is a comprehensive reference book for all surveyors as well as for other professionals and scientists who use electronic distance measurement as a measuring tool. Great emphasis is placed on the understanding of measurement principles and on proper reduction and calibration procedures. It comprises an extensive collection of essential formulae, useful tables and numerous literature references.

After a review of the history of EDM instruments in Chapter 1, some fundamental laws of physics and units relevant to EDM are revised in Chapter 2. Chapter 3 discusses the principles and applications of the pulse method, the phase difference method, the Doppler technique and includes an expanded section on interferometers. The basic working principles of electro-optical and microwave distance meters are presented in Chapter 4, with special emphasis on modulation/demodulation techniques and phase measurement systems. Important properties of infrared emitting and lasing diodes are discussed.

Various aspects of the propagation of electromagnetic waves through the atmosphere are treated in Chapter 5, such as the range of EDM instruments, phase and group refractive indices, coefficient of refraction, measurement of temperature, pressure and humidity and different approaches to the problem of the determination of ambient refractive indices. Chapter 6 continues with the derivation of first velocity corrections for electro-optical and microwave distance meters and concludes with the second velocity correction and a review of more refined methods of velocity corrections.

All equations for the geometrical reduction of electronic distance measurements to the spheroid (or to sea level) are derived in Chapter 7, both, for reductions using station elevations and for reductions using measured zenith angles. Numerical examples are given. Error analyses indicate critical parameters. Additional corrections and computations are discussed in Chapter 8. This chapter includes numerous supplementary reductions which are required in certain cases as well as the computation of height differences from measured zenith angles for EDM tacheometry and EDM height traversing, the derivation of the coefficient of refraction and eye-to-object corrections for distances and zenith angles.

A description of four typical distance meters is given in Chapter 9, together with a discussion of classification criteria and special features of modern electro-optical distance meters. Chapter 10 presents a number of different types of EDM reflectors and many important aspects of reflectors such as the reflector constant (and its computation) and the effect of misaligned reflectors on distance and angle measurements. Chapter 11 discusses the properties of NiCd rechargeable batteries. A review of other suitable power sources has been added.

Chapter 12 gives a comprehensive introduction into major errors of electro-optical distance meters, including additive constant, short periodic errors, scale errors and non-linear distance-dependent errors. The different sources of these errors are indicated, where possible. A mathematical model covering most known error patterns is given. The last chapter provides the neccessary information on how errors can be determined by the user of a distance meter. Included in Chapter 13 is a large section on the calibration of distance meters on EDM baselines. The geometric design of three types of EDM baselines, the physical design, the measurement and analysis procedures and the determination of the baseline lengths are discussed. Other sections describe the calibration procedures on cyclic error testlines and the measurement of the modulation frequency and discuss the accuracy specifications of distance meters.

The appendices include an improved refractive index formula for high precision measurements as well as tables on saturation water vapour pressures (versus temperature), a standard atmosphere (temperature and pressure versus elevation), critical dimensions of reflectors, important data of electro-optical distance meters (for correction and calibration purposes) and technical data of a selection of short range distance meters, pulse distance meters and long range distance meters.

The text uses SI units for all quantities but pressure. Pressures are stated in millibars (mb) rather than the equivalent unit hPa (hectopascal). However, common conversion rates to and from non-SI units are given. Most equations are numbered for easier reference. The definitions of parameters of equations are repeated below the relevant final formulae in order to facilitate the use of the book as a compendium of formulae. The more important symbols are also included in the list of symbols at the beginning of the text.

The writing of this third edition was suggested by F.K. Brunner. His continued encouragement and valuable advice was highly appreciated. The author is further indebted to F.K. Brunner, K. Furuya (Tokyo Optical Co. Ltd.), K. Giger (Wild Leitz Ltd.),

VIII

R. Nünlist and P. Kiefer for their valuable comments on Sections 5.9, 9.2.2, 9.2.3, 9.2.1 and 10.2.5.2, respectively, and to C. Rusu for the preparation of a number of diagrams. The book benefited greatly from the comments made by many readers with respect to the two earlier editions and from the assistance provided by colleagues in editing these earlier editions. The cooperation of manufacturers or their Australian agents with the collection of the technical data for the tables in the appendices is gratefully acknowledged as is the competent support by the staff of the Springer-Verlag.

Sydney, Summer 1989/1990   J.M. RÜEGER

Contents

XII

XVI

List of Symbols

A = amplitude or maximum strength of electromagnetic signal
A = baseline design parameter (Sect. 13.2)
A = parameter of precision formula
A_{ji} = coefficients of instrument correction
B = baseline design parameter (Sect. 13.2)
B = parameter of precision formula
C = baseline design parameter (Sect. 13.2)
C = parameter of first velocity correction
CE = short periodic (cyclic) error
D = baseline design parameter (Sect. 13.2)
D = parameter of first velocity correction
D_s = density factor of dry air (Appendix A)
D_w = density factor of water vapour (Appendix A)
E'_W = saturation water vapour pressure at "wet bulb" temperature over water
E'_{ICE} = saturation water vapour at "wet bulb" temperature over ice
E_W = saturation water vapour pressure at "dry bulb" temperature over water
H_i = spheroidal height of "i"th station
ΔH = height difference = $H_2 - H_1$
H_M = mean height = $0.5(H_1 + H_2)$
I = electric current
IC = instrument correction
I_{th} = threshold current (Sect. 4.1.2.1)
J = radiant intensity at distance d
J_0 = radiant intensity at emitter
K' = first velocity correction
K'' = second velocity correction
K_1 = first arc-to-chord correction (d_1 to d_2)
K_2 = slope correction (d_2 to d_6)
K_3 = sea level correction (d_6 to d_3)
K_4 = second chord-to-arc correction (d_3 to d_4)
K_{23} = chord-to-chord correction (d_2 to d_3)
K_5 = slope correction (d_2 to d_5)
K_6 = sea level correction (d_5 to d_3)
L = fraction of unit length U
N = refractivity
N = number of baseline stations (Sect. 13.2)
P_0 = radiant power output
P_s = partial pressure of dry air (Appendix A)
P_w = partial water vapour pressure (Appendix A)
Q = cofactor
R = mean radius of curvature of spheroid along a line
R_M = mean radius of curvature of spheroid for a specific area
R_i = range of distance meter
R_o = range of distance meter
R_y = range of distance meter to y prisms
S = grid distance (Sect. 8.6.2)
T = atmospheric transmittance

T	$=$	thermodynamic temperature
U	$=$	unit length of an EDM instrument
a	$=$	height of cube corner above front face of prism
b	$=$	distance between front face and vertical/horizontal axis of prism
c	$=$	phase velocity of light in a medium
c	$=$	additive constant (Sect. 13.2)
cy	$=$	cyclic error correction (Sect. 13.3)
c_g	$=$	group velocity of light in a medium
c_0	$=$	velocity of light in a vacuum
d	$=$	measured distance, including first velocity correction
d_G	$=$	slope distance between ground marks
d'	$=$	measured distance (displayed on instrument)
d_1	$=$	wave path length ($= d' + K' + K''$)
d_2	$=$	wave path chord
d_3	$=$	spheroidal chord
d_4	$=$	spheroidal distance
d_5	$=$	horizontal distance at height of EDM instrument station
d_6	$=$	$d_2 + K_2$
$d*$	$=$	distance between satellite stations
d_v	$=$	meteorological range, visibility range
d_{TH}	$=$	distance along zenith angle ray
d_{EDM}	$=$	wave path chord (d_2) (Sect. 8.1.1)
e	$=$	partial water vapour pressure
e	$=$	eccentricity of satellite station (Sect. 8.5)
e	$=$	eccentricity of EDM instrument (Sects. 9.1.3.4, 9.1.3.5)
f	$=$	frequency of signal
f_D	$=$	Doppler frequency
f_{nom}	$=$	nominal modulation frequency
f_{act}	$=$	actual (measured) modulation frequency
h	$=$	relative humidity (Sects. 5.5, 5.8.4)
h_{EDM}	$=$	height of trunnion axis of EDM instrument above survey mark
h_R	$=$	height of reflector above survey mark
h_{TH}	$=$	height of trunnion axis of theodolite above survey mark
h_T	$=$	height of target above survey mark
k	$=$	coefficient of refraction
k_i	$=$	ambiguity increments (Sect. 3.2.1.4)
k_0	$=$	central scale factor (Sect. 8.6)
k_L	$=$	coefficient of refraction of light waves
k_M	$=$	coefficient of refraction of microwaves
l_i	$=$	normalized phase measurement
m	$=$	ambiguities
n	$=$	phase refractive index of a medium
n	$=$	number of observations (Chap. 13)
n_A	$=$	group refractive index of light and IR waves in air
n_G	$=$	group refractive index of light and IR waves in glass
n_{Gph}	$=$	phase refractive index of light and infrared waves in glass
n_L	$=$	group refractive index of light waves for ambient atmospheric conditions
n_M	$=$	refractive index of microwaves for ambient atmospheric conditions
n_{REF}	$=$	reference refractive index of a specific EDM instrument
n_g	$=$	group refractive index of atmosphere for standard conditions
p	$=$	atmospheric pressure
r	$=$	radius of curvature of wave path
s	$=$	chord distance (Sect. 6.5)
s	$=$	standard deviation (Chap. 13)
$s*$	$=$	reduced cyclic error testline data
t	$=$	time (Sect. 2)

t^* = period of sinusoidal signal
Δt = time lead of electromagnetic signal
Δt = time interval (Sect. 3.3)
$\Delta t'$ = flight time of a signal between transmitter and receiver
t = "dry bulb" temperature
t' = "wet bulb" temperature
u = number of unknowns (Chap. 13)
v = speed
v = residual (Chap. 13)
y = sinusoidal signal
z_i = measured zenith angle at station P_i
z_{ij} = zenith angle at station P_i to P_j
z = attenuation or extinction coefficient (Sect. 5.1)
z_G = zenith angle between ground marks
Ω = eye-to-object correction for zenith angle
Φ = phase angle of electromagnetic signal
Φ_B = diameter of return beam
Φ_P = diameter of aperture of reflector
Φ_T = diameter of transmitter optics
$\Delta\Phi$ = phase lead of electromagnetic signal
$\Delta\Phi$ = measured phase difference (Sect. 3.22)
α = coefficient of expansion of air (= 0.003661) (Sect. 5.5)
α = azimuth of the measured line, clockwise through 360° from true north (Sect. 7.1)
α = vertical angle (Sect. 8.3.1)
α = horizontal angle (Sect. 8.5) measured at centre station
α^* = horizontal angle measured at satellite station (Sect. 8.5)
α_A = angle of incidence
α_G = angle of refraction
β = angle of refraction (Sect. 10.2.5.2)
β = angle between the wave path normals through the terminals of a line
δ = refraction angle
$\delta\Phi$ = periodic error
δ ( ) = differential
ε_i = deviation of vertical at station P_i
ζ_i = spheroidal zenith angle at station P_i
ν = angle (Sect. 8.3.1)
λ = wavelength
ν = radius of curvature of the spheroid in the prime vertical
ϱ = radius of curvature of the spheroid in the meridian
σ = vacuum wave number (Appendix A)
σ = standard deviation (Sect. 8.3.2, 8.4.2)
σ = optical length (Sect. 6.5)
ψ = phase lead of contaminated signal (Sect. 12.2.1)
ω = angular velocity

1 History

Historically, the development of electro-optical distance meters evolved from techniques used for the determination of the velocity of light. Fizeau determined the velocity of light in 1849 with his famous cogwheel modulator on a line of 17.2 km length: Light passed through the rotating cogwheel, travelled to a mirror at the other end of the line, was reflected and returned to the wheel where the return light was blocked off by the teeth at high revolutions of the wheel. Fizeau's experiment employed for the first time the principle of distance measurement with modulated light at high frequencies. Later, Foucault employed a rotating mirror in 1862 and Michelson (1927) a rotating prism in 1926 for similar experiments. According to Zetsche (1979), the first electro-optical distance meter was developed by Lebedew, Balakoff and Wafiadi at the Optical Institute of the U.S.S.R. in 1936. In 1940, Hüttel published a technique for the determination of the velocity of light using a Kerr-cell modulator in the transmitter and a phototube in the receiver. This inspired the Swedish Scientist E. Bergstrand to design the first "Geodimeter" (for geodetic distance meter) for the determination of the velocity of light in 1943. The first commercial instrument (Geodimeter NASM-2) was produced by the Swedish company AGA and became available in 1950. With the early Geodimeters, longer distances could only be measured at night. An important development was the introduction of the heterodyne technique to electro-optical distance meters by Bjerhammar in 1954, which enabled the execution of more accurate phase measurements at more convenient low frequencies (Bjerhammar 1971). The first instrument to employ the heterodyne principle was the Geodimeter Model 6A. Subsequently, the principle was employed in distance meters of all makes. The (laser) Geodimeter with the longest range (60 km), the Model 8, was released in 1968. It has been used widely in high order geodetic networks throughout the world.

The use of reflected radiowaves for distance measurements was suggested as early as 1889 by N. Tesla. A first patent application for an electromagnetic distance meter (by Löwy) was made in 1923. The first radiowave distance meter (based on the interference principle) was built in 1926 by Schegolew, Boruschko and Viller in Leningrad, USSR (Zetsche 1979). A radiowave distance meter using the phase measurement principle was developed by T. L. Wadley at the National Institute of Telecommunications Research of South Africa in 1954. It became available under the trade name Tellurometer in 1957 and was immediately employed for long-range traversing in the first-order geodetic control of the Australian continent. Its range exceeded that of the early Geodimeters and it was therefore in much wider use until HeNe lasers were introduced in EDM in the late 1960's. The first light-weight microwave distance meter (Tellurometer CA 1000) became available in 1972.

Prototypes of short-range distance meters incorporating infrared emitting diodes appeared in the mid-1960's (Tellurometer MA100 in 1965; Zeiss SM11 in 1967). Commercial release of these instruments occurred in the late 1960's (Wild/Sercel Distomat DI10 in 1968; Tellurometer MA100 in 1969; Zeiss SM11 in 1970). Further development led to smaller telescope- or theodolite-mounted distance meters (Kern DM500 in 1974; AGA Geodimeter 12 in 1975; Sokkisha SDM-1C in 1976; Topcon DM-C2 in 1979) and smaller semi-electronic tacheometers (Zeiss SM4 in 1976; Topcon GTS-1 in 1980; Sokkisha SDM-3D in 1980; Zeiss SM41 in 1981; Topcon GTS-2 in 1981; Sokkisha SDM-3E in 1982).

More recently, infrared distance meters employing the pulse measurement principle (rather than the phase measurement principle) appeared on the market. Pulsed distance meters for industrial applications were pioneered by Eumig. The first instrument for surveying applications, the Geo-Fennel FEN2000 was released in 1983. The Wild Distomat DI3000 followed as a second type of pulsed IR distance meters in 1985.

The first precision EDM instrument, the Mekometer, was built by K.D. Froome and R.H. Bradsell in 1961 at the National Physical Laboratory, Teddington (U.K.) and became commercially available early in 1973 as the Kern Mekometer ME3000. On short distances, accuracies of 0.2 mm can be achieved. As successors to this instrument, the COM-RAD204DME Geomensor and the Kern Mekometer ME5000 were released in 1984 and 1986, respectively.

The first electronic tacheometer (sometimes termed "total station"), the Zeiss (Oberkochen) Reg ELTA 14, became available in 1970 and featured electronic readout not only of distance but also for the vertical and horizontal circles. The second total station, the AGA Geodimeter 700 followed in 1971. Smaller and lighter second generation instruments entered the market in 1977 and 1978 with the Hewlett-Packard HP3820A, the Wild TC1, the Zeiss ELTA 2 and the Zeiss ELTA 4. By 1985, electronic tacheometers were available from all major surveying equipment manufacturers, as were electronic data storage devices and computer software packages for data processing and plotting.

2 Physical Laws and Units Related to EDM

2.1 Definitions

The frequency f and the wavelength λ of electromagnetic waves are related by the following fundamental equations:

$$\lambda = \frac{c}{f} \; ; \tag{2.1}$$

$$f = \frac{c}{\lambda} \; , \tag{2.2}$$

where c = velocity of electromagnetic waves in a medium, usually referred to as the *velocity of light* in the medium
       f = frequency of signal
      λ = wavelength in the medium.

The mode and velocity of propagation of an electromagnetic wave depend to a certain extent on the frequency of the signal and on the nature of the earth's atmosphere. The distance between two stations can be computed, if the travelling time of the radiation is measured:

$$c\,\Delta t' = d \; , \tag{2.3}$$

where d = distance between the two points
      $\Delta t'$ = time taken by the signal to travel from first to second station ("flight" time)
      c = velocity of light in the medium.

It is assumed in Eqs. (2.1) to (2.3), that the velocity of light in a medium (normally air) is known. This velocity can be calculated if the refractive index of the medium (viz. air) and the velocity of light in a vacuum is known.

$$c = \frac{c_0}{n} \; , \tag{2.4}$$

where n = refractive index of a medium
      c_0 = velocity of light in a vacuum
      c = velocity of light in a medium.

The velocity of light in a vacuum is a physical constant which has to be determined by experiment. The determination of c_0 remains a permanent challenge to physicists and even has led to the development of EDM instruments, as discussed earlier. More information about this natural constant is given in Section 2.3. The

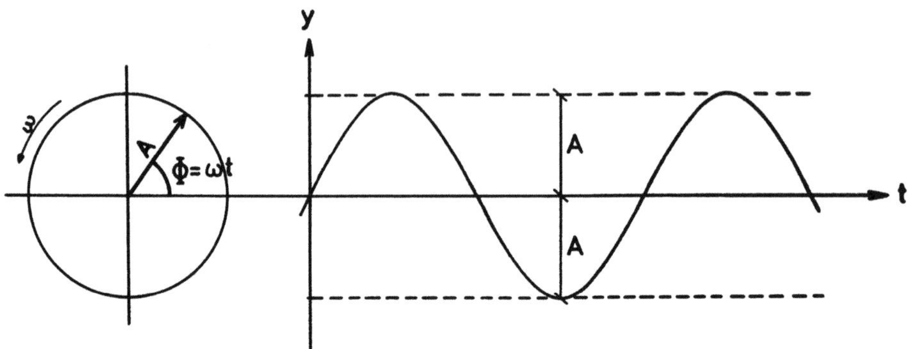

Fig. 2.1. Trigonometric function in a polar and a rectangular coordinate system

refractive index of a medium can be derived from formulae given in Section 5. It is very difficult to accurately assess the refractive index along the wave path, and consequently the accuracy of EDM is often limited by the accuracy to which the integral refractive index is known.

Electromagnetic radiation can be described by the following formulae:

$$y = A \sin(\omega t) \tag{2.5a}$$

$$= A \sin \Phi \, , \tag{2.5b}$$

where A = amplitude or maximum strength
  ω = angular velocity (angular frequency, angular rate of alternation)
  f = frequency of signal
  t = time
  Φ = phase angle,

and where $\Phi = \omega t \tag{2.6a}$

$$\omega = 2\pi f \, . \tag{2.6b}$$

The parameters are further explained in Fig. 2.1.

A signal with a phase lead of $\Delta\Phi$ can be expressed as:

$$y = A \sin(\Phi + \Delta\Phi) \tag{2.7a}$$

$$y = A \sin \omega (t + \Delta t) \, , \tag{2.7b}$$

where Δt = time lead
  $\Delta\Phi$ = phase lead.

The effect of a time lead is depicted in Fig. 2.2.

The phase lead may be written as a function of the time lead:

$$\Delta\Phi = \Delta t \omega \tag{2.8a}$$

$$\Delta t = \frac{\Delta\Phi}{\omega} \, . \tag{2.8b}$$

4

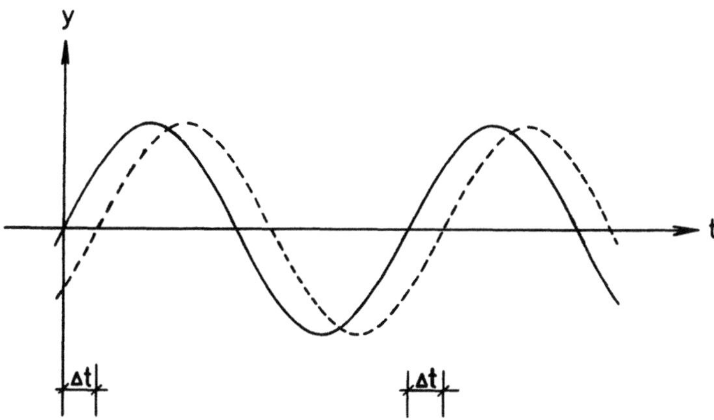

Fig. 2.2. Two sine curves of same amplitude and wavelength but different phase. The *dashed curve* leads the *solid curve* by the time lead Δt

2.2 Frequency Spectrum

The bands of the frequency spectrum are listed below and also depicted in Fig. 2.3, where those used for EDM are hatched.

Radiation	Wavelength λ	Frequency f
X-rays	$1.6\times10^{-11}-6.6\times10^{-8}$ m	$1.9\times10^{19}-4.5\times10^{15}$ Hz
Ultraviolet	$1.4\times10^{-8}-3.6\times10^{-7}$ m	$2.2\times10^{16}-8.3\times10^{14}$ Hz
Visible light	$3.6\times10^{-7}-7.8\times10^{-7}$ m	$8.3\times10^{14}-3.8\times10^{14}$ Hz
Infrared	$7.8\times10^{-7}-3.4\times10^{-4}$ m	$3.8\times10^{14}-8.8\times10^{11}$ Hz
Radiowaves:		
Extra high EHF	$1\times10^{-3}-1\times10^{-2}$ m	$3\times10^{10}-3\times10^{11}$ Hz
Super high SHF	$1\times10^{-2}-1\times10^{-1}$ m	$3\times10^{9}-3\times10^{10}$ Hz
Ultra high UHF (Television)	$1\times10^{-1}-1$ m	$3\times10^{8}-3\times10^{9}$ Hz
Very high VHF	$1-10$ m	$3\times10^{7}-3\times10^{8}$ Hz
(Television, FM Broadcasting)		
High HF (Short waves)	$10-100$ m	$3\times10^{6}-3\times10^{7}$ Hz
Medium MF	$10^{2}-10^{3}$ m	$3\times10^{5}-3\times10^{6}$ Hz
(AM Broadcasting)		
Low LF (Long waves)	$10^{3}-10^{4}$ m	$3\times10^{4}-3\times10^{5}$ Hz
Very low VLF	$10^{4}-10^{5}$ m	$3\times10^{3}-3\times10^{4}$ Hz
Extra low ELF	$10^{5}-10^{6}$ m	$3\times10^{2}-3\times10^{3}$ Hz

Radiowaves belonging to the SHF and UHF bands and parts of the EHF and VHF bands are traditionally called *microwaves*. They may be divided into:

	Wavelength λ	Frequency f
V Band	5.3 mm – 6.5 mm	46 – 56 GHz
Q Band	6.5 mm – 8.3 mm	36 – 46 GHz
K Band	8.3 mm – 27.5 mm	10.9 – 36 GHz
X Band	27.5 mm – 57.7 mm	5.2 – 10.9 GHz
S Band	57.7 mm – 0.194 m	1.55 – 5.2 GHz
L Band	0.194 m – 0.769 m	0.39 – 1.55 GHz
P Band	0.769 m – 1.333 m	0.225 – 0.390 GHz

Microwaves are widely used for telecommunication purposes (including transmissions to and from satellites) and Radar (Radio detection and ranging) systems.

The infrared part of the spectrum is also divided further:

	Wavelength
Near-infrared NIR	0.76 µm – 3.0 µm
Middle-infrared MIR	3.0 µm – 6.0 µm
Far-infrared FIR	6.0 µm – 15.0 µm
Extreme-infrared XIR	15.0 µm – 1 mm

NIR radiation has propagation characteristics which are close to those of visible light and this facilitates its use both in optical communication systems (glass fibres) and EDM.

Electro-optical EDM instruments typcially use NIR radiation of 800 to 900 nm wavelength or red HeNe Laser (Light Amplification by Stimulated Emission of Radiation) light of 633 nm wavelength. Microwave EDM instruments use mainly wavelengths of 30 mm and 8 mm and are therefore classified within the X and K bands and the SHF and EHF bands, respectively. Terrestrial navigation systems make use of a larger range of radiowaves. For example, the Decca system operates in the LF band, Toran and Decca HI-FIX in the HF band and Omega in the VLF band. The satellites of the Global Positioning System (GPS) transmit in the L band at wavelengths of 0.190 m (L 1) and 0.244 m (L 2).

2.3 Velocity of Light in a Vacuum

Over the last 300 years different techniques have been used by astronomers and physicists to determine c_0. The precision of determination has improved during the last 20 years from ± 1 km/s to ± 1 m/s. Some of the instruments and techniques of this development have been applied by EDM instrument manufacturers. Experiments by Essen and Froome (1951) led to a formula for the calculation of the refractive index of microwaves. Some determination methods used a "reverse EDM process", namely to derive the unknown velocity of light from a distance which was measured with precise invar wires or by interferometry.

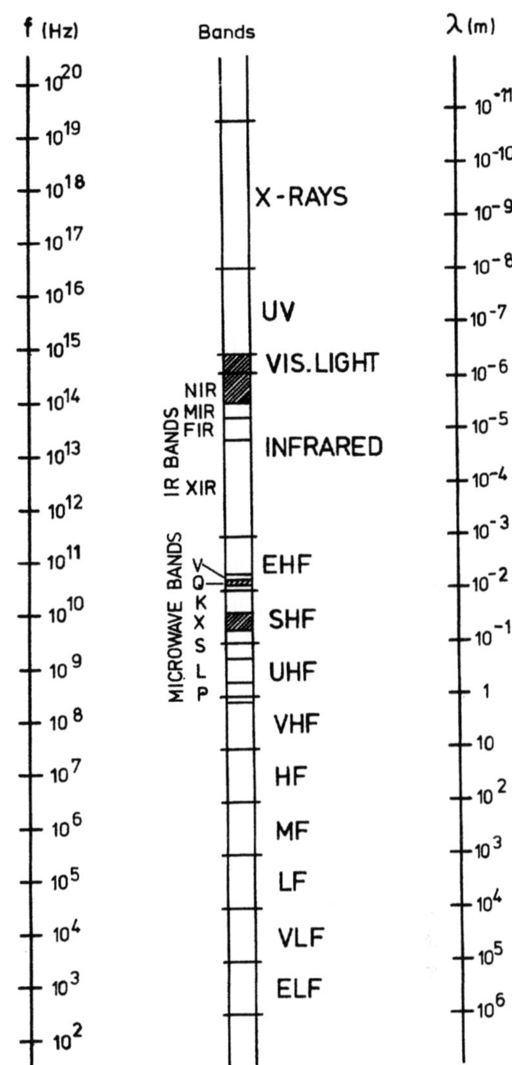

Fig. 2.3. Bands of the frequency spectrum of electromagnetic waves versus wavelength and frequency. *Shaded areas* indicate the bands used in EDM

In 1957 the XIIth General Assembly of the International Scientific Radio Union recommended the value

$$c_0 = 299\,792.5 \pm 0.4 \text{ km/s} \ . \tag{2.9}$$

This value was accepted later by the International Union for Geodesy and Geophysics (IUGG). At the XVIth General Assembly of the IUGG in 1975, the following revised value for c_0 was recommended (IUGG 1975):

$$c_0 = 299\,792\,458 \pm 1.2 \text{ m/s} \ . \tag{2.10}$$

The standard deviation of c_0 quoted above amounts only to 0.004 ppm and is mainly due to the uncertainty in the 1960 definition of the metre. In EDM c_0 can

7

therefore be considered as an error-free value, because the accuracy of EDM is rarely better than one part per million (1 ppm). The change between the 1957 and the 1975 value of c_0 is 42 m/s or 0.14 ppm. The above 1975 definition should be used for the scale definition of distance meters.

2.4 Units and Their Definitions

It is useful to review the definitions of two important units upon which electronic distance measurement is based. The derived frequency unit and one meteorological unit are also mentioned, the latter, because it is not a standard SI unit.

2.4.1 Second of Time

The second was defined by the XIIIth General Conference on Weights and Measures in 1967 as follows:

"The second is the duration of 9 192 631 770 periods of the radiation corresponding to the transition between the two hyperfine levels of the ground state of the caesium-133 atom".

Derived units are

One millisecond $= 1 \text{ ms} = 1 \times 10^{-3} \text{ s}$

One microsecond $= 1 \text{ } \mu\text{s} = 1 \times 10^{-6} \text{ s}$

One nanosecond $= 1 \text{ ns} = 1 \times 10^{-9} \text{ s}$.

This atomic standard of the second can be reproduced to an accuracy of 1 in 10^{-11} according to the National Standards Laboratory (1972). The time standard is maintained by atomic clocks (caesium standards, hydrogen masers).

2.4.2 Metre

An earlier definition of the metre was decided upon by the 11th General Conference on Weights and Measures in 1960 as follows:

"The metre is the length equal to 1 650 763.73 wave lengths in vacuo of the radiation corresponding to the transition between the levels $2 p_{10}$ and $5 d_5$ of the krypton-86 atom."

A new definition of the metre was adopted by the 17th General Conference for Weights and Measures (Conférence Générale des Poids et Mesures, CGPM) on 20 October 1983 in Paris. It reads as follows:

"The metre is the length of the path travelled by light in vacuum during a time interval of 1/299 792 458 of a second."

The new definition will lead to an improved precision in the realization of the unit of length by laser wave lengths and frequency techniques. (Wavelengths from frequency-stabilized lasers can be reproduced to 1 part in 10^{12} or better.) The new definition effectively fixes the value of the velocity of light, thus making c_0 a physical constant. Recommendations on the practical realization of the definition of the metre were adopted by the Comité International des Poids et Mesures during its 72nd Session in October 1983 (Anon 1984).

Derived units are:

$$\text{One kilometre} \quad = 1\,\text{km} \; = 1 \times 10^3\,\text{m}$$

$$\text{One millimetre} \quad = 1\,\text{mm} = 1 \times 10^{-3}\,\text{m}$$

$$\text{One micrometre} = 1\,\mu\text{m} \; = 1 \times 10^{-6}\,\text{m}$$

$$\text{One nanometre} \quad = 1\,\text{nm} \; = 1 \times 10^{-9}\,\text{m} \;.$$

2.4.3 Kelvin

The unit of the thermodynamic temperature T is the Kelvin and is defined as the fraction $1/273.16$ of the thermodynamic temperature of the triple point of water.

Celsius temperatures are, for historical reasons, measured from a thermal state 0.01 Kelvin lower than the triple point of water. A Celsius temperature t is defined by

$$t = T - 273.15\,\text{K} \;. \tag{2.11}$$

The unit of Celsius temperature is the degree Celsius ($°C$). It is equal in magnitude to the Kelvin (K) by definition. The triple point of water is the equilibrium state at which ice, air-saturated water and vapour-saturated air coexist (Fritschen and Gray 1979).

In 1990, the Comité International des Poids et Mesures (CIPM) introduced the International Temperature Scale ITS-90, which closely corresponds to thermodynamic temperature. The difference $(t_{90} - t_{68})$ between ITS-90 and the previous IPTS-68 temperatures is zero at $0\,°C$ and $-0.026\,°C$ at $100\,°C$ (Preston-Thomas 1991).

2.4.4 Other Units in EDM

2.4.4.1 Frequency

Frequency units are not basic units but are based entirely on the definition of time. The name of the frequency unit is taken from the name of the German physicist, Heinrich Hertz (1857−1894), who discovered the existence of electromagnetic waves.

$$\text{One Hertz} \qquad = 1\,\text{Hz} \quad = 1 \text{ cycle per second} = \text{s}^{-1}$$
$$\text{One Kilo Hertz} \; = 1\,\text{kHz} \; = 1 \times 10^3\,\text{Hz}$$
$$\text{One Mega Hertz} = 1\,\text{MHz} = 1 \times 10^6\,\text{Hz}$$
$$\text{One Giga Hertz} \; = 1\,\text{GHz} \; = 1 \times 10^9\,\text{Hz}$$

2.4.4.2 Pressure

In contrast to the derived SI unit for *pressure*, the pascal, the unit of pressure generally used by meteorologists, by international agreement, is the millibar (mb). All formulae in this text involving pressure will be based on millibars. The following relations apply:

$$\text{One millibar} = 1 \text{ mb} = 1 \times 10^{-3} \text{ bar} = 10^2 \text{ Pa} = 10^2 \text{ Nm}^{-2} = 1 \text{ hPa} \; ,$$

where N (newton) is the SI unit for force.

The conversion between mm or inches of mercury (Hg) and mb is as follows:

$$1 \text{ mb} = 0.7500615 \text{ mm Hg} = 0.029530 \text{ inch Hg} \; ; \quad 1 \text{ mm Hg} = 1.333224 \text{ hPa} \; .$$

Example: $1013.250 \text{ mb} = 760 \text{ mm Hg} = 29.9213 \text{ inch Hg} = 101.3250 \text{ kPa}$.

2.4.4.3 Feet

The "International Foot" is used for legal conversion from feet to metres in Australia and in the U.S.A. and is defined as follows:

$$1 \text{ foot} = 0.3048 \text{ metres exactly} \; . \tag{2.12}$$

The above conversion differs by one part in 500000 from the "U.S. Survey Foot" presently used by the surveying and mapping profession in the United States of America. The U.S. Survey Foot was defined in 1893 as

$$1 \text{ foot} \; = 1200/3937 \text{ m} \tag{2.13a}$$

$$1 \text{ foot} \; = 0.30480061 \text{ m} \tag{2.13b}$$

$$1 \text{ metre} = 3.2808333\ldots \text{ feet} \; . \tag{2.14}$$

It should be noted that most (but not all) distance meters, which allow conversion of the display distances from metres to feet, employ the "U.S. Survey Foot" rather than the "International Foot" definition. Instrument-specific information may be found in instrument manuals. Until the introduction of the Australian National Standards Act in 1960 many Australian authorities used "British Feet". The physical length of the British standard yard has changed significantly over the years. Great care is necessary to determine the correct factor to convert old surveys from feet to metres.

2.4.4.4 Fahrenheit

The conversions between degrees Fahrenheit (F) and degrees Celsius (C) are as follows

$$C = (5/9)(F-32) \tag{2.15}$$

$$F = (9/5)C+32 \; . \tag{2.16}$$

Some distance meters have provisions for temperature input in degrees Fahrenheit as well as degrees Celsius.

3 Principles and Applications of EDM

There are many ways to measure distances by electronic means. Four basic principles will be presented although only one will be discussed in greater depth.

3.1 Pulse Method

3.1.1 Principle of the Pulse Method

A short, intensive signal is transmitted by an instrument. It travels to a target point and back and thus covers twice the distance d. Measuring the so-called flight time between transmission and reception of the same pulse, the distance may be calculated as

$$2d = c\,\Delta t'$$

$$= c(t_R - t_E) \, , \tag{3.1}$$

where   d   = distance between instrument and target
        c   = velocity of light in the medium
       $\Delta t'$ = flight time of pulse
       t_E = time of departure of pulse, timed by gate G_E
       t_R = time of arrival of returning pulse, timed by gate G_R.

Figure 3.1 depicts such a instrument. G_E and G_R are gates where the passage of a pulse is monitored and its time recorded.

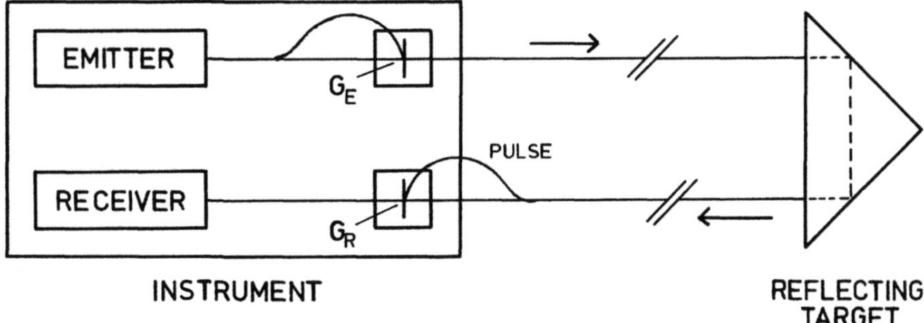

Fig. 3.1. Principle of a pulse distance meter. Timing starts and stops when the light pulse passes the emitter gate GE and the receiver gate GR, respectively

Depending on the nature (light or radio wave) and power of the pulse and on the distance to be measured, suitable natural or artificial features of landscape (or air space) or special retroreflectors may be used as reflecting agents in terrestrial applications. High powered laser systems are used in military laser ranging, thus allowing for the use of the first type of reflecting surfaces. Civil users prefer well-defined, special targets to increase range and precision.

It can be seen from Eq. (3.1) that the accuracy of the distance is dependent on the accuracy of the flight time measurement. An accuracy in the recording of time of 0.1 ns is equivalent to an accuracy of 15 mm.

3.1.2 Applications of the Pulse Method

The pulse method has been used for some time in geodesy, navigation and in military applications, both with light waves and with radiowaves. Powerful pulsed laser systems are employed in geodesy for extremely long distance measurements. These systems require retroreflectors as (cooperative) targets. Military laser rangers measure medium distances by "shooting" at natural or man-made features (non-cooperative targets) visible in the landscape.

Typical light wave applications of the pulse method are:

- Satellite Laser Ranging (SLR)
- Lunar Laser Ranging (LLR)
- Military Laser Rangefinders
- Pulsed Distance Meters for Surveying
- Airborne Laser Terrain Profiler
- Laser Airborne Depth Sounder (LADS)

Satellite laser ranging is carried out for a number of purposes, such as satellite orbit determination (tracking), solid-earth physics studies, polar motion and length of day determination, precise geodetic positioning over long ranges and monitoring of crustal motion. Two satellites have been launched specifically for geodetic and geophysical purposes. Starlette (launched in 1975 by France) carries 60 retroreflectors and orbits 800 km above the earth. LAGEOS (LAser GEOdynamics Satellite) was launched in 1976 by NASA into an orbit 6000 km above the earth. It has a diameter of 0.60 m and carries 422 cube corner reflectors. The apertures of satellite laser ranging telescopes vary from 0.3 to 1.0 m diameter. Pulse energies vary from 0.25 to 3 joules. Typically, green emitting Nd-YAG (neodymium-yttrium-aluminium-garnet) lasers are employed. Ranging accuracies are stated as 1, 0.10 and 0.03 m for first, second and third generation instruments (Wilkins 1980). A global network of fixed stations provides continuous observations. A number of mobile stations, such as NASA's MOBLAS (MOBile LASers) and TLRS (Transportable Laser Ranging Station) provide support for shorter campaigns. The TLRS-2 laser operates at a wavelength of 532 nm (Nd : YAG), at a pulse width of 400 ps, with a power output of 4 MJ and a repetition rate of 10 pulses per second (NASA 1983). Satellite laser rangers have been used (like EDM instruments) for measurements to terrestrial reflectors on rare occasions.

A total of five sets of reflectors were placed on the moon by successive lunar missions by the U.S.A. and the U.S.S.R. The four reflector arrays set-up by the Apollo 11, 14, 15 and the Lunakhod 21 missions are operational and form a quadrilateral with about 1000 km side length. Several laser ranging stations have been measuring to the moon for earth rotation studies and for the determination of the moon's shape, structure and orbit, with the University of Texas McDonald Observatory near Fort Davis, Texas, being the most successful observatory. The flight time of pulses to the moon and back amounts to about 2.6 s. The returns from the moon are received at single photon level and need to be averaged over some time to become recognizable. Single shot accuracy is thought to be at the 0.1 m level for most lunar data (Wilkins 1980).

Battery-operated military-type laser rangefinders may be binocular-shaped for handheld use, tripod-based or attached to sighting periscopes of vehicles, for example. Typically, Q-switched Nd: YAG lasers are employed, operating at 1064 nm and with pulses of 3 to 15 mJ energy, 10 to 20 ns width and repetition rates of 0.5 to 1 Hz. Maximum range varies between 10 and 20 km, minimum range between 50 m and 250 m and ranging accuracy between ± 3 m and ± 5 m.

General purpose pulse distance meters may be divided into three groups. A first group includes instruments developed for use in industry and civil engineering with maximum ranges from 8 to 100 m and 8 to 400 m to matt black and matt white targets, respectively, and resolutions between 10 and 100 mm. The second group includes hand-held or theodolite/tripod-mounted instruments with accuracies of $\pm (0.5$ m $+ 30 - 1000$ ppm) and maximum ranges of 100 m and 3000 m to passive and single prism targets, respectively. The third group is formed by EDM instruments with accuracies of ± 5 mm employing the pulse rather than the phase measurement technique. These EDM instruments have longer range than comparable phase measuring instruments because they can compensate the increased energy emission during a pulse with the idle times between pulses. They permit distance measurements to non-cooperative targets on close range. This third group of pulse distance meters will be discussed later in more details.

Helicopter-mounted laser terrain profilers are routinely used for the determination of longitudinal profiles in the design of roads and transmission lines, typically in connection with inertial surveying systems (Engler 1983). Typically, pulsed GaAs lasing diodes emitting at 904 nm are employed. The single shot accuracies of the systems vary from about ± 0.05 to 0.45 m depending on the signal-to-noise ratio. Discrimination of first and last pulse returns successfully permits mapping terrain as well as ground cover heights (Mamon et al. 1978).

A Laser Airborne Depth Sounder (L.A.D.S.) is being developed for the Hydrographic Service of the Royal Australian Navy. It is designed to measure water depths from 2 to 30 m (up to 50 m at night) for charting purposes on the continental shelf. Two frequencies of a neodymium-doped yttrium-aluminium-garnet (Nd: YAG) laser are employed. The fundamental IR wavelength of 1064 nm is employed for the height measurement to the sea surface and the frequency-doubled green wavelength of 532 nm for the height measurement to the sea floor. Resolution and accuracy are stated as 0.1 and 0.5 m, respectively (Billard 1988; Myres 1983).

Typical radiowave applications of the pulse method are:

- RADAR (RAdio Detection And Ranging)
- DME (Distance Measuring Equipment for aircraft)
- LORAN (LOng RAnge Navigation for aircraft)
- Satellite Radar Altimetry
- Airborne Radar Altimetry

Distance Measuring Equipment (DME) is part of a short-range navigation system for aircraft. Together with VHF omni-range beacons (VOR), it provides bearing and distance (rho-theta) information to pilots. The DME in the aircraft transmits pulse pairs. A ground transponder receives the pulses, delays them for 50 µs and then transmits pulse pairs with a different spacing. The DME measures the time lapse between transmission and reception of the pulse pairs, removes the 50 µs beacon delay and computes the distance to the beacon in nautical miles (Kendal 1979).

Satellite radar altimeters measure continuously the distance between satellites and the surface of the sea. A first instrument was used in the Skylab mission in 1973, a second one in the GEOS-C satellite from 1975. A third operated in SEASAT-1 between June and October 1978 and a fourth in GEOSAT since March 1985. Satellite altimetry is used for ocean geoid determinations and for detailed determinations of the sea surface topography. The accuracy of altimeter data is better than one metre.

3.2 Phase Difference Method

The phase difference method is defined here as the method of measuring phase differences of continuous waves for distance measurement.

3.2.1 Phase Difference Between Transmitted and Received Signal

Most common EDM instruments used in surveying are based on this principle, regardless of whether they use light waves, infrared waves or microwaves as carrier waves. The measuring signal, which is modulated on the carrier wave in the emitter, travels to the reflector (in the case of an optical carrier wave) and back to the EDM instrument, where it is picked up by the receiver (see Fig. 3.2).

In the receiver, the phases of the emitted and the received signals are compared and the phase lead $\Delta\Phi$ is measured.

$$\text{Emitted signal} \quad y_E = A \sin(\omega t) \tag{2.5a}$$

$$= A \sin \Phi \ . \tag{2.5b}$$

$$\text{Received signal} \quad y_R = A \sin \omega(t + \Delta t) \tag{2.7b}$$

$$= A \sin(\Phi + \Delta\Phi) \ . \tag{2.7a}$$

Because a continuous signal is used, the values of y_E and y_R will change with time. But the phase lead $\Delta\Phi$ (as well as the time lead Δt) will remain constant.

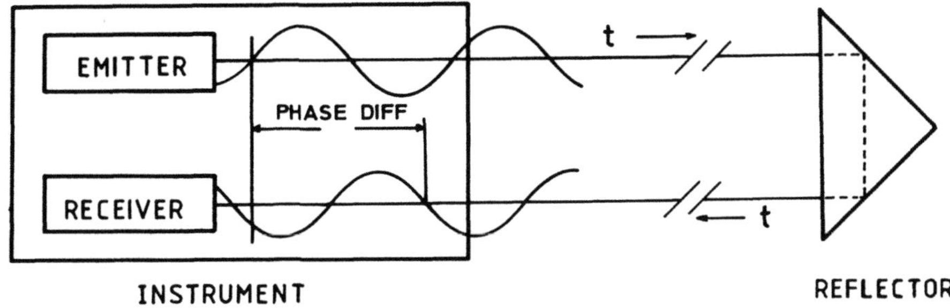

Fig. 3.2. Principle of a phase difference measuring distance meter. The time axis is denoted by t

The instrument can therefore measure a constant phase lead despite the fact that the amplitudes of both signals vary continuously. According to Eq. (3.1), the distance could be computed as follows:

$$d = \frac{c}{2}\,\Delta t' \; . \tag{3.2}$$

Unfortunately, $\Delta t'$ cannot be obtained through phase comparison. Phase comparison provides only a time lead Δt. A time equivalent to the number of full cycles elapsed during the flight of a specific signal has to be added to Δt to obtain the total flight time $\Delta t'$.

$$\Delta t' = m\,t^* + \Delta t \; , \tag{3.3}$$

where $\quad \Delta t' = $ flight time of a specific signal
$\qquad\quad$ m $\;=$ integral number of full wavelengths over the measuring path (ambiguities)
$\qquad\quad$ $t^*\;=$ elapsed time for one full cycle of the modulation signal (period)
$\qquad\quad$ $\Delta t =$ time lead in phase measurement.

In Eq. (3.3), all variables except m and $\Delta t'$ are known. The time lead Δt can be derived from Eq. (2.8a) as a function of the measured phase lead $\Delta\Phi$. Substituting ω and f from Eqs. (2.6b) and (2.2) leads finally to

$$\Delta t = \frac{\Delta\Phi\,\lambda}{2\pi c} \; , \tag{3.4}$$

where λ is the modulation wavelength.

The time interval (t^*) of one full cycle of the modulation wavelength (or period) can be obtained by substituting 2π for $\Delta\Phi$ in Eq. (3.4). This leads to:

$$t^* = \frac{\lambda}{c} \; . \tag{3.5}$$

Equation (3.2) can now be written in a new form, considering also Eqs. (3.3), (3.4) and (3.5):

$$d = \frac{c}{2}\,(m\,t^* + \Delta t)$$

$$= \frac{c}{2}\left(m\,\frac{\lambda}{c} + \frac{\Delta\Phi}{2\pi}\,\frac{\lambda}{c}\right)$$

$$= m\,\frac{\lambda}{2} + \frac{\Delta\Phi}{2\pi}\,\frac{\lambda}{2}\;. \tag{3.6}$$

With the exception of the ambiguities m, all variables of Eq. (3.6) are known.

Usually, the term $\lambda/2$ is replaced by U, which is called the unit length of an EDM instrument. The unit length U is the scale on which the EDM instrument measures a distance.

$$U = \frac{\lambda}{2}\;. \tag{3.7}$$

The second term of Eq. (3.6) is also replaced by a new term, L, indicating the fraction of U.

$$L = \frac{\lambda}{2}\,\frac{\Delta\Phi}{2\pi} = \frac{\Delta\Phi}{2\pi}\,U\;, \tag{3.8}$$

where   L   = fraction of unit length U to be determined by phase measurement,
        $\Delta\Phi$ = (measured) phase lead (in radians)
        U   = unit length of distance meter.

The fundamental Eq. (3.6) now reads as follows:

$$d = m\,U + L\;, \tag{3.9}$$

where m is an integral number of unit lengths and is still unknown. The ambiguity of Eq. (3.9) is solved, not by the determination of m but by the introduction of more than one unit length in an EDM instrument. The procedure may best be explained by some examples. The most important unit length of an instrument is always the smallest, which coincides with the highest frequency. This so-called "main" unit length is used for the fine measurement of distances. The precision of an instrument depends on the choice of this main unit length, because of the limited resolution of the phase measurement.

3.2.1.1 First Example: Hewlett-Packard HP 3800 B

The Hewlett-Packard Distance Meter HP 3800 B used a total of four unit lengths ranging from 10 m to 10 km. The basic principle of operation is explained in the table below.

Step i	Reading $\dfrac{\Delta\Phi_i}{2\pi}$	Unit length U_i	Fraction $L_i = \dfrac{\Delta\Phi_i}{2\pi}\,U_i$
1	0.8250	10 m	8.250 m
2	0.382	100 m	38.2 m
3	0.433	1 000 m	433.0 m
4	0.244	10 000 m	2 440.0 m
		Displayed distance $d = 2438.250$ m	

The underlined figures are transferred mechanically to the distance readout of the instrument. The main unit length of the instrument is 10 metres.

3.2.1.2 Second Example: Kern DM 500/DM 501/DM 502

The Kern Short Range Distance Meters DM 500/501/502 use only two unit lengths of 10 m and 1000 m; the main unit length of the instrument is 10 m.

Step i	Reading $\dfrac{\Delta\Phi_i}{2\pi}$	Unit length U_i	Fraction $L_i = \dfrac{\Delta\Phi_i}{2\pi}\,U_i$
1	0.8253	10 m	8.253 m
2	0.4384	1 000 m	438.4 m
		Displayed distance $d = 438.253$ m	

The underlined figures are on display. The phase measurement is done automatically by digital means. Step one of the procedure is called fine measurement, step two coarse measurement. For distances longer than 1 km, 1 000 m would have to be added to the readout of the instrument.

Later instruments in the 500 series (DM 503, DM 504) employ a third modulation frequency to resolve the multiples of 1 000 m. The third unit length is derived from synthetic frequencies, as shown below in the third example.

3.2.1.3 Third Example: Nikon DTM-1 and NTD-3

The two Nikon distance meters do not use a set of unit lengths which are related by factors of ten, hundred, thousand, etc. They employ unit lengths which are close to each other. The coarse measurements of distance are derived by computation from differences of three fine measurements carried out with slightly different unit lengths. The procedure is similar to techniques used in early Geodimeters, such as the models 4, 6, 6A and 8 (Rüeger 1988).

The following fine modulation frequencies and unit lengths are used:

$$f_1 = 14.973 \text{ MHz } f \rightarrow U_1 = 10.00000 \text{ m} = \frac{c_0}{2 n_{REF} f_1} \qquad (3.10)$$

$$f_2 = 14.935 \text{ MHz } f \rightarrow U_2 = 10.02508 \text{ m} = (400/399) U_1 \qquad (3.11)$$

$$f_3 = 14.224 \text{ MHz } f \rightarrow U_3 = 10.52678 \text{ m} = (20/19) U_1 \; , \qquad (3.12)$$

where $\quad c_0 \quad$ = velocity of light in vacuum
$\qquad n_{REF}$ = reference refractive index.

Subtracting smaller frequencies from larger frequencies (as no negative frequencies can exist) leads to the following synthetic coarse measurement frequencies and unit lengths:

$$f_1 - f_2 = 0.038 \text{ MHz} \rightarrow U_4 = 4000 \text{ m} \qquad (3.13)$$

$$f_1 - f_3 = 0.749 \text{ MHz} \rightarrow U_5 = 200 \text{ m} \; . \qquad (3.14)$$

The distance equations for the two coarse measurements and one fine measurement can be given as

$$d = 4000 \text{ m} (l_1 - l_2) + \text{multiples of } 4000 \text{ metres}$$

$$d = 200 \text{ m} (l_1 - l_3) + \text{multiples of } 200 \text{ metres}$$

$$d = 10.00000 \text{ m} (l_1) + \text{multiples of } 10 \text{ metres} \; ,$$

where $\quad l_i \quad = \dfrac{\Delta \Phi_i}{2 \pi} = \dfrac{L_i}{U_i}$
$\qquad l_i \quad$ = normalized phase measurement $(0 \leq l_i \leq 1.0)$
$\qquad U_i$ = unit length
$\qquad L_i$ = fraction of unit length.

Note: Add 1.0 if differences $l_1 - l_i$ become negative.
A numerical example is given in the table below.

Step	Reading l_i $= \dfrac{\Delta \Phi_i}{2 \pi}$	Composite readings	Unit lengths U_i	Fractions $L_i = \dfrac{\Delta \Phi_i}{2 \pi} \; U_i$
1	0.3658	$l_1 = 0.3658$	10 m	3.658 m
2	0.9096	$(l_1 - l_3) = 0.3658 - 0.2399$	200 m	25.18 m
3	0.2399	$(l_1 - l_2) = 0.3658 - 0.9096 + 1.0$	4000 m	1 824.8 m

Displayed distance d = 1823.658 m

3.2.1.4 Fourth Example: Kern Mekometer ME 5000

The precision distance meter Kern Mekometer ME 5000 operates on a slightly different principle. Rather than using fixed modulation frequencies and measuring

18

phase differences, the modulation frequency is adjusted (within a certain range) until the transmitted and received signals are in phase. This is done at four frequencies, namely at both ends of the tuning range and in the middle. The frequencies are obtained from a frequency synthesizer and are, thus, known. The step interval of the frequency synthesizer is 10 MHz/61824 or 161.749482 Hz (Meier and Loser 1986). All measured frequencies will therefore be multiples of this frequency.

The equations for the four distance measurements can be expressed as

$$d_0 = m_0 U_0 \tag{3.15}$$

$$d_1 = m_1 U_1 \tag{3.16}$$

$$d_2 = m_2 U_2 \tag{3.17}$$

$$d_3 = m_3 U_3 \ , \tag{3.18}$$

where $\quad U_i = \dfrac{\lambda_i}{2} \tag{3.19}$

$\quad\quad m_i =$ multiples of unit length U_i (ambiguities).

With Eq. (2.2) $\lambda_i = \dfrac{c}{f_i}$ the unit length computes as

$$U_i = \dfrac{c}{2 f_i} \ . \tag{3.20}$$

Starting at the lowest frequency, the first phase minimum is tracked and the frequency recorded (f_0). Then the next minimum with higher frequency is tracked and the frequency recorded (f_1). Therefore

$$m_1 = m_0 + 1 \ . \tag{3.21}$$

Solving Eqs. (3.15) and (3.16) for m_i yields

$$m_0 = \dfrac{2 d f_0}{c} \tag{3.22}$$

and

$$m_1 = m_0 + 1 = \dfrac{2 d f_1}{c} \ . \tag{3.23}$$

Substitution of m_0 in m_1 leads to

$$\dfrac{2 d f_1}{c} = 1 + \dfrac{2 d f_0}{c} \ .$$

Rearranging and solution for d yields

$$d = \dfrac{c}{2} \dfrac{1}{(f_1 - f_0)} \ . \tag{3.24}$$

Back substitution in Eq. (3.22) leads to

$$m_0 = \frac{f}{f_1 - f_0} \; . \tag{3.25}$$

Because of the limited accuracy of the difference $(f_1 - f_0)$ the above equation cannot be used directly. It follows from Eq. (3.17)

$$m_2 = m_0 + k_2 = \frac{2 d f_2}{c} \; . \tag{3.26}$$

Substitution of Eq. (3.22) in the above equation yields

$$k_2 = \frac{2d}{c} (f_2 - f_0) \; , \tag{3.27}$$

and with substitution of Eq. (3.24)

$$k_2 = \frac{f_2 - f_0}{f_1 - f_0} \quad \text{(nearest integer value)} \; . \tag{3.28}$$

Solution of Eq. (3.27) for d yields another solution for m_0, after substitution in Eq. (3.22)

$$d = \frac{k_2 c}{2} \frac{1}{(f_2 - f_0)}$$

$$m_0 = \frac{k_2 f_0}{(f_2 - f_0)} \quad \text{(nearest integer value)} \; . \tag{3.29}$$

Similarly m_3 can be computed from $m_0 + k_3$ with

$$k_3 = \left(\frac{f_3 - f_0}{f_1 - f_0} \right) \quad \text{(nearest integer value)} \; . \tag{3.30}$$

The distance can then be computed from three separate equations as follows:

$$d_0 = \frac{c}{2 f_0} \text{INT} \left(\frac{k_2 f_0}{f_2 - f_0} \right) \tag{3.31}$$

$$d_2 = \frac{c}{2 f_2} \text{INT} \left(\frac{k_2 f_0}{f_2 - f_0} + k_2 \right) \tag{3.32}$$

$$d_3 = \frac{c}{2 f_3} \text{INT} \left(\frac{k_2 f_0}{f_2 - f_0} + k_3 \right) \; , \tag{3.33}$$

where
$$c = c_0/n$$
$$c_0 = 299\,792\,458 \text{ m/s}$$
$$= \text{velocity of light in vacuum}$$
$$n = 1.000\,284\,514\,844$$
$$= \text{(reference) refractive index.}$$

It should be noted that the procedure outlined above reflects the principle of the measurement of the Mekometer ME 5000. In reality, the instrument is likely to measure a greater number of phase minima and is likely to use all measurements in the computation of the final distance. The procedure described by Meier and Loser (1988) for a two-colour instrument in fact features more measured minima.

A numerical example is now given to further explain the principle of the ME 5000 measurement.

Example: $f_0 = 474\,005\,888$ Hz $f_2 = 485\,014\,234$ Hz

$\qquad\qquad f_1 = 474\,049\,075$ Hz $f_3 = 479\,488\,548$ Hz

$$k_2 = \frac{f_2 - f_0}{f_1 - f_0} = \frac{11.008\,346}{0.043\,187} = 254.9 \rightarrow 255$$

$$m_0 = \frac{k_2 f_0}{(f_2 - f_0)} = \left(\frac{474.005\,888}{11.008\,346}\right) 255 = 10\,979.9 \rightarrow 10\,980$$

$$k_3 = \frac{(f_3 - f_0)}{(f_1 - f_0)} = \frac{5.482\,660}{0.043\,187} = 126.95 \rightarrow 127$$

$$d_0 = m_0 \frac{c}{2f_0} = 10\,980 \left(\frac{299\,707\,187.1}{2 \times 474\,005\,888}\right) = 3471.2490 \text{ m}$$

$$d_2 = (m_0 + k_2) \frac{c}{2f_2} = 11\,235 \left(\frac{299\,707\,187.1}{2 \times 485\,014\,234}\right) = 3471.2489 \text{ m}$$

$$d_3 = (m_0 + k_3) \frac{c}{2f_3} = 11\,107 \left(\frac{299\,707\,187.1}{2 \times 479\,488\,548}\right) = 3471.2484 \text{ m}$$

The mean value is displayed by the distance meter.

3.2.2 Phase Difference Between Two Received Signals

Another form of distance measurement by phase measurement is adopted in navigation systems such as Toran and Decca which work in the so-called hyperbolic mode.

Two radio transmitters M and S transmit continuous unmodulated signals of equal frequency. The signals are received at a station R of unknown position (see Fig. 3.3).

If Φ_M and Φ_S are the phase angles of the two radiated signals at any instant, their phase difference at station R will be

$$\Delta\Phi = (\Phi_M + \Delta\Phi_M) - (\Phi_S + \Delta\Phi_S)$$

$$= \left(\Phi_M + \frac{\omega}{c}\,\overline{MR}\right) - \left(\Phi_S + \frac{\omega}{c}\,\overline{SR}\right)$$

$$= (\Phi_M - \Phi_S) + \frac{2\pi}{\lambda}\,(\overline{MR} - \overline{SR}) \tag{3.34}$$

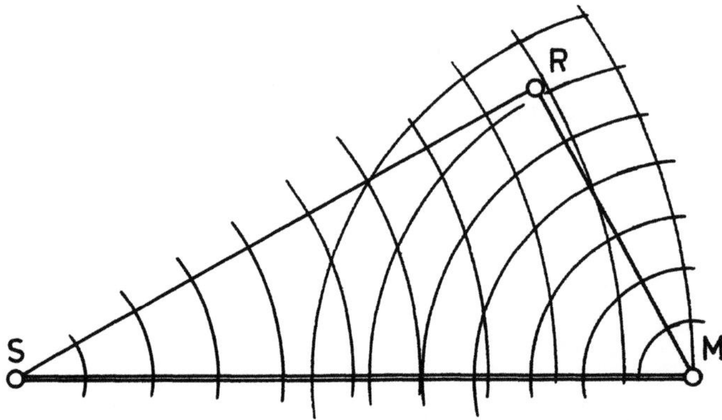

Fig. 3.3. Principle of the measurement of phase differences between two continuous signals (transmitted from *M* and *S*) at a receiver station *R*. *M* and *S* are typically on land, *R* at sea

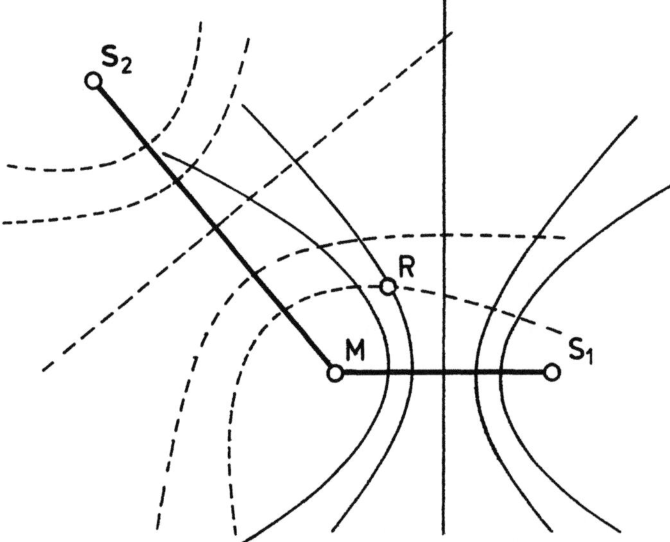

Fig. 3.4. Principle of positioning on the basis of two measured phase (or range) differences. The position is determined by intersecting hyperbolae. The master transmitter and the two slave transmitters are denoted by *M* and S_1/S_2, respectively

where $\quad \Delta\Phi$ = measured phase difference at station R

$\Phi_M, \; \Phi_S$ = phase angles of the M and S transmitters respectively

f = frequency of the M and S transmitters

c = velocity of light in air

$\overline{MR}$ = distance between transmitter M and receiver R

$\overline{SR}$ = distance between transmitter S and receiver R.

The phase angles Φ_M and Φ_S are kept in a constant relation. Both transmitters are phase locked, which produces a constant $(\Phi_M - \Phi_S)$. The second term in

22

Eq. (3.34) is variable and depends on the difference of path lengths from the two transmitters. Lines of equal path difference generate a set of hyperbolae with S and M as foci. A position fix by means of one hyperbola intersecting another requires a second base, S_2M. As shown in Fig. 3.4, this is provided by a third transmitter.

More information about navigation systems may be found in Burnside (1991) and Laurila (1991).

3.3 Doppler Methods

The variation in the pitch of a tone, heard whilst the source of the tone is moved relative to the observer, was first studied by the Austrian physicist Christian Doppler (1803–1853) and later named after him. The *Doppler effect* is observed not only with acoustic waves but also with all electromagnetic waves. It may be explained with the aid of a terrestrial example; its application in satellite geodesy will be discussed later. Figure 3.5 depicts a mobile instrument which consists of a microwave transmitter and receiver. The instrument moves at speed v towards a reflecting surface. The transmitted signal is reflected at this surface and picked up by the receiver.

The emitted frequency is expressed as

$$f_T = \frac{c}{\lambda} , \qquad (3.35)$$

where   c = velocity of light in a medium
       λ = wavelength of emitted radiation.

The following frequency arrives at the reflecting surface:

$$f_S = \frac{c+v}{\lambda} , \qquad (3.36)$$

where   v = speed of instrument relative to reflecting surface.

Back at the receiver, R, the frequency f_R is received.

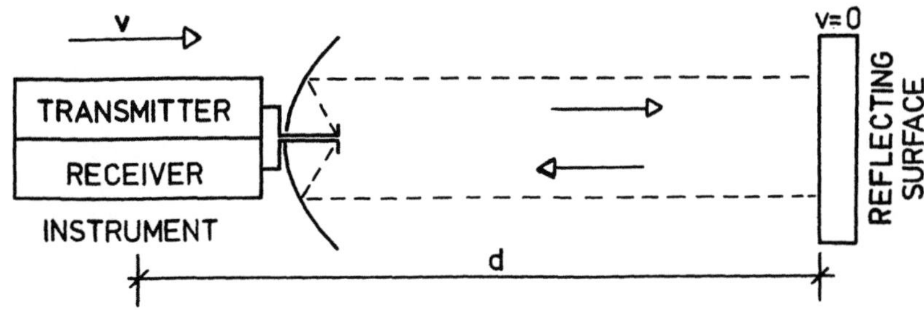

Fig. 3.5. Principle of a microwave distance meter using the Doppler effect. The velocity between instrument and reflecting surface is denoted by v

$$f_R = \frac{c + 2v}{\lambda} \; . \tag{3.37}$$

Mixing of both frequencies f_T and f_R in the instrument produces, amongst other frequencies, the Doppler difference frequency f_D

$$\begin{aligned}
f_D &= f_R - f_T \\
&= \frac{c + 2v}{\lambda} - \frac{c}{\lambda} \\
&= \frac{2v}{\lambda} \; .
\end{aligned} \tag{3.38}$$

The procedure for measuring the Doppler frequency depends on the type of waves used. The Doppler frequency f_D may be obtained by either:

1. Counting the beats per second in the case of sound waves.
2. Counting the bright (or dark) fringes of an optical interference pattern in the case of light waves.
3. Counting the cycles of the Doppler signal ("Doppler counts") per second in the case of radio waves.

Equation (3.38) may be written as:

$$v = \tfrac{1}{2} f_D \lambda \; . \tag{3.39}$$

The distance travelled by the instrument between time t_1 and time t_2 is

$$d_{12} = \int_{t_1}^{t_2} \tfrac{1}{2} f_D \lambda \, dt \; . \tag{3.40}$$

The Doppler method is used in surveying and metrology for distance measurements of highest precision. The Hewlett-Packard HP 5526A and HP 5528A Laser Measurement Systems (laser interferometers) employ the Doppler effect to measure distances travelled by a reflector with a resolution of 10 nm. Because fringe counts need to be obtained, the reflector must travel along the laser beam. This is usually achieved by mounting the reflector on a carriage which travels on a rail.

In geodesy, the Doppler method became well known with the introduction of the U.S. Navy Navigation Satellite System (TRANSIT, NAVSAT) early in 1963. The system was released for public use in 1967 and will be supported until 1992 (Hoskins 1982). (By that time, the new NAVSTAR Global Positionning System (GPS) will have taken over the functions of the TRANSIT (or NAVSAT) system in navigation.) In geodesy, Doppler receivers are employed to track satellite signals (see Fig. 3.6). The most successful instrument, the Magnavox MX 1502 Geoceiver Satellite Surveyor, consists of three portable parts, namely an antenna (which is set-up and centred over the survey mark), the receiver unit with cassette recorder and a power supply (e.g. 12 V battery). The on-board microprocessor permits on-line and off-line processing of data.

The TRANSIT system relies on five to six U.S. navigation satellites in orbit around the earth and provides a world wide navigation system. Setting up a

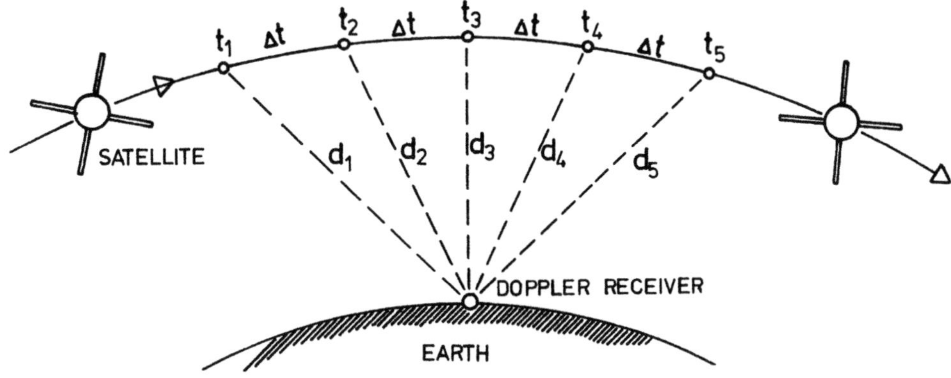

Fig. 3.6. Principle of Doppler satellite positioning

Doppler receiver at a station and recording a number of successive satellite passes allows the computation of the coordinates of the ground station. The Doppler receiver measures the distance differences (see Fig. 3.6)

$$\Delta d_1 = d_1 - d_2$$

$$\Delta d_2 = d_2 - d_3$$

$$\Delta d_3 = d_3 - d_4 \; ,$$

using Doppler counts of the Doppler frequency f_D from interference between the received satellite broadcast frequency and the internal reference frequency.

$$\Delta d_i = \int_{t_i}^{t_{i+1}} f_D \lambda \, dt \tag{3.41}$$

$$= \lambda \, [\text{Doppler counts}]_{t_i}^{t_{i+1}} \; . \tag{3.42}$$

The satellite orbit is known and the position of the ground station can therefore be obtained by intersecting hyperboloids of revolution. Hyperboloids are the loci in space of equal distance differences with respect to two points. Each time interval Δt (Fig. 3.6) produces one pair of hyperboloids of revolution with focal points corresponding to the satellite positions at the beginning and the end of the time interval. The position of the Doppler receiver is obtained in a geocentric coordinate system.

The accuracy of Doppler positioning depends on how long the satellite signals are recorded and on the type of subsequent processing. Assuming observations over 2 days, absolute horizontal positions can be obtained with an accuracy of around ± 10 m. Relative (differential) positions between two stations can be established at the $\pm 0.5 - 1.0$ m level through use of on-board processing. Postprocessing using multi-station (network) solutions may provide accuracies down to ± 0.3 m (Hoar 1981).

3.4 Interferometry

The principle of optical interference is presently used in interferometers for metrology, for high precision distance measurements over short distances and for the definition of the metre. The development of interferometers dates back to 1880, when A. A. Michelson had his first interferometer built in Germany. A first measurement of the metre in terms of light waves followed in 1889. For his work on interferometers, Michelson received the Nobel Prize in Physics in 1907 (Swenson 1987).

3.4.1 Principle of a Michelson Interferometer

The basic principle of an interferometer is depicted in Fig. 3.7. A light source such as a laser produces a light beam which is directed towards a beam splitter. This allows one portion of the beam to pass to the moving reflector and deflects another portion to the fixed reflector. The beams, returned by the reflectors, pass again through the beam splitter and produce an interference pattern there. The interference pattern is recorded by the photodetector and Doppler counts are recorded by the digital counter.

The two waves superimposed in the beam splitter are of equal frequency and amplitude (coherent waves) because they are generated by the same light source. They have, however, a constant phase difference because of the difference in path lengths. They may be described according to Eqs. (2.5a) and (2.7a):

$$y_1 = A \sin (\omega t)$$

$$y_2 = A \sin (\omega t + \Delta \Phi) \ .$$

The superimposed signal is of the form

$$y = y_1 + y_2 = \left[2A \cos \left(\frac{\Delta \Phi}{2} \right) \right] \sin \left(\omega t + \frac{\Delta \Phi}{2} \right) \tag{3.43}$$

and reaches a maximum for a phase lead of $\Delta \Phi = 0$ (constructive interference) and a minimum for $\Delta \Phi = \pi$ (destructive interference). During a displacement of

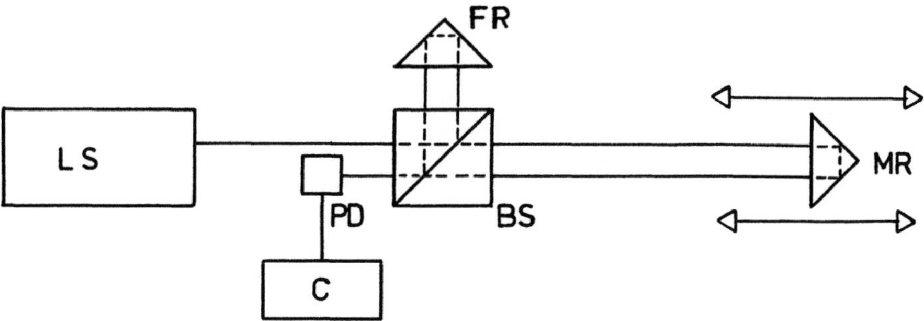

Fig. 3.7. Principal components of a Michelson Interferometer. *LS* light source; *FR* fixed reflector; *BS* beam splitter; *PD* photo detector; *MR* moving reflector; *C* counter

the movable reflector MR, the photodetector counts the number of bright fringes in the interference pattern of the beam splitter.

The distance between the first and the last positions of the reflector is derived from

$$2d = \text{(number of bright fringes)} \times \lambda \,, \tag{3.44}$$

where λ = wave length of light source.

Or in the final form:

$$d = \text{(number of bright fringes)} \times \frac{\lambda}{2} \,. \tag{3.45}$$

The high resolution of interferometers is based on the direct use of the wave length of light waves for measurement. Some EDM instruments use light waves as carrier, but a modulated signal is used for the measurement. Considering a laser interferometer with a HeNe-laser light source, the wave length is $\lambda = 632.8$ nm, thus leading to a least count of about 0.3 µm. The overall accuracy of such systems, however, is limited to about 0.1 ppm by the uncertainty of the refractive index through the limited accuracy of the measurement of ambient temperature and pressure. Common laser interferometers have a maximum range of about 60 m and are mainly used indoors. They are utilized not only for precise length measurements but generally in metrology for measurement of straightness, squareness, parallelism, flatness and angle. For example, the Hewlett-Packard HP 5526 A laser measurement system, as discussed in the next section, is used for such measurements.

3.4.2 Principle of Operation of the HP 5526 A Laser Measurement System

The HP 5526 A uses a two-frequency laser. The two laser frequencies are obtained by so-called Zeeman splitting by applying a weak longitudinal magnetic field to the laser tube (see Fig. 3.8). They are separated by a few MHz. The two frequencies have left and right hand circular polarization. The frequency difference results from different indices of refraction for the two polarizations in the laser tube. The two polarizations are converted to orthogonal linear polarizations by the $\lambda/4$, $\lambda/2$ plates shown in Fig. 3.8. The two-frequency laser is stabilized by balancing the output powers of the two polarizations. Refer to differential amplifier in left bottom corner. Tuning is partly automatic, partly manual.

Both optical frequencies are directed to the polarizing beam splitter (actual interferometer) where the one frequency (f_1) is bent by 90° and the other (f_2) is let through. Both beams then travel to separate reflectors where they are shifted and reflected. The beam separation is 13 mm. The two beams recombine at the beam splitter and then fall on the photodetector. The photodetector senses the difference or beat frequency $f_1 - f_2$.

If, however, the reflectors are moving with velocities v_1 and v_2, the beat frequency will be

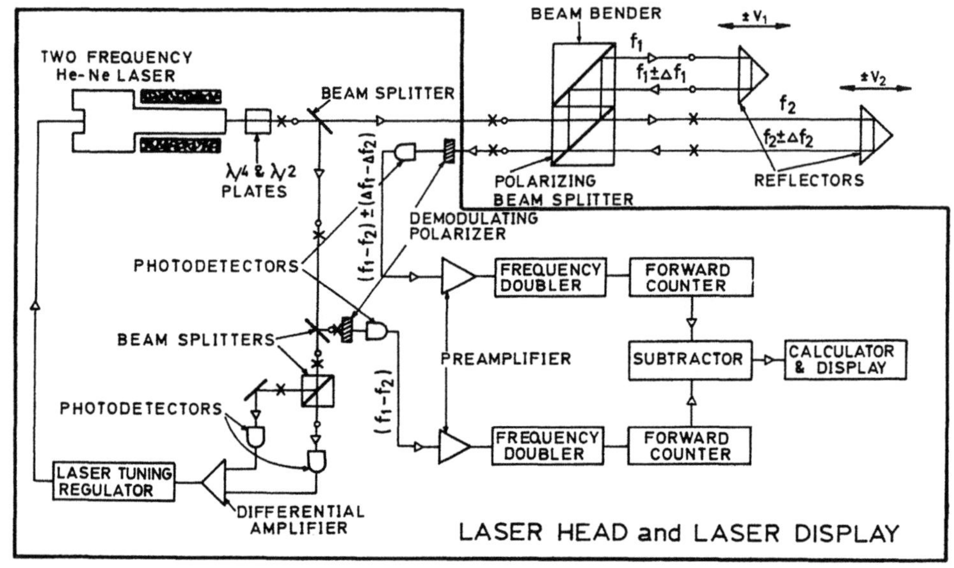

Fig. 3.8. Block diagram of the Hewlett-Packard HP 5526 A laser measurement system. (After Stock 1982; Liu and Klinger 1979)

$$f_1 - f_2 \pm (\Delta f_1 - \Delta f_2) = f_1 - f_2 \pm \Delta f \; , \tag{3.46}$$

where Δf, Δf_1, Δf_2 are the Doppler frequency shifts due to the motion of the respective retroreflectors.

If the difference between Doppler beat frequency from the interferometer (external path) and the reference beat frequency $(f_1 - f_2)$ is counted for a time Δt, the resulting cumulative count corresponds to the differential displacement Δd of the retroreflectors in wavelengths of light, λ, and, by simple conversion, in length units.

$$\Delta f = \frac{2(v_1 - v_2)f}{c} = \frac{2}{\lambda}(\Delta d/\Delta t) \tag{3.47}$$

$$\sum \Delta d = \frac{\lambda}{2} \sum \Delta f \Delta t \; , \tag{3.48}$$

where Δf is the Doppler count per time interval. At long path differences the fringe contrast is degraded because of diffraction and turbulence and may result in the failure of the optoelectronics of the interferometer to recognize fringes. The Hewlett-Packard interferometer displays an error message whenever the Doppler count is lost.

A limitation of the two frequency laser is the low maximum displacement speed, v_{max}, of the reflector. The forward counter in the figure can store only pulses. If $(f_1 - f_2) \pm \Delta f$ approaches zero and crosses zero, the count will increase again, because there is no such thing as a negative frequency. The counter cannot

28

distinguish between Δf larger or smaller than $f_1 - f_2$. To avoid ambiguity, the speed of the reflector is not to exceed

$$f_1 - f_2 \geq \Delta f = 2 v f / c \qquad (3.49)$$

$$v_{max} < \frac{c(f_1 - f_2)}{2f} . \qquad (3.50)$$

For $(f_1 - f_2) = \sim 2\,MHz$, v_{max} amounts to $500\,mm/s$. To be able to detect excessive speed, a maximum speed of $v_{max} = 300\,mm/s$ is specified.

3.4.3 Väisälä Interference Comparator

The Väisälä interference comparator (Väisälä 1923, 1927, 1930) was originally developed to calibrate 24 m invar wires under field conditions. The invar wires were then used, in turn, to determine the length of so-called standard baselines. In 1947, the Finnish Nummela Baseline was measured for the first time directly with the interference comparator with an accuracy of 6 parts in 10^8. This baseline features marks at 0, 6, 24, 72, 216, 432, and 864 m, with the 24 m multiples governed by the earlier determinations by 24 m invar wires.

The working principle of the interference comparator and the observation procedures are described in Kukkamäki (1968), for example. Interference of white light is used to multiply previously determined sections. The fundamental length scale is determined by a 1-metre-long quartz rod, from which the first section (0 m – 6 m) is derived. The length of the standard quartz rods are determined by laboratory techniques. The interval between the first and third mark (0 m – 24 m) is determined by "multiplication" of the first section. The process continues until the 0 m to 864 m interval is derived from the 0 m to 432 m interval.

Subsequently, the IXth General Assembly of IUGG (Brussells 1951) passed a resolution suggesting that baselines be measured by similar techniques in different countries. The following standard baselines were established and measured by interference comparators (Kukkamäki 1978):

- Nummela, Finland (1947, 1952, 1958, 1961, 1966, 1968, 1975, 1977)
- Buenos Aires, Argentina (1953)
- Loenermark, Netherlands (1957, 1969)
- Mata Das Virtudes, Portugal (1962, 1978)
- Potsdam, G.D.R. (1964)
- Mansfield, Ohio, U.S.A. (1966)
- Pienaars River, South Africa (1976)
- Valladolid, Spain (1978)
- Eberberger Forst, Munich, F.R.G. (1958, 1960, 1963)
- Chang Yang, China (1985)
- Gödöllö, Hungary (1987)

The dates in brackets refer to the year(s) of determination(s). Most of these baselines follow the geometrical design of the Nummela baseline. Measurements on the Munich baseline with a Kern Mekometer ME 5000 indicate that the scale of the two determinations is not different. The standard deviation computed from differences between comparator values and ME 5000 values amounts to ±0.2 mm only (Meier and Loser 1986). Väisälä baselines have also been used to field calibrate precision distance meters prior and subsequent to determining a general purpose EDM calibration baseline (Maurer et al. 1988; Humphries 1989).

4 Basic Working Principles of Electronic Distance Meters

EDM instruments are classified according to the type of carrier wave employed. Instruments using light or IR waves are classified as *electro-optical instruments*. Instruments based on radio waves are generally called *microwave instruments*. Because of the different structure of these two types of instruments, they will be discussed separately.

4.1 Electro-optical Instruments

4.1.1 Principle and Components

The basic working principle of an electro-optical distance meter is explained with reference to the type of instrument which uses an analogue phase measuring system. Such instruments display the functions of the different components more clearly.

The *light-source* produces the so-called carrier wave. The wave is described by the carrier wavelength λ_{carr}. The following light sources are presently in use (in order of importance):

1. GaAlAs infrared emitting diode and GaAlAs lasing diode

   $$0.800 < \lambda_{carr} < 0.950 \, \mu m \ .$$

   Gallium aluminium arsenide (GaAlAs) diodes have spectral bandwidths of between 2 and 30 nm between half-power points and are, therefore, almost monochromatic. The degree of coherence of the radiation depends on the structure of the diode and the operating mode. The radiant power output of a single diode could be as high as 30 W but is usually much lower. The effective source size is a few tenth of a millimetre. Gallium aluminium arsenide diodes are used in most short-range and medium-range EDM instruments. They are further discussed in the next section.

2. HeNe-laser

   $$\lambda_{carr} = 0.6328 \, \mu m \ .$$

   The radiant power output of these "red" gas lasers varies between 1 mW and 10 mW. Lasers produce a coherent and monochromatic light of high power density and small divergence. Lasers are used mainly in long range EDM instruments. Distances up to 80 km may be measured, visibility permitting.

3. Xenon flash tube

$$\lambda_{carr} = 0.480\,\mu m \ .$$

High pressure Xenon flash tubes are used in the Kern Mekometer ME 3000 and the COM-RAD Geomensor 204 DME. In the Mekometer ME 3000, flashes of 1.0 μs duration and 0.04 Joule energy are produced at a repetition rate of 100 Hz.

The *modulator* varies the amplitude by intensity modulation of the carrier wave at a modulation frequency produced by an oscillator. The light beam will therefore alternate between bright and dark sequences. The modulation wavelength λ_{MOD} is always much longer than the carrier wavelength, λ_{carr}, or in other words, the modulation frequency is much lower than the carrier frequency. Different modulation techniques are discussed in Section 4.1.2.

The *transmitter lens system*, which may have a fixed or adjustable focus, produces a beam divergence of about 5 min of arc for short range instruments and 20 s of arc for long range laser instruments such as the Geodimeter model 8. Narrow beams produce strong return signals, but because of the pointing precision demanded they may take pointing to distant reflectors a very tedious operation.

The *reflector* consists of a glass prism in a housing. All incident rays are reflected in such a way that the reflected rays are parallel to the incoming ones. The rays are therefore returned to the EDM instrument without the necessity of very accurate orientation of the reflector. Details of reflectors are described in Section 10.

The *receiver lens system* may again be of the fixed or adjustable focus type. It focusses the return signal onto the photo detector.

The *photo detector* transforms the light beam's intensity variations into variations of current. Two devices are commonly used for this purpose:

1. In the range of semiconductors such as Si-(silicon)-photodiodes or Si-avalanche-photodiodes (APD), silicon-photodiodes are preferred to germanium-photodiodes because of their higher response at wavelength $\lambda = 900$ nm. Si-avalanche-photodiodes are preferred to Si-photodiodes because of their better signal-to-noise ratio, and are therefore used in most short-range EDM instruments.
2. Photomultipliers (photo tubes): Light falls onto the cathode which, being coated with a photo-electric substance, emits electrons according to the light's energy. Between anode and cathode, the number of electrons is multiplied by an array of dynodes. Amplification factors up to 10^8 are possible. Such devices are used mainly in long range EDM instruments.

Refer to Burnside (1982) and Saastamoinen (1969) for further details.

The *oscillator* produces the modulation frequency and consists of an oscillator circuit which is locked to the resonant vibration frequency of a quartz crystal. To obtain a unit length of 10 m, the modulation frequency needs to be approximately 15 MHz. The resulting frequency is a function of the shape and size of the quartz crystal employed. Changes in ambient temperature and ageing of the quartz yield changes in the frequencies of oscillators. Differing systems are employed in EDM instruments to minimize temperature effects:

1. Oven-controlled crystal oscillators (OCXO) are used in some precision and long range EDM instruments and are accurate to ± 1 ppm or better. They need a warm-up time of at least 15 min. High performance OCXO's (used in frequency counters and geodetic satellite receivers, for example) feature proportionally controlled double ovens and may exhibit stabilities of better than ± 0.01 ppm between -55 and $+75\,^{\circ}$C.
2. Temperature-compensated crystal oscillators (TCXO) are commonly used in short-range EDM instruments. They need no warm up time and are accurate to ± 1 ppm in the range 0 to $+50\,^{\circ}$C and ± 3 ppm in the range -20 to $+50\,^{\circ}$C. Temperature compensation is typically achieved by analogue compensating networks which employ temperature-sensitive capacitors, thermistors and/or resistors.
3. Non-compensated room temperature crystal oscillators (RTXO) typically exhibit frequency stabilities of ± 5 ppm from -20 to $+70\,^{\circ}$C and ± 2.5 ppm from 0 to $+50\,^{\circ}$C.

The effects of the temperature characteristic of oscillators on distances measured with EDM instruments can be further reduced. Factory calibration of the oscillator's frequency drift with temperature in connection with a built-in temperature sensor permits the application of appropriate corrections to measured distances by the on-board microprocessor.

Typical annual ageing rates of TCXO's and OCXO's are 1 ppm and 0.2 ppm, respectively. The short-term stabilities (one second average) of TCXO's and OCXO's are typically 0.001 ppm and 0.0005 ppm, respectively, with high performance OCXO's reaching 3 parts in 10^{12}. In EDM, the most critical oscillator parameters are the frequency drift with temperature and the frequency drift with time (ageing). More details on oscillators and their performance may be found in Frerking (1978) and Rüeger (1982).

The *resolver* shifts the phase of the reference signal. In Fig. 4.1, a rotation of the phase shift knob simultaneously changes the readout and the resolver angle, and thus the phase of the reference signal.

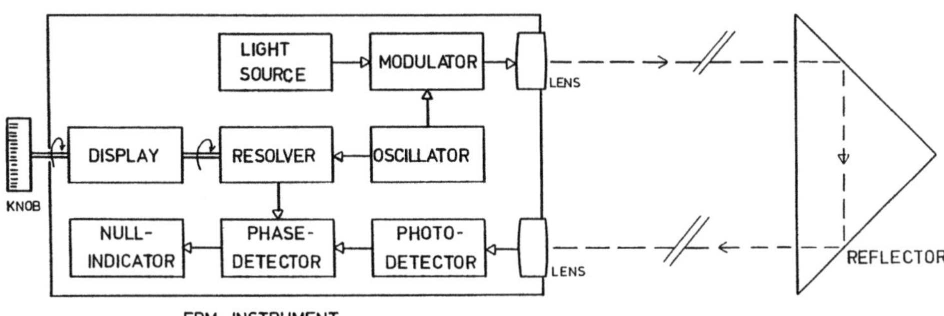

Fig. 4.1. Major components of an electro-optical distance meter using analogue phase measurement

The *phase detector* provides the phase comparison between the return and the reference signal. The result of this phase detection is displayed on the *null indicator*. The null indicator reads zero when both signals are exactly in phase. This is achieved by turning the resolver as explained above. More information on phase measurement is given in Section 4.1.3.

The *display* is directly coupled with the resolver. It indicates the position of the resolver and is therefore a readout of the measured phase difference. Both display and resolver position can be altered by turning the *phase shift knob*.

4.1.2 Methods of Modulation and Demodulation of Light and NIR Waves

4.1.2.1 Infrared Emitting and Lasing Diodes and their Modulation

The first commercial EDM instrument using GaAs infrared emitting diodes was released in 1968 (Wild Distomat DI 10), only 6 years after lasing in GaAs $p-n$ junctions was first observed. Infrared emitting and lasing diodes provide a low cost, light weight, small, low voltage and low current alternative to gas lasers. The former exhibit a larger divergence and inferior spatial and spectral characteristics, but permit much simpler modulation techniques.

Infrared diodes can be classified according to their structure, operation, manufacturing process and direction of emission, to name just a few classification criteria. All devices feature an active region (or laser cavity) which consists typically of undoped low-bandgap material surrounded by higher bandgap n-type (donor) and p-type (acceptor) material. Cross-sections through two typical diodes are depicted in Figs. 4.2 and 4.3.

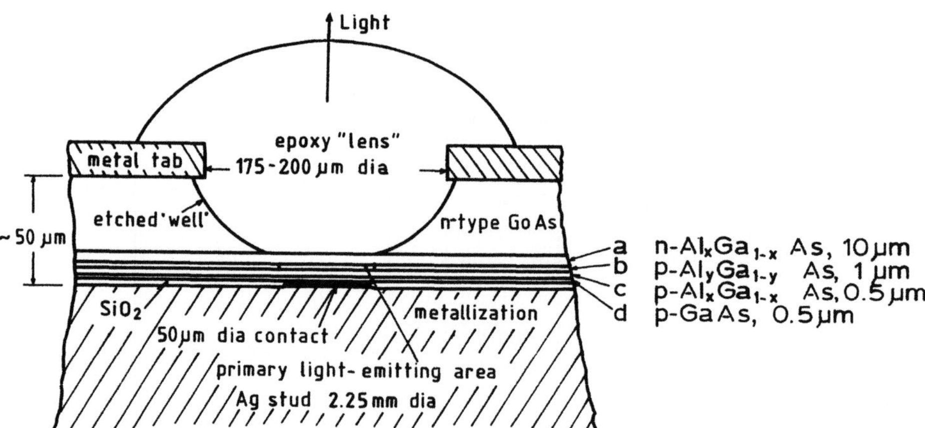

Fig. 4.2. Cross-section through an experimental double heterojunction GaAlAs emitting diode (etched well emitter, after Burrus 1972)

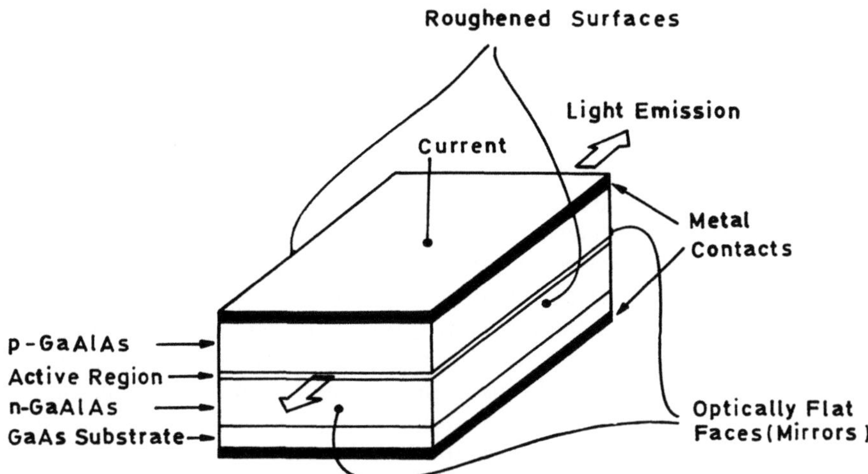

Fig. 4.3. Typical structure of GaAlAs double heterostructure lasing diode with Fabry-Perot cavity. (After Lau 1988)

The principle of GaAlAs emitting diodes is explained with reference to Olsen and Ban (1987). Under forward bias, electrons from the n-region and holes from the p-type region are injected into the active region. The confinement of these carriers to the active region is achieved by the energy barriers and refractive index steps at each heterojunction. This confinement leads to electron-hole recombination in the active layer that generates spontaneous and incoherent infrared emission in all directions. The external quantum efficiency of such LED's (Light Emitting Diodes), namely the ratio of emitted photons to input electrical power, is only a few percent. The high refractive index of the GaAlAs semiconductor material (3.5 to 3.6) causes most of the radiation to be reflected (by total reflection) at the semiconductor/air interfaces and absorbed internally.

Following again Olson and Ban (1987), the working principle of lasing diodes is derived from that of the LED's discussed above: Each photon generated within the active layer can stimulate the recombination of additional electron-hole pairs to emit photons that are coherent (same wavelength and phase) with it. With increased injection current, the gain due to the stimulated emission can approach and then exceed the absorption losses in the active layer. The device becomes an amplifier and exhibits a narrowing of the emitted spectrum as well as an abrupt increase of radiated power (lasing). Amplification is greatest parallel to the active layer [and perpendicular to the mirrored end faces if a Fabry-Perot cavity (see Fig. 4.3) is provided]. This is due to the waveguiding effect of the refractive index steps at the layer boundaries and the gain profile defined by the material parameters. Typically, lasing diodes exhibit higher output, narrower spectral bandwidths and permit higher modulation rates.

In terms of structure, homojunction, single heterojunction and double heterojunction diodes are distinguished. For example, Figs. 4.2 and 4.3 show a double heterojunction surface emitting diode and a heterojunction edge emitting lasing diode, respectively. Increases in the complexity of the layer structure provide thinner and better-defined active layers, faster response times, reduced drive currents and narrower bandwidths of the emitted radiation. The wavelength of the emitted radiation is determined by the structure, doping level, aluminium content and operating temperature of these GaAlAs devices. In principle, emission wavelengths between 710 and 900 nm can be achieved. However, devices with visible wavelengths below 800 nm have been rare so far.

In comparison with lasing diodes infrared emitting diodes (usually referred to as LED's) provide a lower temperature sensitivity, a superior linearity of output power versus input current characteristic, higher reliability, lower cost, greater availability, lower degradation rate, gradual (rather than catastrophic) failure characteristics, permit operation at higher temperatures and require less complicated circuitry. Most EDM instruments have their infrared diodes mounted on printed circuit boards with the output coupled through glass fibres to the transmitting telescope (pig-tailed diodes). In consequence, small area high radiance diodes are favoured. Figure 4.2 depicts an early design of such a surface emitting diode (Burrus 1972). The active area has typically a diameter of 50 micrometre. Small emitting areas can also be achieved by edge emitting LED's.

Different designs are employed for continuous (high duty cycle, low power) lasing diodes and for pulsed (low duty cycle, high power) diodes. The former type is typically used for fibre optics communication systems and EDM and employs double heterojunction structures. It features rise/fall times of about 0.5 ns and modulation bandwidths of 10 GHz (Lau 1988). Figures 4.3 and 4.4 depict the cross-section and the output power versus input current characteristic of a typical laser diode. The transfer characteristic (Fig. 4.4) shows that lasing occurs only when the threshold current is exceeded. Below threshold, the lasing diode operates

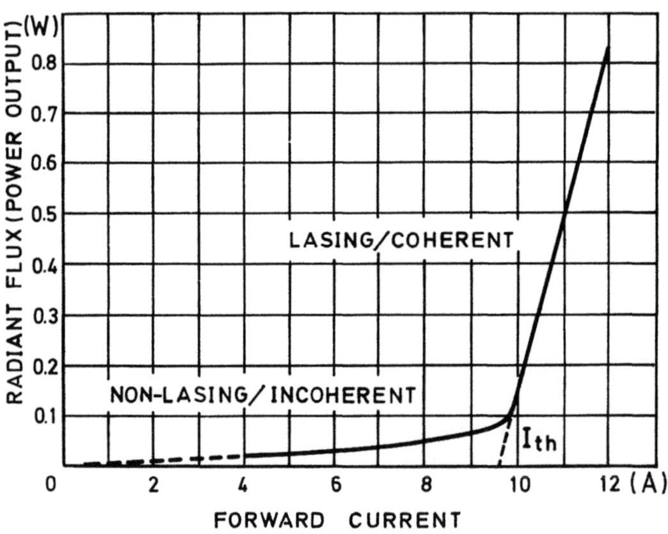

Fig. 4.4. Output power versus input current characteristic for a typical GaAs laser diode. (After Glicksman 1975)

36

Table 4.1. Technical data of some typical infrared emitting and lasing diodes. Values in brackets refer to power output into fibres

	Pulsed single heterojunction laser diode		CW double heterojunction laser diode		High radiance (etched well) emitting diode	
Total peak radiant flux	20	W	7	mW	2.5	mW
			(2	mW)	(600	μW)
Peak injection current	75	A				
Typical current (threshold)	18	A	90	mA	100	mA
Wavelength	904	nm	830	nm	820	nm
Spectral width at 50% power	3.5	nm	2.5	nm	40	nm
Rise time	0.5	ns	0.1	ns	15	ns
Voltage	8	V	2.0	V	1.8	V

as an emitting diode. When using lasing diodes it must be considered that the threshold current as well as slope of the output/input characteristic are strongly temperature-dependent. If follows that the drive current producing the maximum permissible output power is also temperature dependent, and thus must be changed with temperature if no burn-out at low temperatures is to occur. Laser diodes must therefore be temperature-stabilized (typically in interferometers) or their optical output must be monitored and stabilized through an adjustment of the drive current (most non-interferometry applications, including EDM).

The technical data of some typical diodes are listed in Table 4.1, which clearly shows the different operational parameters of pulsed and continuous wave lasing diodes. Gallium-aluminium-arsenide diodes have some further properties which are of particular relevance to EDM applications:

- The emission wavelengths of diodes change with temperature at a typical rate of 0.25 nm/°C, although higher values have been measured (Kopeika et al. 1983: 0.36 nm/°C). In EDM, scale changes of 0.004 ppm/°C result due to uncompensated errors in the first velocity correction. This can lead to an error of 0.1 ppm per 25 °C change from temperature at which wavelength is nominal.
- The actual emission wavelength of specific diode may differ from nominal (as stated in the manual of the EDM instrument) because of manufacturing tolerances of the diodes or change of diode supplier. Typically, these differences are likely to be smaller than 15 nm. The resulting scale errors are thus likely to be smaller than 1.2 ppm.
- The output power of all diodes decreases with increased diode temperature and, thus, ambient temperature. The output power at +70 °C may be 30% to 50% less than at +25 °C. In the case of lasing diodes, larger currents are required at higher temperatures to maintain lasing. This increases power consumption.
- The lifetime of GaAlAs diodes is limited due to gradual degradation of the devices. It is important to note that the lifetime for operation at 25 °C can be 10 times longer than for operation at +50 °C. Typically, the output power drops to 75% after 10000 h of operation at room temperature.

It follows from the above properties that EDM instruments are best operated at the coolest possible temperature. Shading instruments will greatly assist in this matter.

- Because of transit time delays in the GaAlAs material of high refractive index, modulated signals emerging from different parts of the diode surface will typically not be fully in phase. The modulation signal on the infrared carrier will therefore not have a plane wavefront. The magnitude of the effect depends on the transit times of the holes and electrons to the p−n junction as well as on the design, size, shape and location of the electrical contacts. The rise times (time taken for the radiant flux to increase from 10 to 90% of its peak value in response to a step function in the drive current) and fall times of the diodes give an indication of the delays involved. The so-called phase inhomogeneities can be determined by scanning of the diode surface. Figures 4.5 and 4.6 depict scans of power and phase across the emitting surface of a diode (after Leitz 1977; Daino et al. 1976). The maximum variation of phase of 90 mm in Fig. 4.6 refers to a worse than average diode of earlier design. The glass fibres used in most distance meters to connect the diode with the telescope of the instrument have a smoothing effect so that the transmitted beam is likely to exhibit reduced phase inhomogeneities.

Because of the basically linear relationship between input (injection) current and output power (radiant flux) of infrared emitting diodes (over entire operating range) and lasing diodes (specific sections below and above threshold current, only), the infrared output beam can easily be modulated. Considering the input versus output characteristic in Fig. 4.4, it becomes evident that a sinusoidal variation of the drive current between 10.2 A and 11.8 A leads to a sinusoidal modulation of the output power between 0.2 W and 0.75 W.

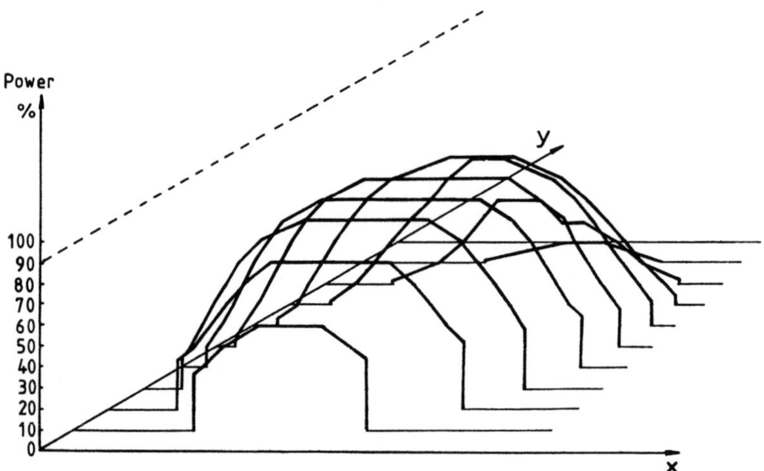

Fig. 4.5. Measured output power (arbitrary units) versus position of an infrared emitting diode of 50−100 μm diameter, as derived from 100 spot measurements. (After Leitz 1977)

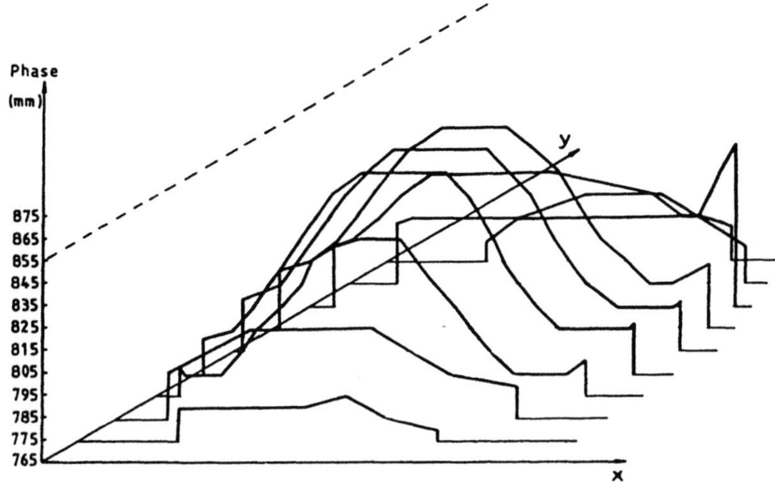

Fig. 4.6. Measured phase (in mm) versus position of an infrared emitting diode of $50-100\,\mu m$ diameter, as derived from 100 spot measurements. (After Leitz 1977)

Further information on diodes, their properties and applications may be found in Hayashi (1984); Kressel and Ettenberg (1982); Arecchi and Schulz-Dubois (1972) and Sze (1969), for example.

4.1.2.2 Direct Demodulation

Photodiodes have the property of transforming radiation into electrical current: the higher the radiation power, the higher the current flow through the diode. The mechanism of photodiodes will not be explained in depth, because they display basically the reverse effect of emitting diodes. EDM instruments use mainly silicon (PIN) photodiodes or Si-avalanche photodiodes (APD). The latter produce a much higher amplification due to the internal multiplication (avalanche) effect. The properties and deficiencies of photodiodes are similar to those of IR emitting and lasing diodes. Please refer, for example, to Webb et al. (1974) and Arecchi and Schulz-Dubois (1972) for further details on photodiodes.

4.1.2.3 Indirect Modulation

Indirect modulation may be achieved by passing a continuous light beam through two polaroid filters of perpendicular polarization planes. Between the two filters the plane of the polarized light is rotated by a special device in phase with a modulation signal. This results in an amplitude modulated light beam emerging from the second filter. The principle is explained in Fig. 4.7. Two devices for the rotation of the polarization plane are commonly used in conjunction with EDM instruments.

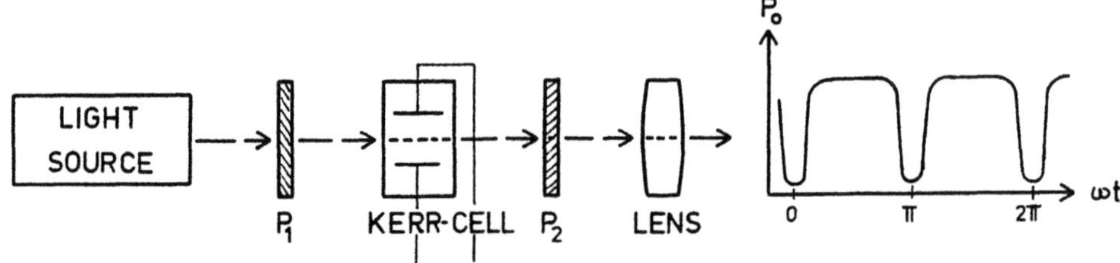

Fig. 4.7. Principle of indirect modulation using Kerr cell. The two polaroid filters P_1 and P_2 are mounted in such a way that their polarization planes are at a right angle to each other. A diagram of output power versus modulation signal is shown on the right and refers to the intensity variations of the light beam after passing through P_2

Kerr Cell. The Kerr cell is a glass tube filled with nitrobenzene and containing two built-in parallel plates, which form the two electrodes of a capacitor. A potential difference across these plates causes a rotation of the plane of polarization of light passing between them. This electro-optical *Kerr-effect* makes it possible by rotating the incident polarization plane of light, to modulate the intensity of the emerging light beam.

In Fig. 4.7, P_1 and P_2 indicate polaroid filters with polarization planes set perpendicular. The Kerr cell's electrodes make an angle of 45° with the polarization planes of P_1 and P_2.

In order to produce the light modulation shown, the modulation frequency is superimposed on the high bias voltage applied to the Kerr cell electrodes. See Saastamoinen (1969) for details of Kerr cells. The Geodimeter models 6, 6A, 6B, 6BL, 700/710, 600 as well as all earlier Geodimeter instruments, were equipped with Kerr cells.

Electro-Optic Crystals. A number of crystals exhibit the linear electro-optical effect called Pockel's effect. From linear polarized light at the input face of the crystal elliptically polarized light is produced at the output face with the semi-axes of the ellipse changing with the applied modulation frequency. After passing a second polarizing filter, the emerging light beam is amplitude-modulated. Crystals exhibiting the Pockel's effect are potassium dihydrogen phosphate (KDP, KH_4PO_4), ammonium dihydrogen phosphate (ADP, NH_4, H_2PO_4), lithium niobate ($LiNbO_3$) and lithium tantalate ($LiTaO_3$), to name just a few. KDP crystals are employed in the AGA Geodimeter Model 8, the Kern Mekometer ME 3000 and the Com-Rad Geomensor 204 DME. Keuffel and Esser did develop lithium tantalite modulating crystals for use in Rangemaster II and III as well as some Ranger models (Erickson 1983).

4.1.2.4 Indirect Demodulation

Indirect demodulation is mostly combined with the subsequent phase measurement. Two devices of this kind are used in present-day instruments.

Photomultiplier. The conversion of light into electric current by means of photomultipliers has already been mentioned in Section 4.1.1. The photomultiplier can be operated with a modulated voltage, the modulation being equal to that of the outgoing light wave. The photomultiplier will then produce a maximum current output when the transmitted and the received measuring signals are in phase. It will produce a minimum current if both signals have a phase difference of 180°. The Geodimeter models 6, 6A, 6B, 600, 8 and 700/710 are equipped with photomultipliers.

Electro-Optic Crystals. Electro-optic crystals can not only be used for modulation but also for demodulation purposes. The same amplitude-modulated voltage applied to the transmitter crystal is applied to the receiver crystal. A maximum light output after the second crystal and its polaroid filter results if there is no phase difference between transmitted and received signal. For a phase difference of 180° the light output would be zero. Such a system is employed in the Mekometer ME 3000 where the light output is detected by a photomultiplier. In the Mekometer ME 5000, a single crystal is used for modulation and demodulation. In this case, the light output is monitored by a photodiode.

4.1.3 Methods of Phase Measurement

4.1.3.1 Optical-Mechanical Phase Measurement

The principle of optical-mechanical phase measurement is depicted in Fig. 4.8. The return signal travels through an internal light path the length of which can be varied by one full unit length. The phase difference can be measured easily and directly in terms of length, by recording that displacement of the moving prisms which is required to null the null-indicator. Such a manual system is employed in the Mekometer ME 3000, where it has been found to be unaffected by cyclic errors. However, linear errors over the 0.3 m unit length were found in some instruments. A similar though less convenient arrangement was already employed in the Geodimeter NASM 2 in 1950.

4.1.3.2 Electric Analogue Phase Measurement

The principle of the analogue phase measurement involves the delay of a reference signal of the same characteristics as the transmitted signal until a zero phase lag with the return signal is obtained. Electrical delay lines are mostly designed as *resolvers*, which are very similar to an electric motor. The phase angle therefore becomes a physical quantity based on the revolution of the resolver. All instruments equipped with resolvers (e.g. Geodimeters 6 and 8, Wild Distomat DI 10, Tellurometer MA 100) may be subject to systematic errors within the phase measurement interval. These so-called cyclic errors are sometimes termed the non-linearity of the phase measurement. The wave length of such cyclic errors is normally equal to half of the unit length $\left(\dfrac{\lambda_{\text{MOD}}}{4}\right)$.

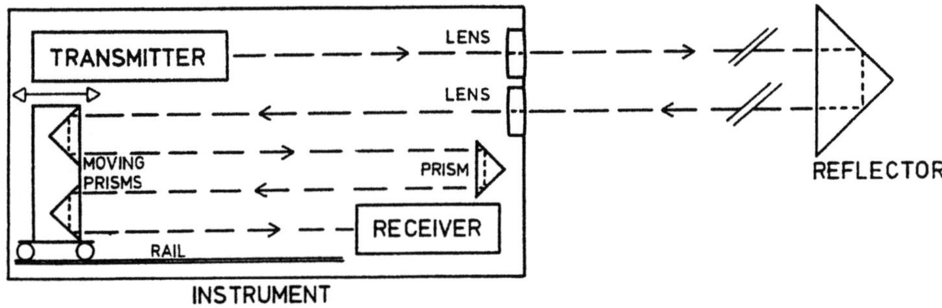

Fig. 4.8. Principle of the optical-mechanical phase measurement system as implemented in the Mekometer ME 3000

The Hewlett-Packard HP 3800 B featured another system of analogue phase measurement where the phase angle was transformed into a direct current. In this process some systematic errors may also occur but will not be of sinusoidal nature.

The testing of analogue phase measurement systems is usually carried out in conjunction with the examination of cyclic errors.

4.1.3.3 Electric Digital Phase Measurement

The digital phase measurement is based on the comparison of two low-frequency sinusoidal signals of equal frequency. One signal is the reference (or transmitted) signal, the other the return signal. Both signals are converted into square waves and operate a gate. The gate is opened when the reference signal begins a new cycle, and closed when the return signal does the same. During the time of the open gate, pulses from a high frequency oscillator are accumulated in a counter (see Fig. 4.9 for the corresponding block diagram).

The phase difference between the two signals can be deduced in two different ways. A first possibility is depicted in Fig. 4.10, which also illustrates the whole measurement: (1) depicts the reference (transmitted) signal, (2) the return signal, (3) the triggered reference signal, (4) the triggered return signal, (5) shows the phase counts "i" between the opening of the gate (GO) and the closure of the gate (GC). In (6), a full wavelength is counted ("j" counts). The phase difference L can be computed according to Fig. 4.10 as:

$$L = \frac{i}{j} \times U \ , \tag{4.1}$$

where U is the unit length which corresponds to the frequency of reference and return signals.

This procedure does not need a constant ratio between the oscillator's high frequency and the low frequency of the return and reference signals.

Another solution is possible if the high frequency pulses are in a fixed ratio to the low frequency of the measurement signals. This can be easily achieved, if

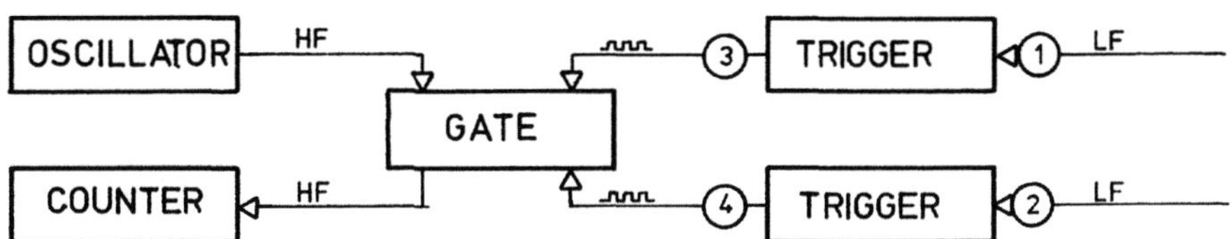

Fig. 4.9. Block diagram of the components of a digital phase measurement system. The low frequency (*LF*) reference signal is numbered *1* and *3* and the low frequency return signal *2* and *4*

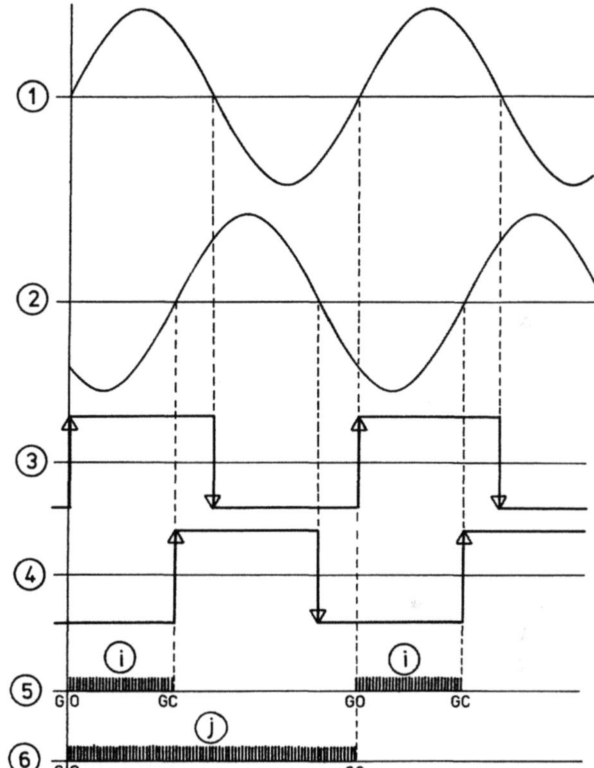

Fig. 4.10. Counting sequences of a digital phase measurement. The curves *1* and *3* represent the reference signal and the curves *2* and *4* the return signal. The counts representing the phase difference between reference and return signals are depicted on line *5* and the counts for a full cycle of the reference signal on line *6*

both frequencies are derived from the same master oscillator in the transmitter. In the Hewlett-Packard HP 3805, for example, the return and reference signals have a low frequency of 3745 Hz whilst the high frequency pulses are at 14 987 103 Hz. There are therefore, during one full low frequency cycle, exactly 4000 pulses of the high frequency. The number of "j" need not be measured but is obtained by calculation as 4000. This leads to the following formula for the phase difference L:

$$L = \frac{i \, f_{LOW}}{f_{HIGH}} \, U \, ,$$

(4.2)

where U is again the corresponding unit length of the distance meter.

Example: HP 3805 A $\quad L = \dfrac{i}{4000}\, U$,

where U is either 2000 m (coarse measurement) or 10 m (fine measurement).

4.2 Microwave Instruments

As microwave distance meters are not discussed in more detail later in this text, some information in addition to the basic working principle is given here. For further details of such instruments, the reader may refer to Burnside (1982); Kahmen (1977); Laurila (1976); Rinner and Benz (1966) and Saastamoinen (1969), for example.

4.2.1 Introduction

Microwave instruments, like optical instruments, measure along the shortest path between the two instruments and therefore need intervisibility between the stations. (It is, however, possible to measure long distances a few metres above the sea without the requirement of intervisibility.) Several carrier wavelengths have been used in microwave EDM, 8 mm (Q band), 18 mm (K band), 30 mm (X band) and 100 mm (S band). S band instruments display very large ground swing effects, while Q band instruments have less power to penetrate haze and cloud and therefore have a reduced range. For these reasons X band and K band instruments have proved to be the most popular.

Microwave instruments are mainly used for the measurement of long distances, up to 150 km, although their all-weather capability may justify even medium or short range applications. The accuracy of microwave distances is mainly dependent on the accuracy of the refractive index. With measurement of atmospherical parameters at the terminals only, an accuracy of 2−3 ppm may be expected. Higher accuracies may be achieved by using better atmospheric models.

4.2.2 Working Principle and Components

A general outline of a microwave instrument using an analogue phase measuring system will now be given. Figure 4.11 depicts the basic design of the Tellurometer, the first microwave instrument which became available in 1957. In Fig. 4.11, the two functions of microwave instruments are clearly distinguished, namely the "master" and the "remote" mode. However, both modes are usually incorporated in the instrument at each terminal of a line, so that the distance can be measured in "forward" and "reverse" directions.

The presence of a parabolic or horn radio antenna and the absence of optical parts are the most obvious differences from an electro-optical EDM instrument. In addition, microwave instruments not only make use of amplitude modulation (AM) but also of frequency modulation (FM) and they provide a built-in phone link between the two stations.

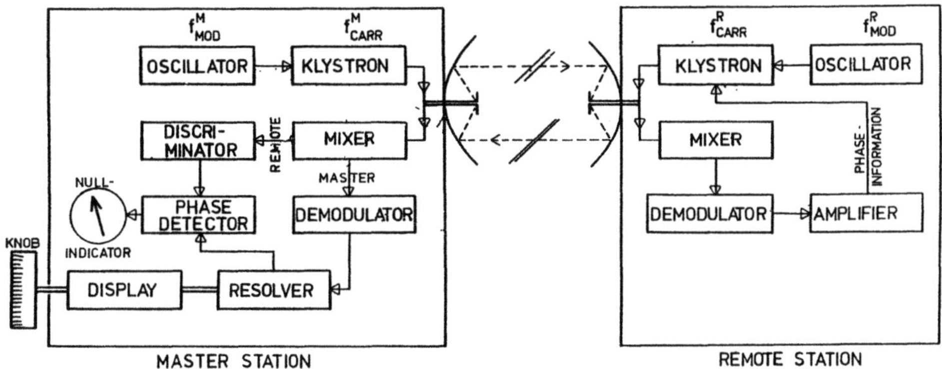

Fig. 4.11. Block diagram of a microwave distance meter using the Tellurometer principle and employing analogue phase measurement techniques

The *oscillator* has the same function as that in electro-optical instruments. It can be switched through different quartz crystals to produce several modulation frequencies. Because microwave instruments are long range instruments, oscillator frequencies have to be temperature-controlled, normally by provision of an oven.

The *klystron* or cavity resonator is an electronic tube producing a microwave. It is operated in such a way that the emerging microwave is frequency modulated by an oscillator frequency.

In newer microwave instruments solid state microwave sources (Gunn Diodes) are used instead of klystrons.

The *antenna* can be of parabolic shape (about 300 mm diameter) and has two small dipoles at its focus. These are perpendicular to each other and set at 45° to the vertical. Transmitted and received signals are polarized in two planes perpendicular to each other.

The *mixer* mixes transmitted and received signals at the antenna.

Demodulators demodulate amplitude modulated (AM) signals into alternating currents (AC).

Discriminators demodulate frequency modulated (FM) signals into alternating currents.

The functions of the resolver, phase detector, null indicator and the diplay have already been discussed.

To obtain a distance measurement, the master and remote stations transmit frequency modulated signals at slightly different frequencies of carrier and modulation. At the time of transmission, the phases of the modulations signals are taken as φ_M and φ_R, respectively. Due to the flight time t_1 of the signal between the stations the two stations receive the following phase informations:

at remote station: $\quad (\varphi_M + t_1 \omega_1)$ $\hfill$ (4.3)

at master station: $\quad (\varphi_R + t_1 \omega_2)$, $\hfill$ (4.4) where the angular velocities of the modulation signals at master and remote stations are denoted by ω_1 and ω_2, respectively.

Through mixing of transmitted and received signals at both stations, the following phase differences can be derived:

at master station: $\quad \varphi'_M = \varphi_M - (\varphi_R + t_1\omega_2)$ $\qquad\qquad$ (4.5)

at remote station: $\quad \varphi'_R = (\varphi_M + t\omega_1) - \varphi_R$. $\qquad\qquad$ (4.6)

The master phase difference information is demodulated by the demodulator and then used as "return" signal in the phase measurement process.

Because both phase differences must be available at the master station for the final phase measurement, the phase difference gained at the remote station is transmitted back to the master instrument by an additional frequency modulation of the remote carrier wave. After demodulation in the discriminator, the remote measurement (with the added time delay of the beat signal) is available at the master station and used as "reference" signal in the phase measurement process:

$$\varphi''_M = (\varphi_M + t\omega_1) - \varphi_R + t_1(\omega_1 - \omega_2) \ . \qquad\qquad (4.7)$$

Subtraction of the master phase difference from the (relayed) remote phase difference leads to the following expression:

$$\varphi''_M - \varphi'_M = \varphi_M + t\omega_1 - \varphi_R + t\omega_1 - (t_1\omega_2) - \varphi_M + \varphi_R + t_1\omega_2 \ , \qquad (4.8)$$

and, after taking note of the cancellation of some if not most terms:

$$\Delta\varphi = 2t_1\omega_1 \ . \qquad\qquad (4.9)$$

The flight time and the distance follow finally as:

$$t_1 = \frac{\Delta\varphi}{2\omega_1} = \frac{\Delta\varphi}{4\pi f} \qquad\qquad (4.10)$$

$$d = \frac{c_0}{n} t_1 \ . \qquad\qquad (4.11)$$

4.2.3 Effects of Reflections in Microwave EDM (Multipath)

Because microwave instruments have a much larger beam divergence than optical instruments, certain parts of the beam may reach the ground or other reflecting surfaces and may be reflected there. The reflected wave travels a longer distance and, being received with the direct wave, has an effect on the measurement of the distance. The effect is usually called *ground swing*. The amplitude of the ground swing is dependent on the reflectivity of the specific ground surface. Water, snow and street surfaces may be highly reflective (Küpfer 1968).

The multipath effect has been expressed in mathematical terms by Poder (Saastamoinen 1969) and Küpfer (1968). Ground swing is mainly a function of the excees path length, the reflectivity of the reflecting surface, the modulation frequency and the carrier frequency. The last two parameters can be chosen by the designer of such instruments, and this has led to two different instrumental designs for the reduction of ground swing:

1. One such design involves the measurement of a distance with up to 20 slightly different *carrier* wavelengths. The resulting distances may then exhibit slight variations, which may be plotted. If a sine curve eventuates, the mean may be taken and this will be free from ground swing effects. For very large and small (< 1.3 m) excess path lengths however, no full "swing" curve is obtained and the mean is hard to estimate (Küpfer 1968). Different carrier waves are, for example, used in the Tellurometers MRA-5, MRA-7 and CA1000.
2. Another method of reducing ground swing is the use of higher modulation frequencies. This method was adopted in the Siemens-Albis MD60 (equivalent to the former Wild Distomat DI60), where a modulation frequency of 150 MHz (unit length $= 1$ m) is employed (Küpfer et al. 1971) and to a lesser extent in the Tellurometers MRA-5 and MRA-7 (unit length $= 1.87$ m).

Ground swing may be further reduced by setting up instruments in positions where the reflected beam is prevented from reaching the antenna by a hump in front of the instrument or by ensuring that the beam cone never reaches any reflecting surfaces (peak to peak, tower to tower) (Rinner and Benz 1966). On flat ground, the effect may be reduced by setting-up very close to the ground (Kahmen 1977), or by variation of the height of instrument. Küpfer (1968, p. 339, Fig. 16) has derived formulae indicating the necessary variation of the height of instrument in order to obtain a full period of the ground swing for various excess path lengths and distances.

5 Propagation of Electromagnetic Waves Through the Atmosphere

5.1 Atmospheric Transmittance

The transmittance of the atmosphere is usually described by the quotient of incident radiant power divided by transmitted radiant power. It is a measure of the attenuation and extinction of wave propagation. The transmittance is a function of numerous variables: wavelength, distance, temperature, barometric pressure, gaseous mixture, rain, snow, dust, aerosols, bacteria and, in more detail, the size of particles of all these constituents. The limitations of atmospheric transmittance are given by the scattering and absorption of the emitted radiation. Scattering by air molecules (Rayleigh Scattering) and scattering by larger aerosol particles (Mie scattering) can be distinguished. Absorption in several spectral regions is mainly caused by water vapour, carbon dioxide and ozone. Figure 5.1 depicts the transmittance of atmosphere as a function of wavelength for a part of the visible and near-infrared (NIR) spectrum under specific conditions. The figure shows that only a limited part of the NIR spectrum is suitable for EDM.

Attenuation of radiation through the atmosphere is the intensity reduction of wave propagation through scattering and absorption [attenuation (in %) = 100 minus transmittance (in %)] and may be expressed by Bouguer's rule

$$J = \frac{J_0}{d^2} \, e^{-zd} \, , \qquad\qquad (5.1)$$

where J = radiant intensity at some distance d from the emitter
   J_0 = radiant intensity at emitter
   d = distance
   z = attenuation coefficient or extinction coefficient.

The first factor of Eq. (5.1) accounts for the purely geometrical propagation. The second, exponential, factor accounts for the transmittance of the atmosphere. The attenuation coefficient is a function of several variables already mentioned at the beginning of this section.

If the "visibility range" or the "meteorological range" is known, the attenuation coefficient can be derived as

$$z_v = \frac{3.912}{d_v} \, , \qquad\qquad (5.2)$$

where z_v is the *attenuation coefficient* for the visible spectrum ($\lambda_{EFF} = 0.555 \, \mu m$) and d_v is the *meteorological range* in kilometres. The meteorological range is defined as the distance over which a large, black target (such as a pine forest) can just be distinguished from the horizon.

48

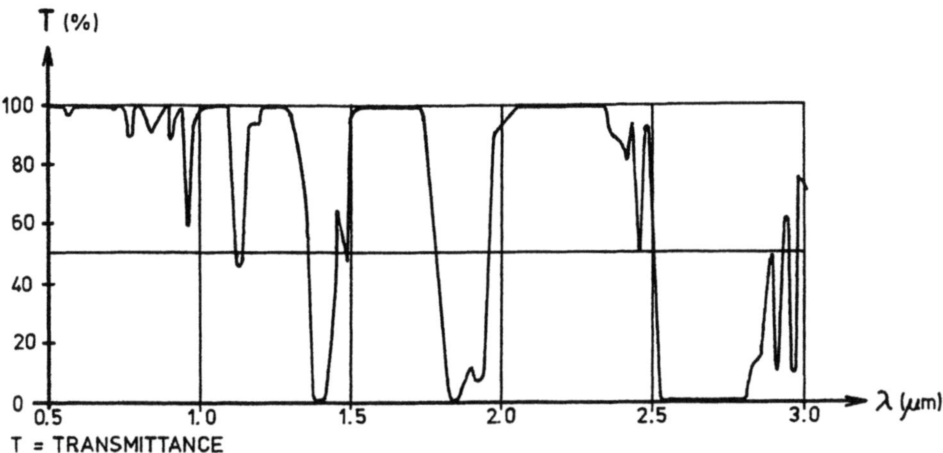

Fig. 5.1. Transmittance of a 300-m-long horizontal air path at sea level, at a temperature of 26 °C and at a partial water vapour pressure of 8 mb. (After RCA 1974). Water molecules (H_2O) cause the absorption effects at 930, 1120, 1450 and 1900 nm, for example. Carbon dioxide is responsible for the absorption effects at 1400 and 2050 nm. Both constituents of air cause the total loss of transmittance between 2550 and 2800 nm

In EDM, the transmittance of the atmosphere affects the range of instruments and the strength of the return signal.

5.2 Range of EDM Instruments

Following Eq. (5.1) and Leitz (1977), the range of a distance meter can be expressed as

$$R_y = (y/a)^{0.5} e^{-zR_y} \, , \tag{5.3}$$

where R_y = range of distance meter with y prisms (in km)
   a   = instrument specific parameter depending on energy density of transmitter, sensitivity of receiver, aperture of telescopes, aperture of prisms (in km^{-2})
   z   = attenuation coefficient (in km^{-1})
   y   = number of prisms used.

An iterative procedure is required for a solution of the above equation and the parameters a and z need to be known. Leitz (1977) quotes $0.2 \, km^{-2}$ and $0.38 \, km^{-1}$ for a and z, respectively, and for a Zeiss Eldi 2 and average weather conditions. The same author uses z values of 0.3 and 0.5 for a visibility of 8 km and overcast conditions and sunny conditions with strong heat shimmer, respectively.

In the case of the instrument specific parameter a not being known, an alternative approach can be chosen. Richter (1970), considering Eqs. (5.1) and (5.2), derived an equation which permits the computation of the range of a distance meter for a particular visibility if the range of the same distance meter is already known for another visibility.

$$R_i = d_{v_i} \left\{ \left(\frac{R_0}{d_{v_0}} + \frac{\log R_0}{X \log e} \right) - \frac{\log R_i}{X \log e} \right\} , \tag{5.4}$$

where R_i = range for visibility range d_{v_i} (1 prism)
       R_0 = range for visibility range d_{v_0} (1 prism)
       $\log e = 0.434294$
       X = 2.5 for $\lambda_{carr} \approx 900$ nm (infrared)
       X = 3.9 for $\lambda_{carr} \approx 550$ nm or 630 nm. (visible)

The attenuation coefficient z can be computed as the ratio X/d_v as in Eq. (5.2). The value of 2.5 for X agrees with the z values suggested by Leitz (1977) for overcast conditions. For sunny conditions with strong heat shimmer, Leitz employs X values between 4.0 and 4.5.

Equation (5.4) allows for the computation of the range R_i at d_{v_i} if d_{v_0} is known for R_0; iteration is required for a solution. It so to say reflects the condition for a single prism observation. With a number y of prisms, the range R_y computes as follows:

$$R_{y_i} = R_i \frac{d_{v_i}}{X \log e} \log \left(\frac{\sqrt{y}\, R_i}{R_{y_i}} \right) . \tag{5.5}$$

A term $(y)^{1/3}$ instead of $(y)^{1/2}$ is employed for the AGA Geodimeter 14; $(y)^{1/2}$ is used for the AGA Geodimeter 600. Equations (5.4) and (5.5) in connection with experimental data permit the computation of range versus visibility diagrams. An example is shown in Fig. 5.2.

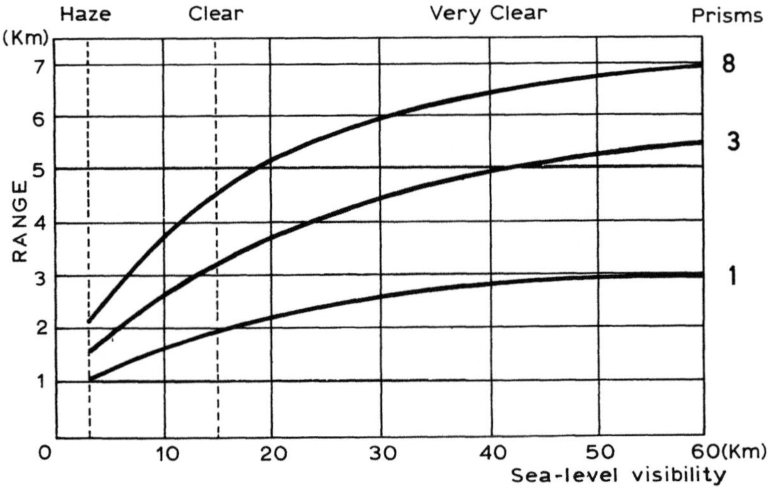

Fig. 5.2. Range versus sea level visibility curves for 1, 3 and 8 prisms for a Geodimeter 220 distance meter. (After Geotronics AB, 1985)

50

The above range equations should be used with caution, as the effective range does not solely depend on the visual range. In practice, the range is also affected by the quality and condition of the prisms (temperature gradients through the prism cause deformations), atmospheric turbulence and heat shimmer, ambient light, atmospheric brightness and ground reflections (Ehbets et al. 1983). These authors define good conditions as overcast, no haze, 30 km visibility, no heat turbulence, and poor conditions as very hazy, 3 km visibility or strong sunlight with substantial heat turbulence. The measuring range differs by a factor of 5 between good and poor conditions. Best range is achieved in overcast conditions, at dawn and, particularly, at night, when air turbulence is minimal and the signal-to-noise ratio of the photodiode improved due to the reduction in ambient light (Ehbets et al. 1983).

5.3 Phase Refractive Index

The phase refractive index n of a medium is defined as

$$n = \frac{c_0}{c} \,, \tag{5.6}$$

where c_0 = velocity of light in a vacuum
c = velocity of light in a medium (phase velocity).

The refractive index of air is a function of:

1. the gaseous composition of the atmosphere, which is very nearly a constant
2. the amount of water vapour pressure in the atmosphere
3. the temperature of the gaseous mixture
4. the pressure of the gaseous mixture
5. the frequency of the radiated signal.

The variation of the refractive index with frequency constitutes the phenomenon of *dispersion.*

5.4 Group Refractive Index of Light and NIR Waves for Standard Conditions

In electro-optical EDM, the refractive index is dependent on the wavelength of the visible or infrared radiation, i.e. a well-defined frequency and its associated frequency band which is formed by several frequencies. The different frequencies would have the same propagation velocity in a vacuum, but not in air, as interference between the different frequencies occurs. The signal resulting from the sum of all frequencies will have the so-called group velocity which is always smaller than the phase velocities of its individual frequencies:

$$c_g = c - \frac{dc}{d\lambda} \lambda \,, \tag{5.7}$$

where c_g and c are the group velocity and phase velocity respectively.

Calling n_g the group refractive index, the following equation may be written:

$$c_0 = c_g n_g \; . \tag{5.8}$$

Substitution of Eqs. (5.8) and (2.4) in Eq. (5.7) leads to

$$\frac{c_0}{n_g} = \frac{c_0}{n} - \frac{d\,(c_0/n)}{d\lambda}\,\lambda \tag{5.9}$$

and, after differentiation of the bracket term, to

$$\frac{c_0}{n_g} = \frac{c_0}{n} - \frac{(-c_0/n^2)\,dn}{d\lambda}\,\lambda \; . \tag{5.10}$$

From an expansion with $(n_g n/c_0)$ follows

$$n = n_g + \frac{n_g}{n}\,\frac{dn}{d\lambda}\,\lambda$$

and after setting $(n_g/n) = 1.0$ and rearranging

$$n_g = n - \frac{dn}{d\lambda}\,\lambda \; , \tag{5.11}$$

where λ is the wavelength in a vacuum.

As soon as an equation for $n = f(\lambda)$ is known, it may be differentiated and the group refractive index derived.

The group refractive index may be calculated according to Barrell and Sears (1939):

$$(n_g - 1) \times 10^6 = 287.604 + 3 \left(\frac{1.6288}{\lambda^2}\right) + 5 \left(\frac{0.0136}{\lambda^4}\right) \; , \tag{5.12}$$

where λ is expressed in μm (micrometres).

This formula is valid for visible light in dry air at $0\,°C$, 1013.25 mb (760 mm Hg) and 0.03 percent CO_2. λ is the "effective" wavelength in a vacuum. Equation (5.12) describes the group refractive index with an accuracy of ± 0.1 ppm or better.

Equation (5.12) was derived from experiments at wavelengths between 436 and 644 nm. It was adopted in 1963 by the XIIIth General Assembly of the International Union of Geodesy and Geophysics (IUGG) in Berkley, together with another formula by Edlen (1953). Both adopted formulae agree within 0.1 ppm for wavelengths of 560 nm as well as 900 nm.

The agreement of Eq. (5.12) with the more recent dispersion formulae by Edlen (1966) and Peck and Reeder (1972) is better than 0.1 ppm at wavelengths between 400 and 1000 nm. The above equation can therefore be used in connection with infrared distance meters. However, for high precision EDM at all visible and infrared wavelengths, it is strongly recommended that the formulae listed in Appendix A be used in place of Eq. (5.12).

Refractive index formulae refer to standard air which is defined as dry air at 15 °C [note that Eq. (5.12) is defined at zero degrees Celsius] and at a total pressure of 1013.25 mb. The molar (or volume) percentages of the major constituents are 78.09% nitrogen (N_2), 20.95% oxygen (O_2), 0.93% argon (Ar) and 0.03% carbon dioxide (CO_2) (Owens 1967). The concentrations of some constant minor constituents are (Iribarne and Cho 1980): neon (Ne) 18 ppm, helium (He) 5 ppm, methane (CH_4) 1.5 ppm, krypton (Kr) 1 ppm, hydrogen (H_2) 0.5 ppm. Typical concentrations of some variable constituents depend on pollution levels: ozone (O_3) 0.005 – 0.5 ppm, nitrogen dioxide (NO_2) 0.001 – 0.1 ppm.

It is known that the amount of carbon dioxide and methane in the air are increasing with time. The carbon dioxide content increased from 316 ppm in 1974 to 346 ppm in 1985 and might reach 600 ppm by volume in 2070. The latter content would lead to an error of 0.05 ppm in refractive index computed with Eq. (5.12). The methane content increased from 1.52 ppm in 1977 to 1.65 ppm by volume in 1985. However, values of 4 ppm have been measured close to the ground in urban areas. The normal concentration of carbon monoxide is 0.1 ppm but can reach 7.5 ppm close to the ground in urban areas (Pelli and Meier-Bukowiecki 1986).

For distances measured in enclosed rooms (factories, laboratories) the refractive index formulae specified for standard air may also not apply in all circumstances. As a worst case example, the effect of a non-standard atmosphere in the car manufacturing industry onto the refractive index was computed as 1.9 ppm (Wilkening 1986). If the composition of a particular non-standard atmosphere is known, the refractive index equations for standard atmosphere [Eq. (5.12), Appendix A] can be corrected accordingly (see Bittel 1985; Landolt-Börnstein 1962).

5.4.1 First Example

The Hewlett-Packard Distance Meter HP 3800 B has a Ga-As diode emitting at a wavelength of 910 nm. Under standard conditions of 0 °C, 1013.25 mb, dry air ($e = 0$) and 0.03% CO_2 in the air, the group refractive index reads:

$$(n_g - 1)\,10^6 = 287.604 + \frac{4.8864}{\lambda^2} + \frac{0.0680}{\lambda^4} \tag{5.13}$$

and for $\lambda = 0.910\ \mu m$,

$$(n_g - 1)\,10^6 = 293.604$$

$$n_g = 1.000\,293\,6\ .$$

5.4.2 Second Example

The distance meter section of the Topcon ET-1 electronic tacheometer employs a GaAlAs diode emitting at 820 nm.

For standard conditions (see above) the group refractive index computes as

$$(n_g - 1)\,10^6 = 287.604 + 7.2671 + 0.1504$$

$$= 295.022$$

$$n_g = 1.0002950 \; .$$

5.4.3 Third Example

Long range instruments such as AGA Geodimeter 8, Geodimeter 600, K+E Rangemaster II as well as Ranger V employ He-Ne lasers as radiation source: $\lambda = 632.8$ nm

$$(n_g - 1)\,10^6 = 287.604 + \frac{4.8864}{(0.6328)^2} + \frac{0.0680}{(0.6328)^4}$$

$$= 287.604 + 12.203 + 0.424$$

$$= 300.231$$

$$n_g = 1.0003002 \quad (\text{for } 0\,°\text{C}, \; 1013.25 \text{ mb, dry air, } 0.03\% \; CO_2) \; .$$

5.4.4 Error Analysis

Differentiating Eq. (5.12) with respect to the wavelength demonstrates that an error of 5 nm in wavelength causes an error of 0.08 ppm in the group refractive index. The carrier wavelength must therefore be known quite accurately. This is the case with stabilized gas lasers such as HeNe lasers, which are monochromatic and feature well-defined wavelengths. Short range distance meters employ typically GaAlAs diodes which may have a centre wavelength 10 to 20 nm off the nominal value (stated in the instrument's handbook) and which may exhibit a drift of the centre wavelength of 0.25 nm per degree Celsius (see Chap. 4.1.2.1 for details). Both effects limit the accuracy with which n_g can be computed for infrared distance meters.

5.5 Group Refractive Index of Light and NIR Waves at Ambient Conditions

The International Union of Geodesy and Geophysics (IUGG) resolved in 1963 at its XIIIth General Assembly in Berkley that the refractive index in electronic distance measurement be reduced to ambient conditions by

$$n_L = 1 + \frac{n_g - 1}{1 + \alpha t} \cdot \frac{p}{1013.25} - \frac{4.125 \times 10^{-8}}{1 + \alpha t}\, e \; , \tag{5.14}$$

where n_L = group refractive index valid for atmospheric conditions described by
                t, p, e.
        t   = "dry bulb" temperature of air (°C)
        p   = atmospheric pressure (mb)
        α = coefficient of expansion of air (= 0.003661 per °C)
        e   = partial water vapour pressure (mb).

The above formula is a simplified version of the interpolation equation given by
Barrell and Sears (1939). It can also be found in Kohlrausch (1955), although with
a slightly different value for the coefficient of expansion of air. Expansion of the
above equation with $273.15(= 1/\alpha)$ leads to the final form of the interpolation
formula.

$$(n_L - 1) = (n_g - 1)\,\frac{273.15\,p}{(273.15 + t)\,1013.25} - \frac{11.27 \times 10^{-6}}{(273.15 + t)}\,e \; . \tag{5.15}$$

Barrel and Sears (1939) state that their equation is valid (at the ± 0.015 ppm level)
for temperature and pressure ranges of $+10$ to $+30\,°C$ and 720 to 800 mm Hg,
respectively. No range of partial water vapour pressures is stated. However,
Edlen's (1966) statement of "not deviating too much from 10 mm" may be taken
as a guide. Edge (1960) stated a maximum error of the above equation of 0.2 ppm
for temperature and pressure ranges of $-40\,°C$ to $+50\,°C$ and 533 to 1067 mb,
respectively.

The numerous deficiencies of Eqs. (5.14) and (5.15) were evaluated by Deichl
(1984) and their systematic errors (the computed refractive index is too large) estimated as about 0.15 ppm at $t = 20\,°C$, $p = 1013.25$ mb and $e = 13.3$ mb. The same
author notes errors of at least 0.4 ppm at $+45\,°C$ and 100% relative humidity.
The humidity term is clearly a weak element in the above equations.

For high precision EDM at all visible and infrared wavelengths and whenever
the systematic errors mentioned above are not tolerable, it is strongly recommended that the formulae listed in Appendix A be used in lieu of Eqs. (5.12), (5.14)
and (5.15).

5.5.1 Error Analysis

The effects of errors in the variables p, t and e on the derived quantity n_L may
be analyzed by their partial differentials. For a temperature of 15 °C, a pressure
of 1007 mb, a partial water vapour pressure of 13 mb and a group refractive index
n_g of 1.0003045 differentiation of Eq. (5.15) yields:

$$dn_L \times 10^6 = -1.00\,dt + 0.28\,dp - 0.04\,de \; , \tag{5.16}$$

where dn_L = differential of the (group) refractive index of light
        dt   = differential of temperature t (°C)
        dp   = differential of pressure p (mb)
        de   = differential of partial water vapour pressure e (mb).

The significance of Eq. (5.16) can be summarized as follows:

1. An error in t of 1 °C affects refractive index and distance by 1 ppm.
2. An error in p of 1.0 mb affects refractive index and distance by 0.3 ppm.
3. An error in e of 1 mb affects the refractive index by 0.04 ppm and e therefore need not be known very accurately.

Temperature is critical for the determination of the refractive index. Temperature may be measured accurately at both terminals of a line and the refractive index calculated for both terminals and the mean taken. Even so, the mean value of the refractive index generally does not represent the prevailing integral value over the wave path to better than 1 ppm. The accuracy of electro-optical distance measurement is therefore normally limited to 1 ppm of the measured distance, although a better accuracy (> 0.1 ppm) can be obtained using special techniques (see Sect. 5.9.3 for more details).

5.5.2 *Omission of Humidity*

Very often, the effect of the partial water vapour pressure e is omitted in formulae provided by manufacturers. Ignoring the following term of Eq. (5.15)

$$-\frac{11.27\times10^{-6}}{(273.15+t)}\,e \qquad (5.16\,a)$$

causes the following errors of the refractive index and, subsequently, of distance.

Temperature t (°C)	Errors in refractive index (ppm)	
	h = 50%	h = 100%
0	0.1	0.2
10	0.2	0.5
20	0.4	0.9
30	0.8	1.6
40	1.4	2.8
50	2.3	4.6

The relative humidity h is related to e. Its definition will be given later.

In summertime, the refractive index could be in error by 1.3 ppm in the Sydney area (30 °C, 80% relative humidity), if the partial water vapour pressure is not taken into account.

Users of reduction formulae or diagrams designed by instrument manufacturers should be aware of this fact. It is recommended that humidity be considered for more precise work and over longer distances.

5.6 Refractive Index of Microwaves

The XIIIth General Assembly (Berkley) of the International Union of Geodesy and Geophysics (IUGG) adopted in 1963 the formula by Essen and Froome

(1951) (as amended) for the determination of the refractive index of microwaves. The equation is the result of experiments carried out at the National Physical Laboratory (U.K.) with the aid of cavity resonators.

$$(n_M - 1)\,10^6 = \frac{77.624}{(273.15 + t)}\,(p - e) + \frac{64.70}{(273.15 + t)}\left(1 + \frac{5748}{(273.15 + t)}\right) e \; , \qquad (5.17)$$

where n_M = refractive index for microwaves
  t = "dry bulb" temperature of air (°C)
  p = pressure of air (mb)
  e = partial water vapour pressure (mb).

According to Edge (1960), the accuracy of Eq. (5.17) is ± 0.1 ppm under normal conditions and better than ± 1.0 ppm under extreme conditions. The above equation is most accurate for wavelengths larger than 8 mm (frequencies smaller than 40 MHz). Deichl (1984) noted that Essen and Froome used 273.00 in the denominator of the above equation. The same author states that the effects of this and other simplifications may cause systematic (refractive index too small) errors in the refractive index of microwaves of about 0.35 ppm.

5.6.1 Error Propagation

The total differential of Eq. (5.17) allows the effects of errors in p, t and e on n_M to be assessed. The coefficients of the total differential are computed for $t = 10\,°C$, $p = 1013.25$ mb and $e = 13$ mb.

$$dn_M \times 10^6 = -1.4\,dt + 0.3\,dp + 4.6\,de \; , \qquad (5.18)$$

where dn_M = differential of refractive index of microwaves n_M
  dt = differential of temperature t (°C)
  dp = differential of pressure p (mb)
  de = differential of partial water vapour pressure e (mb).

Equation (5.18) leads to the following statements:

1. An error of 1 °C in t causes an error of 1.4 ppm in n_M and therefore in distance.
2. An error of 1 mb in the atmospheric pressure p causes an error of 0.3 ppm in n_M or distance.
3. An error of 1 mb in the partial water vapour pressure e causes an error of 4.6 ppm in n_M or distance.

From these values it becomes obvious that the critical parameter in the case of refractive index of microwaves is the partial water vapour pressure. This is the main reason why microwave measurements are likely to be of lower accuracy than the corresponding measurements using light or "near-infrared". An accuracy better than ± 3 ppm in n_M cannot easily be achieved, even if e is measured very precisely at both instrument stations, unless special techniques are employed (Sect. 5.9.3).

Unlike measurements with light wave carriers, microwave EDM requires the partial water vapour pressure e to be measured under all circumstances, if the full precision is required. The measurement of e will be discussed later.

5.7 Coefficient of Refraction

It is known from the theory of trigonometric levelling, that the refractive index of air not only affects the velocity of light but also the geometry of its path. When a wave path passes through regions of differing refractive index n, the wave path will depart slightly from a straight line. The effect will be different for light waves and microwaves because of their different refractive indices.

If R is the mean radius of curvature of the spheroid along a line and r the radius of curvature of the wave path, the coefficient of refraction is defined as:

$$k = \frac{\text{curvature of ray path}}{\text{curvature of spheroid}} = \frac{1/r}{1/R} = \frac{R}{r} \; . \tag{5.19}$$

The curvature of the wave path itself is defined as

$$\frac{1}{r} = -\sin z \, \frac{1}{n} \left(\frac{dn}{dh} \right) , \tag{5.20}$$

where (dn/dh) is the vertical gradient of the refractive index of air, and z is the angle between the direction of the gradient of the refractive index and the tangent to the wave path. The change of n with height h is caused by the vertical density gradient of air.

The following mean values are usually adopted for EDM lines high above the ground:

$$\text{Light waves:} \quad k_L = 0.13 \quad r_L = 8R \; . \tag{5.21}$$

$$\text{Micro waves:} \quad k_M = 0.25 \quad r_M = 4R \; . \tag{5.22}$$

These values may vary considerably under differing atmospheric conditions, for example during the night, at sunrise or at sunset. The variation of k increases the smaller the ground clearance of the lines becomes. Some examples may illustrate the range of values. Höpcke (1964) has reported the following range of refraction coefficients for the layer of air between 40 m and 100 m above ground.

$$k_M = +0.25 \text{ to } +0.50 \; , \quad k_L = +0.13 \text{ to } +0.30 \; .$$

Larger variations were reported for the layer between 2 m and 13 m:

$$k_M = -1.2 \text{ (early morning) to } +3.5 \text{ (early evening) } ;$$

$$k_L = -0.5 \text{ (noon) to } +2.5 \text{ (early morning) } .$$

Brunner (1975) has reported refraction coefficients k_L in the range of -2.3 to $+1.5$ for light paths 1.5 m over ground on clear days. The coefficients of refraction for rays in 3 m height above ground were found to be between -1.0 and $+1.0$ on clear days. Rüeger and Brunner (1981) found coefficients of refraction be-

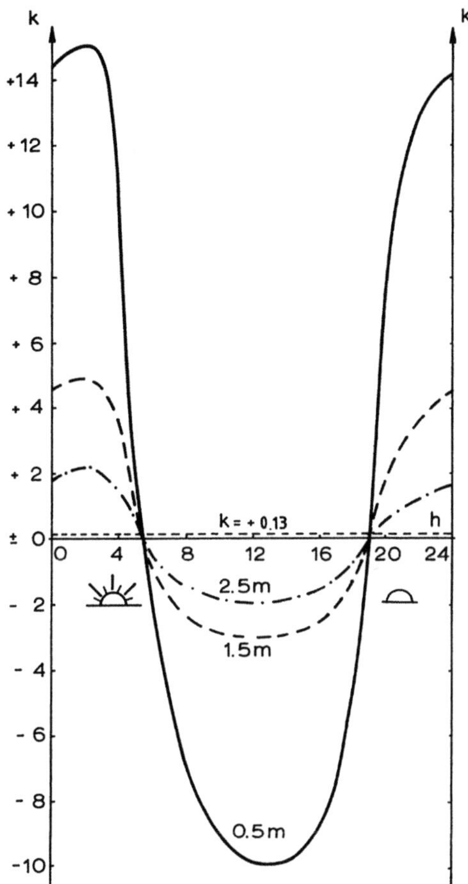

tween -2.3 and $+1.6$ on clear days during an EDM height traversing experiment in Australia featuring line lengths between 190 m and 490 m.

Depending on latitude, season, ground cover and wind speed, the coefficient of refraction will exhibit a daily cycle on clear days (and nights). Figure 5.3 gives an indication of the magnitude of the cycle for different heights of lines above ground.

Different coefficients of refraction are experienced over ice. Lindner and Ritter (1985) report changes of the coefficient of refraction for lines 1.5 to 1.8 m over ice in Antarctica from -0.2 to $+3.0$ to $+5.0$ on clear days between 13 h and 18 h and constant coefficients of $+0.5$ on overcast days during the same time period. Similar values of -0.5 (at 14 h) to $+2.0$ (at 18 h) are given by Ritter (1987) for clear and calm days with temperatures of about $-5\,°C$ and instrument heights of 1.5 m. The same author lists the maximum value measured as $k = 14.9$. A good summary of measured coefficients of refraction over ice may be found in Angus-Leppan (1974).

The mean values given in Eqs. (5.21) and (5.22) should therefore be used with great caution for the reduction of electronic distance measurements. In long range

EDM, the coefficient of refraction is required for the first arc-to-chord correction as well as the second velocity correction. Where highest precision is required (such as in geodetic control networks), simultaneous reciprocal zenith angle measurements should be carried out for the sole purpose of determination of k. In short range EDM, the coefficient of refraction is often required for the computation of horizontal distances and height differences. Considering Fig. 5.3, a value of $k = 0.0$ for the coefficient of refraction is as good an assumption as $k = 0.13$ in these cases. Where required, the uncertainty in the value of k can be greatly reduced by direct determination of the coefficient of refraction by simultaneous reciprocal zenith angle measurements.

5.8 Measurement of Atmospheric Parameters

The extent and precision required in the measurement of atmospheric parameters depends largely on the EDM instrument type, its internal accuracy, the distance accuracy required and the length to be measured. Temperature measurements for short range instruments with an internal accuracy of ± 5 mm require a simple mercury thermometer for distances of about 1000 m; an aneroid will suffice for the measurement of pressure. For shorter distances of a few 100 m length, a good estimate of temperature (based on experience or weather information broadcast by radio/tv stations) and standard pressure at field elevation (refer to Appendix C – do not employ pressures reported in weather bulletins/broadcasts as they are reduced to sea level!) may be sufficient. Actual atmospheric pressures are unlikely to deviate by more than 30 mb from the standard pressure at the field elevation (see Appendix C) and, thus, unlikely to cause errors in excess of 1 mm over 100 m. An error of 10 °C in temperature, for example, would cause an error in a 100 m distance of only 1 mm, which can be neglected in relation to the instrument's internal accuracy of ± 5 mm. Thermometers and barometers used for these purposes would not required precise calibration. However, regular comparison of the instruments with calibrated instruments or standards should be made.

Medium and long distance measurements with microwave instruments, long range measurements with light or infrared instruments and precision short range measurements require precise determination of atmospheric temperature, pressure and humidity at least at the terminals of the EDM line. Details of accurate meteorological measurements and some related computations follow in the next sections. More refined and alternative methods of refractive index determination are discussed in Section 5.9 and 6.5.

5.8.1 Measurement of Atmospheric Pressure

It has been demonstrated in Sections 5.5 and 5.6 that an error of 1 mb affects distances by only 0.3 ppm. Depending on the length of the EDM lines and the resolution of the distance meter used, good quality aneroid barometers are generally sufficient. Such instruments are:

1. Pocket barometers with accuracies of about ± 1.0 mb
   - with mechanical pick-up and display
   - with electronic pick-up and display (least count: 1.0 mb)
2. Handheld (precision) barometers with accuracies of about ± 0.5 mb
   - with mechanical pick-up and display
   - with electronic pick-up and display (least count: 0.1 mb).

Aneroid barometers may require corrections for temperature, scale errors and zero errors. They should be checked against a reference mercury barometer before and after field trips. Barometers can be submitted to National Standard Laboratories, local Bureaus of Meteorology and registered Testing Laboratories for calibration. A standard test would include the determination of the hysteresis effect, corrections at a number of points over the full range of the barometer and possibly, an evaluation of the temperature compensation (National Measurement Laboratory 1975).

Aneroids should be handled with great care and should always be set up level and in the shade. They should be allowed time to settle and be tapped gently before reading.

5.8.2 Measurement of Atmospheric Temperature

In connection with precision or long range EDM, the temperature should be measured with an instrument which provides an accuracy of $\pm 0.2\,°$C. Such instruments are:

1. mercury-in-glass thermometers
2. platinum resistance thermometers
3. electronic thermistor thermometers.

The advantage of equipment (2) and (3) is that the actual measuring device (probe, sensor) is separated from the display unit, and from the thermal effects of the observer. The probe can be put on a mast, enabling a continuous, electronic recording of data. Temperatures should be measured in the shade, exposed to wind and well above the ground. The distance to any object, including the ground or people, should be more than 1.5 m. At low temperatures, mercury thermometers should be shielded from the heat radiation of the human body by a transparent plastic plate. During distance measurement, temperature readings should be taken at a rate of at least one reading per minute for a minimum period of 5 min, to account for the fluctuations of meteorological parameters. Electronic devices have a much faster response to temperature changes than mercury thermometers. More observations should therefore be taken to provide a reliable mean value if no automatic recording device is used. Thermometers should be calibrated from time to time, at least for the zero error. Graduation errors need to be checked at least once. Calibration services are usually provided by the laboratories listed in the previous section.

5.8.3 Measurement of Atmospheric Humidity

Three types of instruments and procedures for the measurement of atmospheric humidity are discussed below, although only the latter two are currently employed in EDM work.

1. Hygrometer

The hair hygrometer was originally based on the change in length of human hair as a function of relative humidity. The present day hygrometers may also use other materials for the same purpose. Hair hygrometers provide the relative humidity directly in percent. The accuracy is about 3% which is not always accurate enough for EDM purposes. Hygrometers have a very slow response to humidity changes and thus provide a mean value over a period of time.

2. Aspiration Psychrometer

Designed by R. Assmann, the aspiration psychrometer consists basically of two thermometers, one of which is covered with a cotton wick which must be kept wet with distilled water. The "dry bulb" and "wet bulb" thermometers are enclosed in tubes to protect them from radiation. An air current of at least 3 m/s is used to aspirate the thermometer bulbs. The psychrometer provides readings of the dry bulb and the wet bulb temperature from which the partial water vapour pressure is derived.

Two types of *aspiration psychrometers* are commercially available:

a) Mercury-in-glass psychrometer. Originally designed by Assmann, these comprise a dry bulb and wet bulb thermometer and a small spring driven fan and are accurate to 0.2 °C.

b) Thermistor psychrometer. The mercury thermometers are replaced by thermistors, which allow temperature readings at a distance from the probes. The fan is operated electrically. An accuracy of about 0.2 °C can be achieved.

In addition to the operating instructions for simple thermometers, given in Section 5.8.2, further care is required in using (mercury) psychrometers. After wetting the wet bulb wick, repeated readings must be taken to confirm that the wet bulb has reached its equilibrium. An increase in wet bulb temperatures may indicate that the wick has run dry. The wick must be clean and in contact with the thermometer. No water should bridge the gap between wick and shield. The observer should avoid breathing on the air intake, especially at low temperatures, and should stand clear of the instrument. More advice may be found in Bomford (1975, page 54). Some of the problems are eliminated if thermistor instruments are used.

A psychrometer should be calibrated periodically and needs some maintenance with regard to the polished shields and thermometers and the replacement of the wick. The internal air velocity also must be checked and should be at least 3 m/s to ensure proper functioning.

3. Humidity Sensors

With the emergence of hand-held solid state thin film humidity sensors, an alternative to psychrometer measurements has become available in EDM in cases where humidity (and partial water vapour pressure) is required, typically in precision EDM and long range EDM.

The humidity is sensed by a hygroscopic dielectric between two electrodes. By absorbing or desorbing water, the element puts itself in equilibrium with the ambient conditions and alters the capacitance of the device at the same time. The capacitance change is used to derive the humidity reading. The thin film reacts very fast, e.g. $5-10\,s$ for 90% response. As the temperature is required to correct the humidity reading, humidity sensors used in surveying are typically combined with (platinum resistance) temperature sensors.

The accuracy of humidity sensors is typically specified as $\pm 2\%$ relative humidity (R.H.) between 0 and 80% R.H. and $\pm 3\%$ between 80 and 100% R.H. The temperature sensitivity of the humidity reading is typically 0.05% R.H./°C. Frequent calibration is essential to maintain these accuracies.

5.8.4 Computation of Partial Water Vapour Pressure from Psychrometer Measurements

The partial water vapour pressure e, which is required for the determination of the refractive index (see Sects. 5.6 and 5.5), can be derived from dry bulb and wet bulb readings of an aspiration psychrometer. The psychrometric equation by A. Sprung is (Rinner and Benz 1966, p. 228):

$$e = E'_W - 0.000662\,p\,(t-t') \ , \tag{5.23}$$

where E'_W = saturation water vapour pressure for temperature t' (wet bulb temperature) in mb over water
    t = dry bulb temperature (°C)
    t' = wet bulb temperature (°C)
    p = atmospheric pressure (mb)
    e = partial water vapour pressure (mb).

This formula is accurate to $\pm 1\%$ and valid for $t > 0\,°C$ and $t' > 0\,°C$. In case of a frozen wet bulb wick, the following equation must be used:

$$e = E'_{ICE} - 0.000583\,p\,(t-t') \ , \tag{5.24}$$

where E'_{ICE} is the saturation water vapour pressure over ice at the temperature t' (wet bulb temperature) in millibar.

The saturation water vapour pressure over water and over ice can be obtained from the table in Appendix B. Alternatively the saturation water vapour pressure may be calculated according to the equation by Magnus-Tetens (Rinner and Benz 1966, p. 228):

$$E'_W \ = 10^{[(7.5t'/(237.3+t'))+0.7858]} \tag{5.25}$$

$$E'_{ICE} = 10^{[(9.5t'/(265.5+t'))+0.7858]} \tag{5.26}$$

or its revised form by Murray (1967):

$$E'_W = 6.1078 \exp\left(\frac{17.269\,t'}{237.30 + t'}\right) \qquad\qquad (5.25\,\text{a})$$

$$E'_{ICE} = 6.1078 \exp\left(\frac{21.875\,t'}{265.50 + t'}\right), \qquad\qquad (5.26\,\text{a})$$

where E'_W and E'_{ICE} are the saturation water vapour pressures (in mb) over water and ice, and t' is the wet bulb temperature (°C). Equations (5.25), (5.25 a) are valid for the temperature range -70 °C to $+50$ °C. Equations (5.25) and (5.25 a) are accurate to 0.1 mb between 0 °C and 40 °C.

Improved formulae for the saturation water vapour pressures have been published by Buck (1981). These formulae are more accurate than those by Murray [Eqs. (5.25 a) and (5.26 a)], Magnus-Tetens [Eqs. (5.25) and (5.26)] and Goff-Gratch (on which the tables in Appendix B are based).

$$E'_W = [1.0007 + (3.46 \times 10^{-6}\,p)] \times 6.1121 \exp\left[\frac{17.502\,t'}{240.97 + t'}\right] \qquad (5.27)$$

$$E'_{ICE} = [1.0003 + (4.18 \times 10^{-6}\,p)] \times 6.1115 \exp\left[\frac{22.452\,t'}{272.55 + t'}\right], \qquad (5.28)$$

where p is the ambient atmospheric pressure (in mb). The maximum errors of Eq. (5.27) occur at -20 °C and $+50$ °C and amount to 0.20%. The maximum errors of Eq. (5.28) do not exceed 0.02% at -50 °C and 0 °C (Buck 1981).

Example: $\quad t = 21.3$ °C $\quad t' = 17.9$ °C $\quad p = 1010.6$ mb

$$E'_W = 10^{[(7.5 \times 17.9/(237.3 + 17.9)) + 0.7858]} = 20.50 \text{ mb} \qquad \text{[from Eq. (5.25)]}$$

$$e = 20.50 - 0.000\,662\,(1010.6)\,(21.3 - 17.9)$$

$$= 20.50 - 2.28 = 18.22 \text{ mb}.$$

Effect of the Precision of t, t' on n_M. If the refractive index n_M of microwaves is computed from a value of the partial water vapour pressure obtained by psychrometer observations of t and t', it can be shown that an error of 0.22 °C in the dry bulb temperature t or an error of 0.14 °C in the wet bulb temperature t' causes an error of 1 ppm in n_M. Psychrometer observations must therefore be carried out with the highest possible care. On long distances, the accuracy of a distance is limited mainly by the accuracy of the psychrometric data.

Effect of the Precision of t, t' on n_L. In the same manner, it can be shown that an error of 1.0 °C in the dry bulb and an error of 12.5 °C in the wet bulb reading each cause an error of 1 ppm in the refractive index of light and NIR waves. The wet bulb reading is significantly less critical than in the case of n_M.

5.8.5 Computation of Partial Water Vapour Pressure from Relative Humidity

The partial water vapour pressure e may also be derived from known values of relative humidity, h. Relative humidity can be measured with humidity probes and hygrometers or may be available from daily weather information. As will be shown below, partial water vapour pressures computed from relative humidity values may not be accurate enough for all EDM applications.

The partial water vapour pressure is computed from

$$e = \frac{Eh}{100} , \tag{5.29}$$

where   e = partial water vapour pressure (mb)
      E = saturation water vapour pressure (mb) at temperature t (dry bulb temperature)
      h = relative humidity in percent.

The values for E may be taken from the table in Appendix B or computed from Eqs. (5.25) to (5.28). Replace E' by E and t' by t.

Effect of the Precision of h (and t) on n_L and n_M. It can be shown that a precision of $\pm 3\%$ and $\pm 0.2\,°C$ of measured relative humidity h and measured temperature, respectively, lead to the following precisions of the humidity term of the refractive index formulae at 50 and 100% relative humidity.

Temperature	Uncertainty at 50% R.H. of		Uncertainty at 100% R.H. of	
	n_L	n_M	n_L	n_M
(°C)	(ppm)	(ppm)	(ppm)	(ppm)
+10	±0.02	±1.7	±0.02	±1.9
+20	±0.03	±3.3	±0.03	±3.5
+30	±0.05	±5.9	±0.05	±6.2
+40	±0.09	±10.4	±0.09	±10.8

The above table indicates that humidity probes permit the measurement of humidity with sufficient accuracy for all but the most accurate distance measurements with light and NIR waves. In the case of microwave distance measurements, the use of measured relative humidity causes a significant loss of accuracy in most cases. Psychrometer measurements are therefore the preferred humidity measurement option for microwave distance measurement.

Computation of e from Weather Reports in Newspapers. Daily average values of the partial water vapour pressure can be recovered from weather reports of the past day in newspapers as long as the temperatures and humidities published refer to the same time and location.

For example, the following data were published by the Sydney Morning Herald on 30 March 1988 for the 29 March 1988 under the heading "Yesterdays Weather": 9 am: 19.4 °C, 74% R.H.; 3 pm: 23.8 °C, 63% R.H.; 8 pm: 22.3 °C, 61% R.H. Using Eq. (5.29), the following values of e are obtained: 9 am 16.7 mb, 3 pm 18.6 mb, 8 pm 16.4 mb.

Such information may be used to correct electro-optical distance measurements carried out on the specified day in the local area covered by the weather report in cases where no humidity was measured in the field and where complete omission of the humidity term in the refractive index equation is not acceptable.

Mean Values of Partial Water Vapour Pressure. In some cases it is warranted to use yearly mean values for the partial water vapour pressure. This applies to cases where the field measurement of humidity is not warranted but where the total omission of the humidity correction [see Eq. (5.16a) in Sect. 5.5.2] would cause unacceptable systematic errors. Annual mean values of partial water vapour pressure may be derived from data held by the local Bureau of Meteorology or from retrospective weather reports in newspapers (see above).

As a first example, the 10-year mean (1962−1971) of partial water vapour pressure at Richmond Aerodrome (near Sydney, Australia) at 15.00 hours Australian Eastern Standard Time may be mentioned. From 3358 days of measurements, a mean value of 12.8 mb is obtained. The measured partial water vapour pressures vary betwen 3.3 and 31.0 mb and have a standard deviation of ±5.8 mb about the above mean.

As a second example, some values given by Kneissl (1956) for Western Europe are quoted. Averaging over 30 meteorological stations, mean values of 6.1 and 14.8 mb are given for January and July, respectively.

5.9 Determination of the Refractive Index

In principle, electronic distance measurements should be corrected for the integral refractive index over the entire wave path. (In later sections, the correction for the refractive index will be referred to as first and second velocity correction). This could involve the measurement of temperature, pressure and humidity along the wave path to calculate the refractive index and the subsequent integration of these refractive index values along the wave path. In practice, simplified methods based on endpoint measurements are usually adopted. These methods, their limitations and some alternative methods are discussed below.

5.9.1 Normal Procedures

The representative refractive index of a line is computed from meteorological observations taken at one or both terminals of the line. For lines in excess of a few kilometres and whenever high accuracy is required, t, p and t' or h are measured at both endpoints and the refractive indices are computed separately for both terminals. The mean refractive index is then used to correct the measured distance. This procedure provides the best refractive index corrections in cases where solely meteorological measurements at the endpoints are considered.

For distances of a few hundred metres to a few kilometres, t, p and, possibly, t' or h are measured at both terminals of the line. In this case, the above procedure may be simplified by taking the mean of the endpoint meteorological measurements before calculating the refractive index once. Small errors are introduced in the process due to the non-linearity of the refractive index equations. These errors are sufficiently small to be ignored in most cases, as a numerical example may demonstrate: Considering an infrared distance meter and a height difference of 1000 m with changes of 6.5 °C and 114.4 mb in t and p between the terminals, the simplified computation incurs an error of 0.2 ppm in comparison with the rigorous method above.

For short distances (such as in EDM tacheometry), t and p (and, rarely, t' or h) are only measured (or estimated) at the instrument station. The meteorological measurements taken at the instrument station are then used to calculate the refractive index. As a result of this, the temperature and pressure variations over the lines are ignored. Because of the change of pressure with height, systematic errors are incurred which amount to 1.6 ppm per 100 m height difference between instrument and reflector.

Numerical examples of such computations are given later in Chapter 6, on velocity corrections.

5.9.2 *Limitations of Normal Procedures*

Refractive indices evaluated from temperature, pressure and humidity readings taken at endpoints do represent the integral value over the wave path reasonably well as long as the EDM wave path is parallel to the ground. This case occurs most likely on close range. For lines of more than a few hundred metres length, the average ground clearance is likely to exceed the instrument and reflector heights significantly. In these cases, the refractive indices computed from meteorological observations at the terminals will no longer closely approach the integral value over the wave path because the temperature and humidity measurements at the endpoints are strongly influenced by the closeness of the ground (ground proximity effects).

Figure 5.4 depicts some typical temperature variations with height on clear (sunny) days. It becomes evident that temperatures measured at instrument height (e.g. 1.5 m above ground) are not representative for temperatures at a height of, say, 100 m above ground. Temperatures measured during day light hours are too high and those measured at night too low when compared with temperatures at 100 m.

To illustrate the ground proximity effects on endpoint measurements of atmospheric parameters (and EDM corrections derived from such) further, it is assumed that the temperature distributions of Fig. 5.4 apply and that distances with an average ground clearance of 100 m were measured at 04.00 h and at 14.00 h. It can be shown that the first velocity corrections make the distances at 04.00 h and 14.00 h 4 ppm too short and 2 ppm too long, respectively. It should

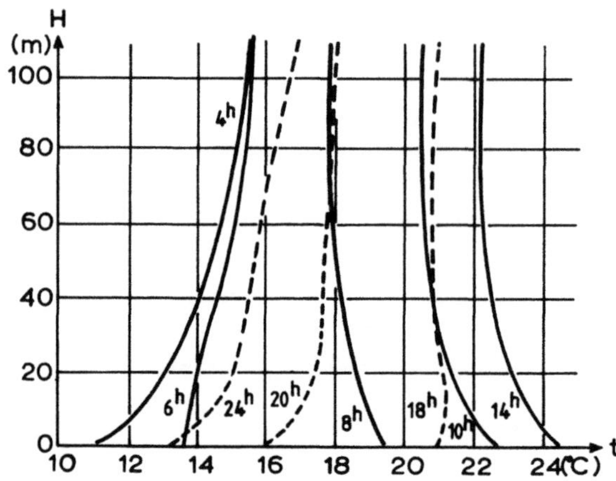

Fig. 5.4. Diurnal temperature variations in the lowest 100 m of the atmosphere on clear (sunny) summer days in Southeastern England. (After Jones 1971; Best et al. 1952)

be noted that these errors apply to clear nights and sunny days as defined above. The magnitude of these effects will be considerably reduced for similar lines on heavily overcast and windy days.

The ground proximity effects on endpoint measurements of temperature and humidity also explain why it is a common experience to find light wave instruments giving longer distances than microwave instruments during simultaneous measurements in day time. The difference amounts usually to 2 – 5 ppm. It could be easily explained by the ground level partial water vapour pressure exceeding the integral value by 0.7 mb. The diagrams of e versus height above ground given by Jones (1971) and Best et al. (1952) support such an assumption. Based on an analysis of a large sample of Austrian radio sonde and ground meteorological data, Brettenbauer (1975) concludes that temperatures and partial water vapour pressures measured at terminals of lines at 1500 m above sea level are typically too high by 3.0 °C and 1.3 mb, respectively. The temperature error leads to light wave and microwave distances being 3 ppm and 4.2 ppm too long, respectively. The humidity error leads to microwave distances being 6 ppm too short. In consequence, the light wave minus microwave difference becomes +4.8 ppm. A similar difference was reported for long distance measurements in Switzerland (Fischer 1971). The temperature and humidity structure in the lower atmosphere has been reviewed by Webb (1984).

It becomes evident from the above discussions that refractive indices (and first velocity corrections) computed from measurements of atmospheric parameters at the terminals of lines can severely bias the corrected distances. The ground proximity effects are worst during clear sunny days (unstable stratification with temperature decreasing with height and upward heat flux) and during clear nights (stable stratification, inversion layer, with temperature increasing with height and downward heat flux).

Reduced ground proximity effects are experienced during neutral stratification with no heat transfer between ground and air (zero heat flux) and zero temperature gradient. Neutral and near-neutral conditions are usually restricted to the transition periods between stable and unstable conditions, typcially 1.5 h after sunrise and 1.5 h before sunset or whenever the heat flux does not exceed 10 W/m^2 (Rüeger 1984). Near-neutral conditions also prevail in heavily overcast weather with moderate to strong wind. Generally speaking, cloud cover over EDM lines and/or moderate to strong winds will reduce the ground proximity effects on meteorological endpoint measurements. With respect to microwave distance measurements, Brunner and Groenhout (1979) reported on experiments where measurements taken during periods of rainfall exhibited drastically reduced ground proximity effects.

As it is not practical to restrict EDM observations to heavily overcast conditions with strong winds or to times around sunrise or sunset, methods have been developed which overcome some of the limitations of the endpoint measurements of atmospheric parameters. Some of these techniques are briefly discussed in the next section.

5.9.3 Special Procedures

The special procedures developed to overcome the limitations of the endpoint measurements of atmospheric parameters can involve instrumental solutions, modelling of the atmosphere (rectification models) and modified field and computing procedures (operational models).

5.9.3.1 Atmospheric Measurements Along the Wave Path

With the aid of suitably equipped planes, motor gliders or helicopters, continuous recordings of temperature and humidity along the wave path can be obtained. Pressures need to be extrapolated from measurements at the terminals of the EDM lines. This technique is used by the U.S. Geological Survey on a routine basis for distance measurements carried out in connection with the monitoring of crustal movements in California. The accuracy of the Geodolite measurements, as corrected with the aid of airborne meteorological observations, is generally given as $\pm(3\text{ mm} + 0.2\text{ ppm})$ (Savage and Prescott 1973; Savage et al. 1979, 1987).

5.9.3.2 Multiple Wavelength EDM Instruments

Simultaneous distance measurements with two or three different carrier waves take advantage of the dispersion effect and permit the determination of the differential refractive indices between the different carrier waves. This method leads to a greatly reduced dependency on accurate determinations of atmospheric parameters. In the case of "two-colour" instruments, the distance is derived as follows (Rüeger 1980b):

$$D = d_{red} - A(d_{blue} - d_{red}) \, , \tag{5.30}$$

where d_{red} = distance measurement with red laser (HeNe)

        d_{blue} = distance measurement with blue laser (HeCd)

           A = coefficient, dependent on wavelength, pressure and water vapour pressure

            = approximately 21 for red and blue light.

Being the largest residual effect of atmospheric conditions on distance, an error of 1 mb in water vapour pressure leads to 1 part in 10^7 error in distance. To obtain the water vapour pressure to 1 mb, one would need to know the relative humidity to 5%, which seems feasible.

Although the two-colour method eliminates a considerable part of the uncertainty of the refractive index, the distance difference has to be measured with at least 20 times higher resolution than required in the final result due to the factor A in Eq. (5.30).

One two-colour instrument (Terrameter LDM-2) has been produced in small commercial quantities (Huggett 1981). The accuracy of the instrument is quoted by Langbein et al. (1987) as $\pm(0.3 \text{ mm} + 0.12 \text{ ppm})$ and by Gervaise (1984) as ± 0.08 ppm in a network with distances between 3.5 and 13.6 km. Other development work on two-colour instruments were reported by Bradsell (1976); Querzola (1979); Kasser (1982); Levine (1984) and Meier and Loser (1988). Accuracies of 0.15 ppm (and better) were obtained using the prototype of a "three-colour" [HeNe (red), HeCd (blue) and microwave] instrument (Huggett and Slater 1975).

5.9.3.3 Atmospheric Rectification Models

It has been mentioned in Section 5.9.2 that the measurements of atmospheric parameters at the endpoints of EDM lines do often not approximate the integral values of the wave path. Brunner (1984) notes that the use of the endpoint measurements "limits the accuracy of EDM to about 2 to 4 ppm, even under favourable conditions". The endpoint measurements of temperature, pressure and humidity are representative for the atmospheric surface layer close to the ground. To make these measurements representative for a wave path high above the ground, they need to be transformed to the boundary between the surface layer and the adjoining adiabatic (convection) layer.

Brunner (1984) summarizes the large number of atmospheric rectification models developed for EDM applications. Models for short range applications were developed by Angus-Leppan and Brunner (1980). They are restricted to lines of maximum lengths of 3 km and a ground clearance of 100 m and aim at an accuracy of ± 0.5 ppm for the reduced distances. Different equations are proposed for unstable (sunny days) and stable (nights) conditions. Three additional parameters must be evaluated in the field on the basis of diagrams. Another two must be measured or estimated: wind speed and the elevation angle of the sun. A rough profile of the line must also be known. The fictitious examples given by the two authors indicate clearly that the additional corrections on clear days are negative (e.g. -2 ppm) and on clear nights positive (e.g. $+4$ ppm). They confirm the evaluations in Section 5.9.2. A partial field test of the short range model was reported by Rüeger (1984).

For long range lightwave/infrared and microwave measurements, Brunner and Fraser (1978) developed the atmospheric Turbulent Tranfer Model (TTM). The method is restricted to periods of free convection during which a turbulent regime applies, namely to clear sunny days between about 10.00 and 15.00 h. The TTM requires the determination of some additional parameters, namely net radiation, heat flux into the ground, evaporation and wind speed. The former three can be

70

estimated from empirical formulae. Successful applications of the TTM to microwave and light wave distance observations were reported by Brunner and Fraser (1977) and Fraser (1984).

5.9.3.4 Refractometers

Refractometers are devices which determine the ambient refractive index directly without the requirement for the measurement of atmospheric parameters. In most cases, refractometers are designed as dual-beam laser interferometers, with the first beam travelling in vacuum and the second beam in ambient air. During the evacuation of the chamber of the second beam, the change in the optical length can be measured by interferometric techniques. Wilkening (1986) claims uncertainties of 0.2 ppm for such a device.

Refractometers cannot overcome the limitations of atmospheric measurements at endpoints of EDM lines, as their measurements also reflect the conditions of the atmospheric surface layer rather than integral values along the wave path. However, they are ideally suited for indoor precision EDM and interferometry in general and for measurements in an industrial or laboratory environment in particular, where temperature, pressure and humidity are usually uniform and the composition of the atmosphere is often non-standard. The refractive index equations given in Sections 5.4 and 5.5 are only valid for air of standard composition. For example, errors of 1.9 ppm were reported when using the standard refractive index formulae for the correction of distances measured in car manufacturing halls featuring a non-standard composition of air (Wilkening 1986).

Historically it is of interest to note that the refractometer principle was used by the designers of the Mekometer ME 3000 for the direct application of the refractive index correction within the instrument (Froome 1971). Although the internal microwave refractometer (standard cavity) took care of most of the first velocity correction, additional corrections were required, for example to account for the conditions at the reflector stations (Meier and Rüeger 1984).

5.9.3.5 Ratio Method

The length ratio method is an operational model (Brunner 1984) and makes use of the fact that all distance measurements made from one station to a number of reflector stations in a short time interval exhibit proportional atmospheric (refractive index) effects. The principles and applications of line ratios were first discussed by Baarda (1962, 1967); Allan (1967); Vincenty (1969) and Robertson (1971). Angus-Leppan (1979) gives a summary of early applications of the method. Originally, the ratio of line pairs was formed and introduced as observations into least-squares adjustments. It is, however, of advantage to employ the technique in the form of a "local scale parameter model" (Vincenty 1979), where all distances (or groups of distances) measured on a station are assigned a common but unknown scale parameter. When compared with the pure ratio method of line pairs, a much improved degree of freedom results for the network adjustment.

In principle, no measurements of atmospheric parameters are required when using this operational model, nor is there any need for the scale calibration of the distance meter. This greatly simplifies the field operations of EDM. The model has, however, two disadvantages. Firstly, most lines in a geodetic network have to be measured both ways in order to maintain a suitable degree of freedom of the adjustment (Brunner 1984). This increases the fieldwork considerably. Secondly, the absolute scale of the distances is lost as no first velocity corrections are applied. To permit network solutions, the scale must be defined by fixing the coordinates of two stations, for example. The resulting coordinates of the unknown stations are biased by the selection of the fixed points. Depending on the functions of the network, this may not be a serious disadvantage.

The ratio method works best for EDM lines of similar length and similar profiles. Slater (1979) quotes precisions of ± 0.2 ppm and ± 0.5 ppm of length ratios of lines of similar and dissimilar profiles, respectively. Typically, precisions of ± 0.5 ppm are reported for lines of a few kilometres length (Rüeger 1984). On shorter lines, the effects of the (phase) resolution of the distance meter may limit the gains of the ratio method to a certain extent. Because of the pressure (and temperature) gradient with height, the measured distances will exhibit different scales whenever the reflector stations contributing to a unknown scale parameter are not at the same height above sea level (Jäger 1985). In addition to this correction for the height difference, distances must be reduced to a common horizon, if a two-dimensional network adjustment program is used. Successful applications of the method in engineering networks were reported by Jäger (1985) and Gründig and Teskey (1984). An improvement of about 30% and 70% in the accuracy of point positions and in the external reliability, respectively, was achieved in comparison with the use of distances reduced by standard methods.

The local scale parameter model is extremely powerful in terms of achievable accuracies and is extremely user-friendly as far as the meteorological field operations are concerned. It is very likely the cheapest and easiest special procedure which may be used to overcome the limitations of endpoint measurements of atmospheric parameters. This operational model is particularly suited for applications where a network has to be measured repeatedly, such as in monitoring networks. The loss of the absolute scale is not critical in these cases.

6 Velocity Corrections to Measured Distances

The basic formula for the computation of d was introduced in Sections 2.1 and 3.1.1. Equation (3.1) may be written, after substitution of Eq. (2.4) for c, to describe the distance value d′ actually displayed on a distance meter.

$$d' = \frac{c_0}{n_{REF}} \frac{\Delta t'}{2} , \tag{6.1}$$

where d' = the distance displayed on the EDM instrument,
            c_0 = the velocity of light in a vacuum,
            $\Delta t'$ = the measured "flight" time of the signal to the reflector and back,
           n_{REF} = the reference refractive index of the instrument.

6.1 Reference Refractive Index

The reference refractive index is instrument-specific. It defines that refractive index for which the distance meter provides directly a correct readout. No refractive index (or first velocity) corrections are therefore required whenever the integral refractive index along the wave path equals the reference refractive index.

The instrument constant n_{REF} is defined by the following equations:

$$n_{REF} = \frac{c_0}{\lambda_{MOD} f_{MOD}} \tag{6.2}$$

$$= \frac{c_0}{2 U f_{MOD}} , \tag{6.3}$$

where λ_{MOD} = the constant modulation wavelength of the fine measurement for which the instrument is designed,
          f_{MOD} = the constant modulation frequency of the fine measurement,
          U = the exact half of λ_{MOD}, called the unit length of the instrument.

It may be seen from Appendix E that 10 m is presently the most common unit length.

The reference refractive index n_{REF} of an instrument is fixed by the manufacturer by adopting a suitable unit length and by adjusting the main oscillator to such a modulation frequency f_{MOD} that the then fixed value n_{REF} corresponds more or less to an average refractive index encountered under field conditions. Reference refractive indices for some electro-optical EDM instruments are given in Appendix D.

If such details as λ_{MOD} and f_{MOD} are not available for a specific instrument, a reference pressure p_{REF} and a reference temperature t_{REF} may be available instead. Substitution of the reference meteorological conditions into Eqs. (5.15) or (5.17) provides the reference refractive index in these cases.

Example: For the Kern DM501, the manufacturer specifies meteorological reference data of $+12\,°C$ and 1013 mb as well as a carrier wavelength of 900 nm.

Application of Eq. (5.13) leads to a group refractive index n_g of 1.00029374. The reference refractive index follows then from the substitution of n_g as well as $t = 12\,°C$ and $p = 1013$ mb into Eq. (5.15) and assuming a partial water vapour pressure of $e = 0.0$ mb:

$$n_{REF} = 1.00028238 \; .$$

Use of the above Eq. (6.3) in connection with

$$c_0 = 299792458 \text{ m/s}$$

$$f_{MOD} = 14985400 \text{ Hz}$$

$$U = 10 \text{ m exactly}$$

provides the following value for the reference refractive index:

$$n_{REF} = 1.00028180 \; .$$

The differences between the two values are insignificant and due to round off errors in the specified reference atmospheric parameters.

6.2 First Velocity Correction

The length of the wave path d is

$$d = \frac{c_0}{n} \frac{\Delta t'}{2} \; , \tag{6.4}$$

where n is the actual refractive index which ideally is the mean value over the distance.

In order to obtain the (true) wave path length d from the read-out distance d', the term $c_0 \Delta t'/2 = n_{REF} d'$ can be derived from Eq. (6.1) and substituted in Eq. (6.4). This provides the following rigorous correction equation:

$$d = \left(\frac{n_{REF}}{n} \right) d' \; . \tag{6.5}$$

However, it is usually preferred to work with a differential equation. The first velocity correction K' is now derived by subtracting d' [Eq. (6.1)] from d [Eq. (6.4)]:

$$K' = d - d' = c_0 \frac{\Delta t'}{2} \left(\frac{1}{n} - \frac{1}{n_{REF}} \right)$$

$$= c_0 \frac{\Delta t'}{2} \left(\frac{n_{REF} - n}{n \times n_{REF}} \right)$$

$$= \frac{c_0 \Delta t'}{2 n_{REF}} \left(\frac{n_{REF} - n}{n} \right) , \tag{6.6}$$

and, after substitution of Eq. (6.1):

$$K' = \left(\frac{n_{REF} - n}{n} \right) d' . \tag{6.7}$$

The above equation is rigorous. For simplicity, the denominator in Eq. (6.6) is usually set equal to 1. It can be shown that this simplification does not introduce errors in excess of 0.02 ppm. The first velocity correction K' thus may be written with sufficient accuracy as

$$K' = d' (n_{REF} - n) \tag{6.8}$$

and the corrected distance d as

$$d = d' + d' (n_{REF} - n) \tag{6.9}$$

$$= d' + K' . \tag{6.10}$$

First velocity corrections for electro-optical short range distance meters are generally given in a form like

$$K' = \left[C - \frac{Dp}{(273.15 + t)} + \frac{11.27 e}{(273.15 + t)} \right] 10^{-6} d' , \tag{6.11}$$

where the coefficients C and D can be computed from

$$C = (n_{REF} - 1) 10^6 = N_{REF} \tag{6.12}$$

$$D = (n_g - 1) 10^6 \frac{273.15}{1013.25} = \frac{273.15}{1013.25} N_g , \tag{6.13}$$

where the reference refractivity and the group refractivity are denoted by N_{REF} and N_g, respectively. A list of C and D values is given in Appendix D for a number of instruments. Eqs. (6.11) and (6.13) are based on Eqs. (5.14) and (5.15). The overall accuracy of these equations has been discussed in Section 5.5 and is unlikely to be better than 0.5 ppm. This is, however, entirely sufficient for most EDM applications.

For high precision EDM at all visible and infrared carrier wavelengths and whenever the systematic errors mentioned above are not tolerable, it is strongly recommended to use the formulae listed in Appendix A together with the rigorous Eqs. (6.5) or (6.7) with (6.10).

6.2.1 Derivation of First Velocity Correction for the Infrared Distance Meter Kern DM501

Given are
$$c_0 = 299\,792\,458 \text{ m/s}$$
$$f_{MOD} = 14\,985\,400 \text{ Hz}$$
$$\lambda_{MOD} = 20 \text{ m exactly}$$
$$\lambda = 0.900 \,\mu\text{m} \ .$$

The modulation frequency f_{MOD} and the unit length U refer to the fine measurement and were laid down by the manufacturer. The reference refractive index n_{REF} is the only unknown in Eq. (6.3) and may be computed as follows:

$$n_{REF} = \frac{c_0}{\lambda_{MOD} f_{MOD}} = \frac{299\,792\,458 \text{ m/s}}{20 \times 14\,985\,400 \text{ m/s}} = 1.000\,281\,8 \ . \tag{6.14}$$

Using Eq. (6.8) leads to:

$$K' = d'(n_{REF} - n)$$

$$= d'(1.000\,281\,8 - n)$$

$$= d'\{(1.000\,281\,8 - 1) - (n - 1)\} \ .$$

Equation (5.15) in Section 5.5 gives the following value for $(n_L - 1)$:

$$(n_L - 1) = (n_g - 1)\,\frac{273.15\,p}{(273.15 + t)\,1013.25} - \frac{11.27\,e\,10^{-6}}{(273.15 + t)} \ .$$

With $n_g = 1.000\,293\,74$ from Eq. (5.12) for $\lambda = 0.900\,\mu\text{m}$:

$$K' = d'\left\{281.8 \times 10^{-6} - (293.74)\,10^{-6}\,\frac{273.15\,p}{(273.15 + t)\,1013.25} + \frac{11.27\,e\,10^{-6}}{(273.15 + t)}\right\}$$

$$K' = \left(281.8 - \frac{79.186\,p}{(273.15 + t)} + \frac{11.27\,e}{(273.15 + t)}\right)\,10^{-6}d' \ , \tag{6.15}$$

where
- K' = first velocity correction
- d' = measured distance displayed on EDM instrument
- t = dry bulb temperature (°C)
- p = atmospheric pressure (mb)
- e = partial water vapour pressure (mb) (e.g. computed from dry and wet bulb readings on a psychrometer).

6.2.2 Derivation of First Velocity Correction for the Infrared Distance Meter Pentax PM-81

The Pentax PM-81 features an on-board compensation facility for the first velocity correction based on refractivity rather than on temperature and pressure or ppm correction input. The equations supplied by the manufacturer can be converted to the form given in Eq. (6.11) for post-processing as shown below.

The following correction input is stated by the manufacturer:

$$(PPM) = \frac{293.6\,p\,(mm\,Hg)}{760 + 2.78236\,t\,(^\circ C)} \; . \tag{6.16}$$

Expansion with $\dfrac{273.15}{760}$ and conversion to mb leads to:

$$(PPM) = \frac{79.145\,p^{(mb)}}{[273.15 + t\,(^\circ C)]} = N_L \; .$$

The PM-81 incorporates a default value for the refractivity of $(PPM) = 278$, corresponding to $15\,^\circ C$, $760\,mm\,Hg$. Therefore, to apply the first velocity correction by computation:

1. Set (PPM) in instrument always to 278 (default value)
2. Apply by computation:

$$K' = \left[278 - \frac{79.145\,p}{(273.15 + t)} + \frac{11.27\,e}{(273.15 + t)} \right] 10^{-6} D \; . \tag{6.17}$$

This follows directly from Eqs. (6.11) to (6.13).

6.2.3 Derivation of First Velocity Correction for the Pulse Distance Meter Distomat Wild DI 3000

The Distomat Wild DI 3000 is a pulse distance meter. Equations (6.2) and (6.3) for the computation of the reference refractive index therefore do not apply. As the operators' manual does not state reference atmospheric parameters, the first velocity correction for post-processing in the form of Eq. (6.11) can be easily derived from the correction equation given by the manufacturer:

$$\Delta D_1 = 281.5 - \frac{0.29035\,p}{1 + 0.00366\,t} + \frac{11.27\,h}{100\,(273.15 + t)} \times 10^x \; , \tag{6.18}$$

where
ΔD_1 = atmospheric correction in ppm
p = pressure in mb
t = temperature in $^\circ C$
h = relative humidity in %
x = $\dfrac{7.5\,t}{237.3 + t} + 0.7857$.

Considering Eqs. (5.25) and (5.29) and $\dfrac{1}{0.003661} = 273.15$ and expansion of second term with 273.15 gives the following equation in the form of Eq. (6.11)

$$K' = 281.5 - \frac{79.3091\,p}{(273.15 + t)} + \frac{11.27\,e}{(273.15 + t)} \; . \tag{6.19}$$

This equation may be used for post-processing if the ppm correction is set to 0.0 ppm during the distance measurements.

6.2.4 Derivation of First Velocity Correction for the Microwave Distance Meter Siemens-Albis SIAL MD 60

The following data are given by the manufacturer:

$$c_0 = 299\,792\,500 \text{ m/s}$$

$$f_{MOD} = 149\,848\,300 \text{ Hz}$$

$$\lambda_{MOD} = 2 \text{ m exactly}$$

$$U = 1 \text{ m exactly} .$$

The modulation frequency f_{MOD}, the modulation wavelength λ_{MOD} and the unit length U refer to the fine measurement of the distance. Using Eq. (6.2) the reference refractive index of the SIAL MD 60 may be computed as follows:

$$n_{REF} = \frac{c_0}{\lambda_{MOD} f_{MOD}} = \frac{299\,792\,500 \text{ m/s}}{2 \times 149\,848\,300 \text{ m/s}} = 1.0003200 . \tag{6.20}$$

Should the refractive index of the ambient atmosphere coincide with n_{REF}, then the displayed readout would be the correct distance.

The first velocity correction K' is obtained according to Eq. (6.8) as:

$$K' = (n_{REF} - n)\,d'$$

$$= [(n_{REF} - 1) - (n - 1)]\,d' .$$

Substituting Eq. (5.17) (see Sect. 5.6) for $(n-1)$ gives for the SIAL MD 60:

$$K' = \left\{ 320.0 - \frac{77.624\,(p-e)}{(273.15+t)} - \left(1 + \frac{5748}{(273.15+t)}\right) \frac{64.70\,e}{(273.15+t)} \right\} 10^{-6} \times d', \tag{6.21}$$

where p = air pressure (mb)
t = temperature (°C)
e = partial water vapour pressure (mb)
d' = measured distance (reading on display of instrument).

6.2.5 Derivation of First Velocity Correction for the Microwave Distance Meter Tellurometer CA 1000

The following data are known:

Given are $c_0 = 299\,792\,458 \text{ m/s}$
$f_{MOD} = 49\,949\,180 \text{ Hz}$
$\lambda_{MOD} = 6 \text{ m exactly}$
$U = 3 \text{ m exactly} .$

Using Eq. (6.2) the reference refractive index of the CA 1000 may be computed as

$$n_{REF} = \frac{c_0}{\lambda_{MOD}\, f_{MOD}} = \frac{299\,792\,458 \text{ m/s}}{6 \times 49\,949\,180 \text{ m/s}} = 1.000\,3249 \ . \tag{6.22}$$

The first velocity correction K′ is obtained according to Eq. (6.8)

$$K' = (n_{REF} - n)\, d'$$

$$= [(n_{REF} - 1) - (n - 1)]\, d' \ .$$

Substituting Eq. (5.17)

$$K' = \left\{ 324.9 \times 10^{-6} - \frac{77.624 \times 10^{-6}}{(273.15 + t)}\, (p - e) \right.$$

$$\left. - \frac{(64.70)\, 10^{-6}}{(273.15 + t)} \left[1 + \frac{5748}{(273.15 + t)} \right] e \right\} d'$$

or, in final form:

$$K' = \left\{ 324.9 - \frac{77.624\,(p - e)}{(273.15 + t)} - \left[1 + \frac{5748}{(273.15 + t)} \right] \frac{64.70\, e}{(273.15 + t)} \right\} 10^{-6} d' \ , \tag{6.23}$$

where p, e in mb, t in °C, K′ in units of d′ .

6.3 Real-Time Application of First Velocity Correction by EDM Instrument

Most short-range distance meters permit the application of the first velocity correction in real-time in one form or another. Originally, this required the input of a ppm (or environmental) correction value. More recently, keyboard entry of temperature, pressure (and humidity) became available. The key-board entry of refractivity (see Sect. 6.2.2, for example) is unlikely to gain wide acceptance. Two precision distance meters apply the first velocity correction based on internal measurements without external input. The Kern Mekometer ME 3000 employs, in principle, a microwave refractometer. Some models of the Com-Rad Geomensor 204 DME feature built-in temperature and pressure sensors and apply the first velocity correction by computation. It should be noted that the real-time application of the first velocity is really only practicable as long as the measurements of the atmospheric parameters are carried out at the instrument station only and the errors incurred in the process (see Sects. 5.8 and 5.9.1) are tolerable.

For instruments with ppm input, manufacturers provide either correction charts or slide rules for the evaluation of the appropriate setting of the ppm (or environmental) correction dial (or switches, key-board entry). The ppm correction charts are entered with temperature and pressure or with temperature and elevation above sea-level. (Humidity is generally ignored, see Sect. 5.5.2 for an estimate of the errors incurred.) In the latter case, the conversion from sea level elevation to pressure is based on a standard atmosphere such as the one described in Appendix C. Differences between ambient and standard pressures lead to errors in the first velocity correction, and thus in the distances displayed by the instrument at the rate of 0.28 ppm/mb [refer to Eq. (5.16)].

The errors incurred by the use of standard pressures may be assessed with the aid of the following table of pressure variations (Anthes et al. 1975):

1084 mb	Highest recorded sea level pressure
1050 mb	Very strong anticyclone
1033 mb	Moderate anticyclone
1013.25 mb	Mean sea level pressure
996 mb	Moderate cyclone
976 mb	Very strong cyclone
908 mb	Hurricane Camille (1969)
877 mb	Lowest recorded sea level pressure

The highest and lowest sea level pressures recorded in Sydney (Australia) are 1039.7 mb and 982.2 mb, respectively. Based on the above information, the likely maximum difference between ambient and standard pressures under surveying weather conditions may be estimated as about 20 mb. This is equivalent to a maximum error of 6 ppm in the first velocity correction. In cases where such systematic errors cannot be tolerated, the first velocity correction should be computed from measured atmospheric pressure.

The direct input of measured temperatures, pressures (and, possibly, humidity) is likely to replace the traditional ppm input. This procedure is likely to be more accurate, as no diagrams or nomograms need to be entered and read. Most instruments ignore the effect of humidity and do not offer the entry of elevations instead of pressure.

Users should be aware of the first velocity equations employed by the instrument manufacturer before using ppm inputs or temperature and pressure inputs. Instrument users should always have the option of by-passing the real-time first velocity correction by either entering zero ppm or the reference temperature and reference pressure.

6.4 Second Velocity Correction

As mentioned earlier, pressure and psychrometer observations are taken normally only at instrument and reflector stations or, for microwave instruments, at the master and remote stations. The first velocity corrections are computed for both terminals of a line. The mean of both values is subsequently used to correct the distance. However, an error is introduced if the mean first velocity correction of both stations is used without a further small correction, called the second velocity correction.

In Fig. 6.1, the mean refractive index $n = \frac{1}{2}(n_1 + n_2)$ would be correct for the upper curve with a radius R but not for the actual wave path below it having radius r, assuming a spherically layered atmosphere on a sphere. For the actual wave path, the correct refractive index would be, assuming linear vertical gradients of temperature and pressure (Höpcke 1964; Saastamoinen 1964):

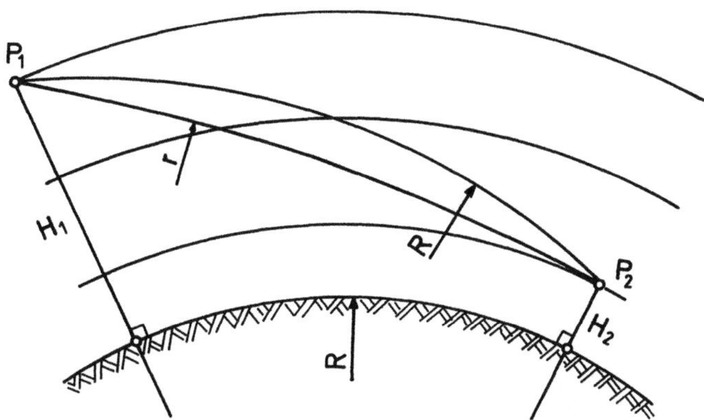

Fig. 6.1. EDM wave paths in a spherically layered atmosphere. The *upper curve* indicates a wave path featuring a coefficient of refraction of k = 1.0 and a radius equivalent to the earth's radius *R*. The *lower curve* depicts a wave path with a coefficient of refraction of less than 1.0 (but larger than 0.0) and, thus, a radius of curvature *r* larger than *R*. The midpoint of the upper curve is at the mean height $(H_1 + H_2)/2$ of the terminals P_1 and P_2 of the line

$$n = \frac{n_1 + n_2}{2} + \Delta n \; , \tag{6.24}$$

where

$$\Delta n = (k - k^2)\, \frac{d'^2}{12R^2} \; . \tag{6.25}$$

The second velocity correction may be written as

$$K'' = -d'\Delta n \tag{6.26}$$

or

$$K'' = -(k - k^2)\, \frac{d'^3}{12R^2} \; , \tag{6.27}$$

where K'' = second velocity correction
        k = coefficient of refraction (see Sect. 5.7)
        d' = measured distance, displayed on instrument
        R = mean radius of curvature of the spheroid along the line.

The second velocity correction K'' is more important for microwaves than for light waves and more important on long distances than on short distances. For example, K'' would become $-50\,\text{mm}$ for a distance d' of 50 km, measured by a microwave instrument ($k_M = 0.25$).

The wave path length d_1 is obtained as

$$d_1 = d' + K' + K'' \; , \tag{6.28}$$

where d' = measured distance, displayed on instrument
        K' = first velocity correction (according to Sect. 6.1.2)
        K'' = second velocity correction.

6.5 Refined Method of Reduction of Measured Distance to Wave Path Chord

Using basic theorems of geometrical optics a formula can be developed which gives the direct relation between the measured optical path and the straight line between two stations. This formula replaces not only the first and second velocity correction, but also the first arc-to-chord correction, which is discussed later in Chapter 7. A few basic theorems are listed here. A comprehensive treatise of the principles of optics and electromagnetic waves is given by Born and Wolf (1970). The development of appropriate EDM reduction formulae has been demonstrated by Brunner and Fraser (1978) and Brunner and Angus-Leppan (1976).

The principle of Fermat (Principle of shortest optical path or principle of least time) states that the optical length σ of an actual ray between any two points, P_1 and P_2, is shorter than the optical length of any other curve through the same two points. In mathematical form:

$$\sigma = \int_{P_1}^{P_2} d\sigma = \int_{P_1}^{P_2} n\,ds = c \int_{P_1}^{P_2} dt \ , \tag{6.29}$$

where ds is an incremental chord element: $ds^2 = dx^2 + dy^2 + dz^2$. Since EDM is based on the measurement of time, the optical length σ can be derived from the direct readout distance d' as follows (Fraser 1984):

$$\sigma = n_{REF} d' \ . \tag{6.30}$$

The basic equation of geometrical optics can be derived from Maxwell's equations assuming very small wavelengths ($\lambda \rightarrow 0$)

$$(\text{grad } \sigma)^2 = \left(\frac{\partial \sigma}{\partial x}\right)^2 + \left(\frac{\partial \sigma}{\partial y}\right)^2 + \left(\frac{\partial \sigma}{\partial z}\right)^2 = n^2(x, y, z) \ . \tag{6.31}$$

The first order partial differential equation of the optical length σ is also known as the *eikonal equation* and describes the effect of vertical and horizontal refraction for EDM. The solution of the eikonal equation corresponds to the conformal projection of the geodesic and was developed by Moritz (1967). In the form of Brunner and Angus-Leppan (1976), the correction of the optical length to the wave path chord becomes:

$$\Delta S = S - \sigma = -10^{-6} \int_0^S N\,dx + \frac{1}{2} 10^{-12} \int_0^S \left[\left(\int_0^x \frac{dN}{dy} \xi\,d\xi \right)^2 \right.$$
$$\left. + \left(\int_0^x \frac{dN}{dz} \xi\,d\xi \right)^2 \right] \frac{dx}{x^2} \ , \tag{6.32}$$

where σ is the optical length ($\sigma = d' n_{REF}$), s is the wave path chord, N is the *refractivity* $[N = (n-1) \times 10^6]$ and ξ is an integration variable along x.

The origin of the Cartesian coordinate system x, y, z is defined at one endpoint of the line. The x-axis is colinear with the chord. The z-axis is perpendicular to the x-axis and on a vertical plane through the origin. The y-axis is in a horizontal plane through the origin. All integrations are carried out along the x-axis. Equation (6.32) is mathematically accurate to 2×10^{-11} s assuming that the laws of geometrical optics are valid for describing the wave propagation. However, the practical accuracy of Δs will largely depend on the accuracy of the refractivity N and its gradients. The gradients can be derived from a combination of measurements of atmospheric parameters (see Sect. 5.8) and atmospheric models. Different atmospheric models apply depending on the specific conditions of the atmosphere.

Refined methods of reducing an optical length (or measured distance) σ to its chord length s will not be considered further in this text, because their application requires both field observations of additional atmospheric parameters as well as more extensive computations. Some atmospheric rectification models have been discussed in Section 5.9.3, together with alternative methods. For high precision and long range EDM, the application of refined methods should be considered.

With respect to the following, the correction $\Delta s = s - \sigma$ will be approximated in the traditional way by the sum of the first and the second velocity correction and the first arc-to-chord correction.

7 Geometrical Corrections

In principle, the wave path d_1 is obtained after applying the two velocity corrections to an actual observation of the length. Unless the length is subsequently processed in a three-dimensional network adjustment, it has to be reduced to the equivalent spheroidal distance on a geodetic reference surface, for example the Australian Geodetic Datum. Selected methods of reduction to the spheroid will now be reviewed, prior to a discussion of additional corrections and computations.

There are basically two different methods of reduction. The first method uses known or measured heights, the second measured zenith angles. The first method is employed in conjunction with medium to long range distance measurements. The second method is used for short ranges where the heights of target stations are usually not available.

In connection with three-dimensional networks, the wave path d_1 only needs to be reduced to the wave path chord d_2 with the aid of the first arc-to-chord correction discussed below. This provides the space distance between the two terminals of the line.

It should be noted that some additional corrections may be necessary after the application of the velocity corrections before the correct wave path length is obtained. Such additional corrections may be necessary for telescope- and theodolite-mounted EDM instruments and when measured zenith angles are used in the reduction process. These corrections are discussed later in detail in Sections 8.1 and 8.2.

7.1 Reduction to the Spheroid Using Station Heights

The reduction problem is illustrated in Fig. 7.1; d_1 stands for the wave path length, d_2 for the wave path chord, d_3 for the spheroidal chord and d_4 for the spheroidal distance. The spheroidal distance corresponds to the length of a normal section and differs only slightly from the length of the geodesic. The difference is less than 20 mm for a distance of 3000 km and may therefore be ignored (Bomford 1975). The radius of curvature of the wave path is denoted by r and the mean radius of curvature of the spheroid along the line by R. H_1 and H_2 are the spheroidal heights of stations P_1 and P_2, measured along the normal to the spheroid. The spheroidal height is the sum of the orthometric (geoidal) height and the corresponding height of the geoid above the spheroid (geoid-spheroid separation). γ is the angle between the normals to a sphere of radius R through P_1 and P_2, and β is the angle between the tangents to the wave path through P_1 and P_2 and also the angle between the wave path normals through P_1 and P_2.

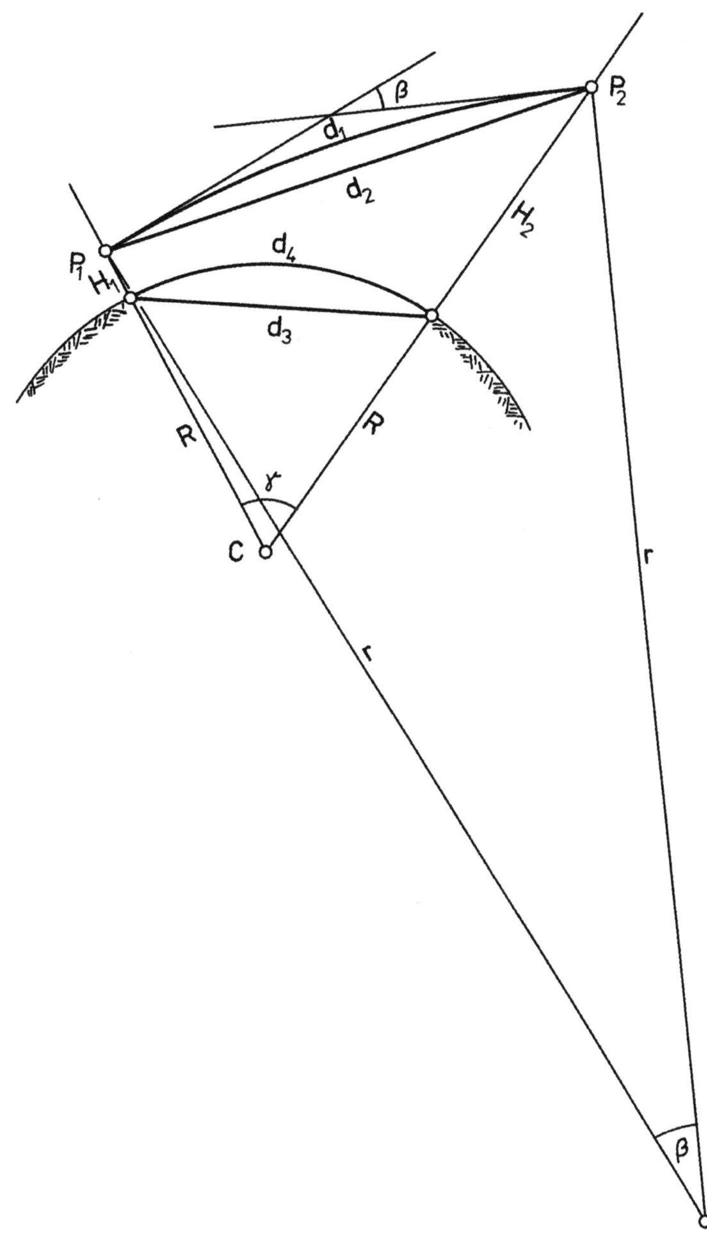

Fig. 7.1. Reduction of the wave path length d_1 to the spheroidal arc length d_4 with the aid of the elevations (H_1 and H_2) of the terminals P_1 and P_2. The wave path chord and the spheroidal chord lengths are denoted by d_2 and d_3, respectively. The angle γ is subtended by the normals to the spheroid of radius R and through the terminals of the line. Similarly, the angle β is subtended by the normals to the wave path with radius r

The value R is computed from

$$R = \frac{\nu \varrho}{\nu \cos^2 \alpha + \varrho \sin^2 \alpha} \,, \qquad (7.1)$$

where ν = radius of curvature of the spheroid in the prime vertical
ϱ = radius of curvature of the spheroid in the meridian
α = azimuth of the measured line, clockwise through $360°$ from the north.

The two radii of curvature follow from (e.g. Cooper 1987):

$$\varrho = \frac{a(1-e^2)}{(1-e^2 \sin^2 \varphi)^{3/2}} \qquad (7.2)$$

$$\nu = \frac{a}{(1-e^2 \sin^2 \varphi)^{1/2}} \,, \qquad (7.3)$$

where φ is the latitude and where the eccentricity e and the flattening f of the spheroid are given by

$$e^2 = (a^2 - b^2)/a^2 \qquad (7.4)$$

$$f = (a-b)/a \ . \qquad (7.5)$$

The semi-major and semi-minor axes of the spheroid are denoted by a and b, respectively. National (and international) reference spheroids are usually defined by a and e. For example, the Australian National Spheroid is defined by $a = 6378160$ m and $f = 1/298.25$ exactly and the ellipsoid of the World Geodetic System WGS 84 is based on $a = 6378137$ m and $f = 1/298.257223563$.

The spheroid between the terminals of a line is always replaced by a sphere of radius R for the reduction of distances. This leads to an intersection of the spheroidal normals which were originally skew. Sometimes, a mean radius R_M is adopted for a certain area to avoid repeated computations of R. The Integrated Survey Grid in the State of New South Wales (Australia) refers for example to a mean radius of $R_M = 6370100$ m causing errors of less than 1 ppm in reduced distances for N.S.W. latitudes.

The average radius of curvature R_M at a point on the spheroid can be expressed by

$$R_M = (\nu \rho)^{1/2} \ . \qquad .6)$$

Spheroidal heights are often replaced by orthometric (geoidal) heights because the former are not readily available. Orthometric heights, for example heights based on the Australian Height Datum (A.H.D.), may be obtained from spirit and trigonometric levelling. The use of orthometric instead of spheroidal heights for the reduction of distances may lead to errors in the spheroidal distance depending on the height of the geoid above the spheroid at the two terminals of a line (Vincenty 1975). Two cases may be distinguished:

1. Equal heights of geoid above spheroid at the terminals of a line introduce errors of 1 ppm per 6 m geoid-spheroid separation.

2. Unequal heights of geoid above spheroid at the terminals produce an additional error which may be estimated using the second term of Eq. (7.40) in Section 7.1.3. For extreme cases in alpine areas, errors of up to 0.15 m for 2.5 km lines have been reported (Elmiger 1977).

Spheroidal heights should be used in high order geodetic surveys. However, the difference between spheroidal and orthometric heights is often ignored in low order and short range surveys. The effects of this omission are not always negligible. With the introduction of surveying techniques based on satellite positioning systems such as the Global Positioning System (GPS), the determination of the geoid-spheroid separation has found renewed interest and may lead, eventually, to more accurate and more readily available local geoid maps.

Two calculation methods are now presented. The first executes the reduction in several steps. The second gives a closed solution.

7.1.1 First Method: Step-by-Step Solution

7.1.1.1 First Arc-to-Chord Correction $K_1(d_1$ to $d_2)$

d_1 stands for the measured distance along a wave path with radius of curvature r and already includes the first and second velocity correction explained in Chapter 6.

Thus:

$$d_2 = 2r \sin\frac{\beta}{2} = 2r \sin\frac{d_1}{2r} \; . \tag{7.7}$$

Developing $\sin\dfrac{d_1}{2r}$ by its series expansion:

$$d_2 = d_1 - \frac{d_1^3}{24r^2} + \frac{d_1^5}{1920r^4} - \cdots \; . \tag{7.8}$$

The third term amounts to 1 mm for $d = 1000\,\text{km}$ and $r = 4R$, and hence may be safely neglected.

Therefore

$$d_2 = d_1 - \frac{d_1^3}{24r^2} \; . \tag{7.9}$$

Substituting Eq. (5.19)

$$r = \frac{R}{k} \tag{5.19}$$

in Eq. (7.9) leads to

$$d_2 = d_1 + K_1 \tag{7.10}$$

and

$$K_1 = -\frac{d_1^3}{24\,r^2} = -k^2\,\frac{d_1^3}{24\,R^2} \; . \tag{7.11}$$

7.1.1.2 Chord-to-Chord Correction K_{23} (d_2 to d_3)

Applying the cosine rule to the triangle $P_1 P_2 C$ (Fig. 7.1) yields:

$$d_2^2 = (R+H_1)^2 + (R+H_2)^2 - 2(R+H_1)(R+H_2)\cos\gamma \; . \tag{7.12}$$

$\cos\gamma$ can now be expressed by known values. The basic relationship

$$\cos\gamma = 1 - 2\sin^2\frac{\gamma}{2} \tag{7.13}$$

may be used, and by substituting $\sin\dfrac{\gamma}{2} = \dfrac{d_3}{2R}$ (see Fig. 7.1), the following equation is obtained:

$$\cos\gamma = 1 - \frac{d_3^2}{2R^2} \; . \tag{7.14}$$

Substituting Eq. (7.14) in Eq. (7.12) leads to:

$$d_2^2 = R^2 + 2RH_1 + H_1^2 + R^2 + 2RH_2 + H_2^2 - 2R^2 - 2RH_1 - 2RH_2 - 2H_2H_1$$

$$+ \frac{d_3^2}{R^2}(R+H_1)(R+H_2) = (H_2-H_1)^2 + d_3^2(R+H_1)(R+H_2)\,\frac{1}{R^2} \; . \tag{7.15}$$

Solving for d_3 yields:

$$d_3 = +\sqrt{\frac{d_2^2 - (H_2-H_1)^2}{[1+(H_1/R)]\,[1+(H_2/R)]}} \; , \tag{7.16}$$

where
$\quad d_2 \quad$ = wave path chord
$\quad d_3 \quad$ = spheroidal chord
$\quad H_1, H_2$ = spheroidal heights
$\quad R \quad\;$ = radius of curvature of the spheroid along the line [see Eq. (7.1)].

Equation (7.16) is a rigorous formula for a spherical reference surface of radius R. The spheroidal heights H_1 and H_2 refer, in the case of electro-optical EDM, to the trunnion axes of EDM instrument and reflector, respectively. For an EDM instrument mounted on the telescope of a theodolite, H_1 refers to the trunnion axis height of the theodolite.

The chord-to-chord correction is sometimes split into two separate corrections, namely

1. the *slope correction* K_2 (d_2 to d_6) and
2. the *sea level correction* K_3 (d_6 to d_3) leading to:

$$d_6 = d_2 + K_2 \tag{7.17}$$

$$d_3 = d_2 + K_2 + K_3 \; . \tag{7.18}$$

Both corrections may be derived from the rigorous Eq. (7.16), which may be written as follows:

$$d_3 = d_2 \left[1 - \left(\frac{H_2 - H_1}{d_2} \right)^2 \right]^{1/2} \left[1 + \left(\frac{H_1 + H_2}{R} + \frac{H_1 H_2}{R^2} \right) \right]^{-1/2} . \tag{7.19}$$

Substituting $\Delta H = H_2 - H_1$ and $H_M = \frac{1}{2}(H_1 + H_2)$ and expanding the terms in the second brackets leads to:

$$d_3 = d_2 \left[1 - \left(\frac{\Delta H}{d_2} \right)^2 \right]^{1/2} \left[1 - \frac{H_M}{R} - \frac{H_1 H_2}{2R^2} + \frac{3}{2} \frac{(H_M)^2}{R^2} \right.$$
$$\left. + \frac{3}{2} \frac{H_M H_1 H_2}{R^3} + \frac{3}{8} \frac{(H_1 H_2)^2}{4R^4} - \cdots \right] . \tag{7.20}$$

The magnitude of the terms in the second bracket may be estimated by substituting maximum values (applicable in Australia) for $H_1 = H_2 = H_M = 2.5$ km and assuming $R = 6370.1$ km:

$$\frac{H_M}{R} = 4 \times 10^{-4} , \quad \frac{H_1 H_2}{2R^2} = 8 \times 10^{-8} , \quad \frac{3 H_M^2}{2R^2} = 2.4 \times 10^{-7} , \ldots .$$

Also, all terms with R^2, R^3 etc. in the denominator may be safely disregarded. The maximum error resulting from this omission of terms will not exceed 0.2 ppm for heights below 2500 m.

Equation (7.20) may therefore be written in a reduced form

$$d_3 = d_2 \left[1 - \left(\frac{\Delta H}{d_2} \right)^2 \right]^{1/2} \left(1 - \frac{H_M}{R} \right) . \tag{7.21}$$

Multiplication of the first two factors in Eq. (7.21) leads to a "horizontal" distance d_6 computed in a right-angled triangle. The correction which has to be applied to the "slope" distance to obtain the "horizontal" distance d_6 is traditionally called the slope correction. Considering the first two factors of Eq. (7.21) and Eq. (7.18) the *slope correction* K_2 (d_2 to d_6) may now be defined as

$$K_2 = d_2 \left[1 - \left(\frac{\Delta H}{d_2} \right)^2 \right]^{1/2} - d_2 \tag{7.22}$$

$$= (d_2^2 - \Delta H^2)^{1/2} - d_2 \tag{7.23}$$

or, using a series expansion of the square root term in Eq. (7.22), as

$$K_2 = -\frac{\Delta H^2}{2 d_2} - \frac{\Delta H^4}{8 d_2^3} - \frac{\Delta H^6}{16 d_2^5} - \cdots . \tag{7.24}$$

Considering Eqs. (7.18), (7.21) and (7.22), the so-called *sea level correction* K_3 (d_6 to d_3) yields:

$$K_3 = -\frac{H_M}{R} (d_2^2 - \Delta H^2)^{1/2} \tag{7.25}$$

or with regard to Eq. (7.23):

$$K_3 = -\frac{H_M}{R}(d_2 + K_2) \tag{7.26}$$

$$= -\frac{H_M}{R} d_6 \; . \tag{7.27}$$

Substitution of K_2 in Eq. (7.26) by Eq. (7.24) leads to a third equation for K_3:

$$K_3 = -\frac{H_M}{R} d_2 + \frac{H_M \Delta H^2}{2 d_2 R} + \frac{H_M \Delta H^4}{8 d_2^3 R} + \frac{H_M \Delta H^6}{16 d_2^5 R} + \dots \; . \tag{7.28}$$

The first term is the conventional value for the sea level correction which is not only used in EDM but also in reductions of horizontal distances to sea level and vice versa. The first term of Eq. (7.28) is not always sufficient for reducing EDM slope distances accurately to the spheroid because the slope correction K_2 does not lead to the chord at mean height H_M, which would be the true horizontal distance. The terms of higher order in Eq. (7.28) may be estimated by inserting extreme values of $\Delta H = 1$ km, $d_2 = 1.5$ km and $H_M = 1.5$ km. The second term then becomes 78 mm, the third 9 mm and the fourth 2 mm. The use of the first term of Eq. (7.28) alone is therefore insufficient. It is advisable to use the rigorous formulae (7.25) or (7.26) for the sea level correction, in conjunction with slope corrections according to Eqs. (7.22) and (7.23).

7.1.1.3 Second Chord-to-Arc Correction K_4 (d_3 to d_4)

From Fig. 7.1 follows:

$$d_4 = R\gamma = 2R \left(\frac{\gamma}{2}\right) = 2R \, \text{arc sin} \left(\frac{d_3}{2R}\right) . \tag{7.29}$$

Replacing arc sin by its series

$$d_4 = d_3 + \frac{d_3^3}{24 R^2} + \frac{3 d_3^5}{640 R^4} + \dots \; . \tag{7.30}$$

The third term in (7.30) amounts to only 1 mm for a distance d_3 of 200 km. It may be safely neglected.

$$d_4 = d_3 + K_4 \; , \tag{7.31}$$

where

$$K_4 = +\frac{d_3^3}{24 R^2} \; . \tag{7.32}$$

7.1.1.4 Combined Correction for K'', K_1, K_4

A combined correction for the second velocity correction K'', for the first arc-to-chord correction K_1 and the second chord-to-arc correction K_4 was derived by Saastamoinen (1964):

90

$$K'' + K_1 + K_4 = -(k-k^2)\,\frac{d'^3}{12R^2} - k^2\,\frac{d_1^3}{24R^2} + \frac{d_3^3}{24R^2}\,. \qquad (7.33)$$

For long lines, the above equation may be simplified by assuming $d' = d_1$ and $d_3 = d_1$:

$$K'' + K_1 + K_4 = \frac{d_1^3}{24R^2}\,(-2k + 2k^2 - k^2 + 1) \qquad (7.34)$$

$$= \frac{d_1^3}{24R^2}\,(k^2 - 2k + 1)$$

$$= (1-k)^2\,\frac{d_1^3}{24R^2} \qquad (7.35)$$

7.1.2 Second Method: Closed Solution

The reduction of the wave path length d_1 to the spheroidal distance d_4 can be done in one step, if several formulae of Section 7.1.1 are combined. Considering the equation

$$d_4 = 2R \arcsin\left(\frac{d_3}{2R}\right) \qquad (7.29)$$

and substituting Eq. (7.16) for d_3 leads to:

$$d_4 = 2R \arcsin \sqrt{\frac{d_2^2 - (H_2 - H_1)^2}{4(R+H_1)(R+H_2)}}\,. \qquad (7.36)$$

Substituting d_2 from Eq. (7.7) and considering Eq. (5.19) yields after some simple computation,

$$d_4 = 2R \arcsin \sqrt{\frac{R^2 \sin^2(d_1 k/2R) - k^2(H_2 - H_1)^2/4}{k^2(R+H_1)(R+H_2)}}\,. \qquad (7.37)$$

The above formula is rigorous and reduces d_1 to d_4 without any approximations

where R = mean radius of curvature of the spheroid [along the line, see Eq. (7.1)]
d_1 = wave path length
d_4 = spheroidal distance
H_1, H_2 = spheroidal heights of stations P_1, P_2
k = coefficient of refraction, as defined by Eq. (5.19).

The spheroidal heights represent again the trunnion axis elevations of EDM instrument and reflector at the terminals P_1 and P_2. For an EDM instrument, which is mounted on the telescope of a theodolite, the height refers to the trunnion axis height of the theodolite.

Note on the Use of Pocket Calculators and Computers. Before Eq. (7.37) is used in conjunction with pocket calculators the accuracy of the pocket calculator's trigonometric functions for small angles must be known. If the number of accurate significant digits is smaller than the number required, both trigonometric functions in Eq. (7.37) should be replaced by series expansions. When programming Eq. (7.37) on personal and other computers, it is essential to declare all parameters involved as double precision variables, as squares of large numbers are involved.

7.1.3 Analysis of Errors

To test the effect of the parameters in Eq. (7.37) on the spheroidal distance d_4, the propagation of real errors is assessed on the basis of Eqs. (7.11), (7.16) and (7.32).

Defining

$$\Delta H = H_1 - H_1 \tag{7.38}$$

$$H_M = \tfrac{1}{2}(H_1 + H_2) , \tag{7.39}$$

and assuming

$$(R + H_1)(R + H_2) = (R + H_M)^2$$

$$R^2/(R + H_M)^2 = 1.0 ,$$

the following total differential may be derived:

$$\delta(d_4) = \frac{d_2}{d_3}\,\delta(d_1) - \frac{\Delta H}{d}\,\delta(\Delta H) - \frac{d}{R + H_M}\,\delta(R + H_M) + \frac{d}{R}\,\delta(R)$$
$$- \frac{k\,d^3}{12 R^2}\,\delta(k) , \tag{7.40}$$

where d may be taken as d_4 and where the partial derivatives of the variables d_4, d_1, ΔH, $(R + H_M)$, R and k are denoted by δ. It should be noted that any error $\delta(R)$ will create an error $\delta(H_M)$ of equal magnitude but opposite sign and that the error $\delta(R + H_M)$ becomes zero.

If an error δd_4 in d_4 should not exceed a certain limit, the maximum allowable errors in each of d_1, ΔH, $(R + H_M)$, R and k can be calculated from the total differential above.

7.1.3.1 Example of a Long Line

Given are the following approximate quantities:

$$d_1 = 20\,\text{km} , \quad \Delta H = 1000\,\text{m} , \quad H_M = 500\,\text{m} ,$$

$$R = 6370\,\text{km} \quad \text{and} \quad k = 0.13 .$$

How accurately must the spheroidal height difference ΔH, the mean spheroidal height H_M, the radius of curvature of the spheroid R and the coefficient of refraction k be known so that the effect of each parameter on the spheroidal distance d_4 does not exceed 5 mm?

With $\delta(d_4) = 5$ mm, the individual errors can be solved for as

$$\delta(\Delta H) = (d/\Delta H)\,\delta(d_4) = 100\ \text{mm}$$

$$\delta(R+H_M) = (1/d)(R+H_M)\,\delta(d_4) = 1.6\ \text{m}$$

$$\delta(R) = (R/d)\,\delta(d_4) = 1.6\ \text{m}$$

$$\delta(k) = (12R^2/k\,d^3)\,\delta(d_4) = 2.3\ .$$

It becomes evident that the height difference must be known with much higher accuracy (0.1 m) than the spheroidal mean height of the line (1.6 m). It should be noted that the former value must also include the uncertainty of the change of the geoid-spheroid separation between the endpoints of the line and that the latter must include the uncertainty of the mean geoid-spheroid separation of the line (refer to Sect. 7.1). No state-wide mean radius of the spheroid can be adopted if the above accuracy has to be achieved. Equations (7.2), (7.3) and (7.6) should be used instead. The coefficient of refraction is not critical.

7.1.3.2 Example of a Short and Steep Line

Considering high precision EDM on close range, such as for the purpose of the monitoring of structures, the following parameters are assumed:

$$d_1 = 140\ \text{m}\ ,\quad \Delta H = 100\ \text{m}\ .$$

How accurately must ΔH be known, if the effects of its uncertainty on the reduced distance should not exceed 0.1 mm?

Solving for $\delta(\Delta H)$ in the second term of Eq. (7.40) and omission of the negative sign leads to

$$\delta(\Delta H) = (d/\Delta H)\,\delta(d_4)$$

$$= (140/100)0.1\ \text{mm}$$

$$= 0.14\ \text{mm}\ .$$

It follows that the use of given height differences for the reduction of relatively short distances may cause considerable errors in the reduced distances, if the height differences are not known with the necessary precision. In this case, the height difference would have to be established by precise geodetic levelling or, possibly, by simultaneous reciprocal trigonometric heighting. The measurements of the heights of instrument and reflector would also need to be accurate to 0.1 mm.

7.2 Reduction to the Spheroid, Using Measured Zenith Angles

7.2.1 Introduction

The problem is depicted in Fig. 7.2, where

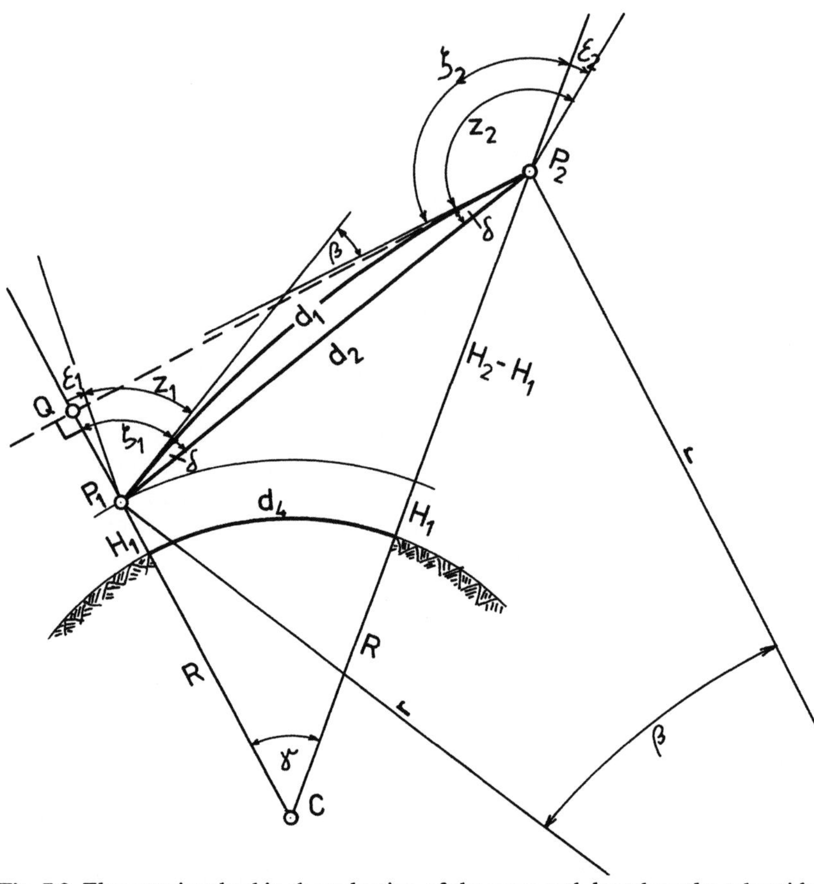

Fig. 7.2. Elements involved in the reduction of the wave path length to the spheroidal arc length using measured zenith angles. (Explanation see text)

z_1, z_2 = measured zenith angles at stations P_1 and P_2 respectively
δ = refraction angle, assumed equal at P_1 and P_2
$\varepsilon_1, \varepsilon_2$ = deviations of vertical at points P_1 and P_2 in azimuth α_{12} of the line
ζ_1, ζ_2 = spheroidal zenith angles at P_1, P_2 respectively
r = radius of curvature of wave path
β = angle between wave path normals through P_1 and P_2
R = radius of curvature of the spheroid along the line
H_1, H_2 = spheroidal heights
d_1 = wave path length
d_2 = wave path chord
d_4 = spheroidal distance
ΔH = spheroidal height difference = $H_2 - H_1$
γ = angle between the spheroidal normals through P_1 and P_2.

Figure 7.2 clearly assumes that the two zenith angle measurements as well as the distance measurement were collinear, and that the trunnion axis heights (above ground marks) of theodolites, EDM instruments, reflectors, traversing targets,

94

were thus the same on any one station. The spheroidal heights at the endpoints refer to these common trunnion axis heights of all equipment used. Additional corrections to measured zenith angles and/or distances may apply, if the conditions are not fulfilled (see Sect. 8.2).

A rigorous reduction of the wave path length d_1 to the spheroidal distance d_4 is only possible if the spheroidal heights and the deviations of the vertical ε are available. For practical purposes, the spheroidal zenith angles ζ are usually replaced by the measured zenith angles z, the spheroidal heights by the orthometric (geoidal) heights and R by the mean radius of spheroid for a specific area R_M. For example, R is usually replaced by $R_M = 6370100$ m for computations in the Integrated Survey Grid of the State of New South Wales (Australia).

The impact of such approximations is described in Section 7.1 and in Vincenty (1975). The height difference ΔH may be computed from either spheroidal ζ or measured zenith angles z. In the former case, *spheroidal height differences* are obtained, in the latter case, *orthometric height differences*.

The relationship between the spheroidal zenith angle ζ and the measured zenith angle z is given by

$$\zeta = z + \varepsilon \ . \tag{7.41}$$

Again, two different calculation methods are presented.

7.2.2 Reduction to the Spheroid: Closed Solution

Following a solution given by Dabrowski and Maier (1977) and referring to Fig. 7.2, the distances QP_1 and QP_2 may be computed in the right-angled triangle P_1P_2Q as

$$QP_1 = d_2 \cos (z_1 + \varepsilon_1 + \delta) \tag{7.42}$$

$$QP_2 = d_2 \sin (z_1 + \varepsilon_1 + \delta) \ . \tag{7.43}$$

Considering the right-angled triangle QCP_2 the angle γ may be derived as follows

$$\gamma = \arctan \left[\frac{d_2 \sin (z_1 + \varepsilon_1 + \delta)}{R + H_1 + d_2 \cos (z_1 + \varepsilon_1 + \delta)} \right] \ . \tag{7.44}$$

The spheroidal distance d_4 may now easily be computed as

$$d_4 = R \arctan \left[\frac{d_2 \sin (z_1 + \varepsilon_1 + \delta)}{R + H_1 + d_2 \cos (z_1 + \varepsilon_1 + \delta)} \right] \ . \tag{7.45}$$

The computation of the refraction angle δ follows from Fig. 7.2:

$$\delta = \frac{\beta}{2} = \frac{d_1}{2r} = \frac{d_1 k}{2R} \ . \tag{7.46}$$

The wave path chord d_2 is computed from Eq. (7.7) of Section 7.1.1.1 and Eq. (5.19) of Section 5.9:

$$d_2 = \frac{2R}{k} \sin\left(\frac{d_1 k}{2R}\right) \simeq d_1 \;. \tag{7.47}$$

Equations (7.45), (7.46) and (7.47) permit a rigorous computation of the spheroidal distance d_4. For $k = 0.13$ and $R = 6370$ km, the difference between d_2 and d_1 (the first arc-to-chord correction according to Sect. 7.1.1.1) amounts to only 0.02 mm for $d = 10$ km, and 0.47 mm for $d = 30$ km. The wave path chord d_2 may therefore be safely replaced by the wave path length d_1 in all practical cases. Equation (7.45) may subsequently be written as

$$d_4 \simeq R \arctan \left[\frac{d_1 \sin(z_1 + \varepsilon_1 + d_1 k/2R)}{R + H_1 + d_1 \cos(z_1 + \varepsilon_1 + d_1 k/2R)} \right] , \tag{7.48}$$

where d_4 = spheroidal distance
      d_1 = wave path length
      H_1 = spheroidal height of station P_1
      z_1 = measured zenith angle at P_1 (in radian)
      k = coefficient of refraction of light ($= 0.13$)
      R = radius of curvature of the spheroid along the line $P_1 P_2$
      ε_1 = deviation of vertical at P_1 in azimuth of $P_1 P_2$, derived from additional observations (e.g. astronomical observations) (in radian).

The deviation of the vertical is usually unknown, which means that ε is very often ignored in computing the spheroidal distance d_4. The effect of ε on d_4 can be derived by taking the total differential of Eq. (7.48). Replacing spheroidal heights by orthometric heights (geoidal heights) leads to additional errors which have already been discussed in Section 7.1.

Note on the Use of Pocket Calculators and Computers. The use of arc tan in Eqs. (7.45) and (7.48) may be critical because the term in brackets is always very small. The accuracy limitations of a particular calculator should be checked for small arguments of this trigonometric function. If necessary, arc tan may be replaced by its series expansion. Pocket calculators should be switched to radian entry of arguments of trigonometric functions before using Eqs. (7.45) and (7.48). When programming personal or other computers for the computation of Eqs. (7.45) and (7.48), all parameters involved should be defined as double precision variables.

7.2.3 Reduction to the Spheroid: Step-by-Step Solution

Apart from the one-step solution of the problem, given in Section 7.2.2, the reduction of distances can also be evaluated by a similar method as shown previously in Section 7.1.1.

7.2.3.1 First Arc-to-Chord Correction K_1 (d_1 to d_2)

This correction has already been discussed in Section 7.1.1. The wave path chord d_2 has been obtained as follows

$$d_2 = d_1 + K_1 \ , \quad \text{where} \quad K_1 = -k^2 \frac{d_1^3}{24\,R^2} \ . \tag{7.10}$$

The size of K_1 may be evaluated for $R = 6370$ km and a coefficient of refraction $k_L = 0.13$ (light waves). K_1 then amounts to -0.02 mm for $d = 10000$ m and to -0.47 mm for $d = 30000$ m.

The first arc-to-chord correction may therefore be ignored in almost all cases.

7.2.3.2 Slope Correction K_5 (d_2 to d_5)

The slope correction reduces the wave path chord d_2 to the horizontal distance d_5 at the height H_1. All necessary parameters are depicted in Fig. 7.3. Applying the sine rule to triangle $P_1 P_2 P_2'$ yields

$$d_5 = d_2 \frac{\sin\,[z_1 - \gamma(2-k)/2]}{\sin\,(\pi/2 + \gamma/2)} = d_2 \frac{\sin\,[z_1 - \gamma(2-k)/2]}{\cos\,\gamma/2} \ . \tag{7.49}$$

Equation (7.49) is rigorous as long as $\gamma/2$ is correct; it is, however, usually approximated by

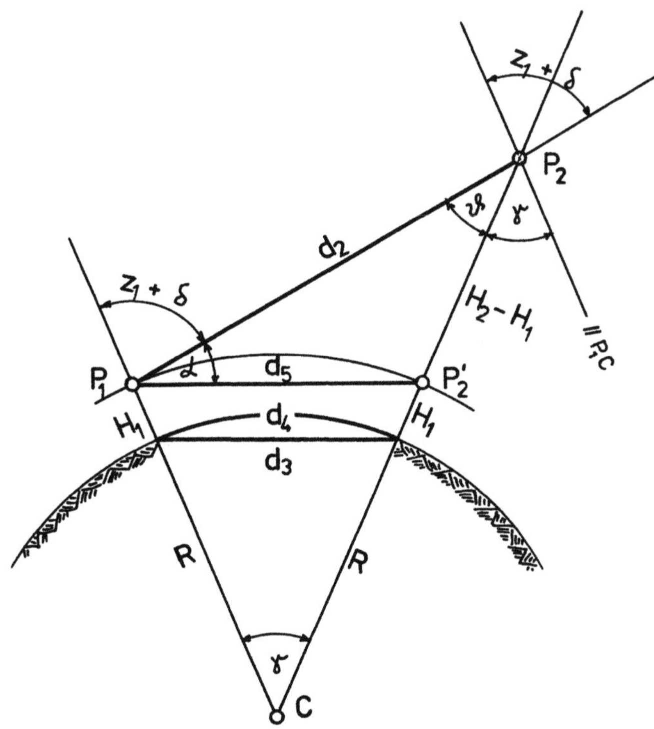

Fig. 7.3. Elements involved in the step-by-step reduction of the wave path chord length to the spheroidal arc length using measured zenith angles. The horizontal chord length at the height H_1 is denoted by d_5. The spheroidal height difference is given by $H_2 - H_1$. The deviations of the vertical are not shown

$$\frac{\gamma}{2} \approx \frac{d_2 \sin z}{2(R+H_1)} \; . \tag{7.50}$$

Using the series expansion of cosine and Eq. (7.50), it can be shown that the denominator in Eq. (7.49) may be safely taken as unity for all practical cases. The approximation causes an error of 1 part in 10^8 over 2 km of horizontal distance of 3 parts in 10^7 over 10 km.

Rearrangement of the numerator in Eq. (7.49) leads to

$$d_5 = d_2 \sin z_1 \cos \left[\frac{\gamma}{2}(2-k) \right] - d_2 \cos z_1 \sin \left[\frac{\gamma}{2}(2-k) \right] \; . \tag{7.51}$$

Because $\dfrac{\gamma}{2}$ is small, the term $\left[\dfrac{\gamma}{2}(2-k) \right]$ will also be small.

Substituting $\cos \left[\dfrac{\gamma}{2}(2-k) \right]$ by unity and $\sin \left[\dfrac{\gamma}{2}(2-k) \right]$ by $\dfrac{\gamma}{2}(2-k)$ and considering Eq. (7.50) yields

$$d_5 = d_2 \left[\sin z_1 - \frac{d_2(2-k)}{2(R+H_1)} \sin z_1 \cos z_1 \right] \; . \tag{7.52}$$

Finally, using a theorem of trigonometric functions and neglecting H_1 leads to

$$d_5 = d_2 \sin z_1 - \frac{d_2^2(2-k)}{4R} \sin 2z_1 \; , \tag{7.53}$$

where d_5 = horizontal distance at height H_1
        d_2 = wave path chord
        z_1 = zenith angle in P_1
        R = mean radius of curvature of spheroid along the line
        H_1 = height of station P_1.

The effect on d_5 of neglecting H_1 can be evaluated by multiplying the figures in the table below by the factor H_1/R. The maximum error owing to the omission of H_1 in d_5 is smaller than 0.4 mm, if the slope distances d_2 are smaller than 5000 m and the zenith angles between 70° and 110°.

Note: To reduce distances correctly to the spheroid, spheroidal zenith angles and spheroidal heights need to be used (see Sect. 7.2).

The magnitude of the second term in Eq. (7.53) is demonstrated with the aid of a table. It is seen that the term should usually be taken into account.

Slope distance d_2	$z_1 = 80°$	$z_1 = 70°$
100 m	− 0.3 mm	− 0.5 mm
300 m	− 2.3 mm	− 4.2 mm
500 m	− 6.3 mm	− 11.8 mm
1 000 m	− 25.2 mm	− 47.2 mm
2 000 m	− 100.8 mm	− 188.8 mm

For zenith angles of 90°, the second term is zero. The above table assumes $H_1 = 0$ and $k = 0.13$ (light waves).

7.2.3.3 Sea Level Correction K_6 (d_5 to d_3)

The sea level correction K_6 will be slightly different from the solution in Section 7.1.1.2 because d_5 is defined at the station height H_1 of P_1 and not at the mean height H_m as before.

$$K_6 = -\frac{H_1}{R} d_5 \tag{7.54}$$

$$d_3 = d_5 + K_6 \ . \tag{7.55}$$

Equations (7.54) and (7.55) are rigorous and can be easily derived from the triangle $P_1 P_2' C$ of Fig. 7.3:

$$
\begin{aligned}
d_3 &= \frac{d_5 R}{R + H_1} \\
&= \frac{d_5}{(1 + H_1/R)} \\
&\approx d_5 \left(1 - \frac{H_1}{R}\right) \ .
\end{aligned}
\tag{7.56}
$$

7.2.3.4 Second Chord-to-Arc Correction K_4 (d_3 to d_4)

This correction has already been discussed in Section 7.1.1.3. It can be shown, by using Eq. (7.32) and assuming $R = 6370.1$ km and $k = 0.13$, that the omission of this correction causes errors of only 0.1 mm and 1.0 mm on lines of 5 km and 10 km length. K_4 may therefore be ignored in almost all cases.

7.2.4 Analysis of Errors

The effects of errors in the parameters of Eq. (7.45) may be analyzed by the total differential of Eqs. (7.45) and (7.48). After some substitutions and with a radius $R = 6370000$ m the total differential of Eq. (7.45) yields (Dabrowski and Maier 1977):

$$\delta(d_4) = \sin z\, \delta(d_1) - 1.6 \times 10^{-7} d_4\, \delta(H_1) + \Delta H\, \delta(z) + 7.8 \times 10^{-8} d_1 \Delta H\, \delta(k) \ , \tag{7.57}$$

where ΔH is defined by $(H_2 - H_1)$ and all length parameters are taken in metres.

The first two terms will not be discussed further. The third term evaluates the effect of errors in the measured zenith angle or of the omission of the deviation of the vertical. An error of 2 seconds of arc leads to the following errors $\delta(d_4)$:

for $\quad \Delta H = 50$ m $\qquad$ 100 m $\qquad$ 200 m $\qquad$ 1 000 m

$\qquad \delta(d_4) = 0.5$ mm $\qquad$ 1.0 mm $\qquad$ 1.9 mm $\qquad$ 9.7 mm .

These effects are critical for precise distance meters such as, for example, the Mekometer, which has an internal accuracy of 0.2 mm. This may require precise levelling for the determination of the height differences. The knowledge of the deviations of the vertical is essential in this case if the accuracy in measurement is not to be spoiled by inaccurate reduction due to errors in, or omission of, the deviations of the vertical.

The fourth term in Eq. (7.57) demonstrates the effect of the difference between the actual coefficient of refraction and the assumed value of k = 0.0, which is typically used for EDM reduction. It has been shown in Section 5.7 that the coefficient of refraction of "grazing rays" can easily vary between k = 0 at sunrise and sunset and k = −2 or 3 in the middle of the day. Based on an (optimistic) error of $\delta(k) = 1.0$ in the assumed value of k, the following errors $\delta(d_4)$ may be computed for the spheroidal distance:

	$d_1 = 100$ m	300 m	600 m	1000 m	2000 m	3000 m
$\Delta H = 100$ m	0.8 mm	2.3 mm	4.7 mm	7.8 mm	15.6 mm	23.4 mm
300 m		7.0 mm	14.0 mm	23.4 mm	46.8 mm	70.2 mm
1000 m					156.0 mm	234.0 mm

The resulting errors $\delta(d_4)$ are quite large, due to the large error in k. Smaller $\delta(k)$ may be expected, if distances are measured well above the ground ($\delta(k) = 0.1-0.4$). Two or three times larger $\delta(k)$ may be expected on clear, sunny days in summer, as indicated in Section 5.7. For precise EDM, large height differences should be avoided or the coefficient of refraction should be determined, through reciprocal trigonometric levelling for example.

8 Miscellaneous Corrections, Computations and Numerical Examples

Occasionally, some additional EDM corrections are required in order to achieve the collinearity requirements between EDM wave path and zenith angle ray paths which were assumed during the derivation of the geometrical corrections in Sect. 7. A number of approaches to the problem are discussed below. These derivations are followed by the discussion of three EDM-related computational problems, namely the computation of height differences from measured zenith angles and slope distances (EDM-Trigonometric Heighting), the computation of the coefficient of refraction from reciprocal zenith angle (and slope distance) measurements and the centring of measured zenith angles and slope distances from satellite stations to their respective centre stations. This section then concludes with fully worked examples for the reduction to the spheroid of the measurements of a short and a long EDM line.

8.1 Correction of Measured Distance to Zenith Angle Ray Path

Whenever measured distances are to be reduced with the aid of measured zenith angles (see Sect. 7.2) although they were not measured along the zenith angle's ray path, a correction from the EDM wave path (d_{EDM}) to the zenith angle ray path (d_{TH}) is required. The resulting distance (d_{TH}) refers then to the trunnion axis heights of theodolite and theodolite target applicable during the zenith angle measurement. The correction equations depend on the degree of integration between the EDM instruments and theodolites concerned.

Such corrections are not necessary in the case of electronic tacheometers (total stations) with co-axial EDM and visual telescope axes measuring zenith angles and distances to the same reflector and in cases where the vertical offset of the distance meter from the theodolite is compensated by the same vertical offset of a special reflector-target assembly. This aspect is discussed later in more detail.

8.1.1 Correction for Unequal Heights of Theodolite, EDM Instrument, Target and Reflector

The correction derived below is typically required when zenith angles and distance measurements are executed independently with separate station occupancies.

The correction is best applied after the application of velocity corrections (see Sect. 6.2 and 6.4) and the first arc-to-chord correction (see Sect. 7.2.3.1) but before the slope correction. The situation is depicted in Fig. 8.1, where the following elements are defined:

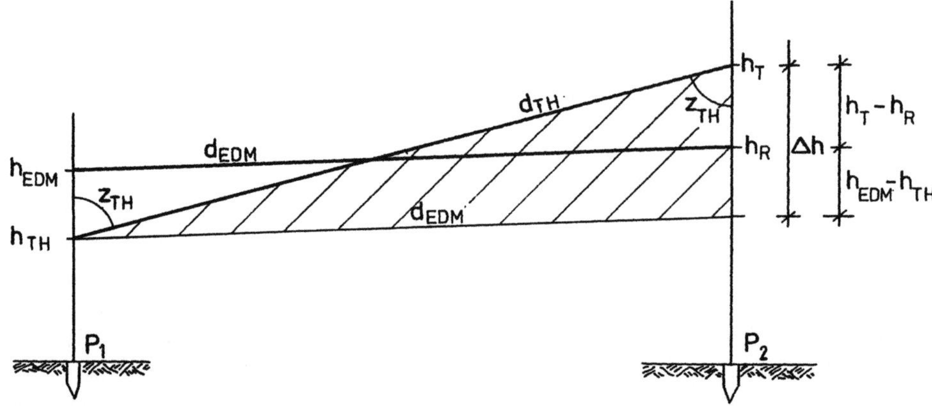

Fig. 8.1. Principle of the reduction of the measured distance (d_{TH}) to the zenith angle ray path distance (d_{TH}) using the measured zenith angle (z_{TH}) at P_1 and the measured heights above ground marks of theodolite (TH), EDM instrument (EDM), reflector (R) and traversing target (T)

d_{EDM} = wave path chord (d_2)
d_{TH} = distance along zenith angle ray
h_{EDM} = height of trunnion axis of EDM instrument (above survey mark P_1)
h_{TH} = height of trunnion axis of theodolite (above survey mark P_1)
h_T = height of target (above survey mark P_2)
h_R = height of reflector (above survey mark P_2)
z_{TH} = measured zenith angle at station P_1 between theodolite and target.

Application of the cosine rule in the hatched triangle together with the equation for Δh:

$$\Delta h = h_{EDM} - h_{TH} + h_T - h_R \tag{8.1}$$

leads to

$$d_{EDM}^2 = d_{TH}^2 + \Delta h^2 - 2 d_{TH} \Delta h \cos z_{TH} \tag{8.2}$$

and

$$d_{TH}^2 = d_{EDM}^2 + 2 d_{TH} \Delta h \cos z_{TH} - \Delta h^2 \; . \tag{8.3}$$

The distance d_{TH} may be replaced, with sufficient accuracy, by d_{EDM} in the second term of Eq. (8.3), because the value of the term is very small and because the difference between d_{TH} and d_{EDM} is small.

$$d_{TH} = d_{EDM} \left(1 + \frac{2 \Delta h \cos z_{TH}}{d_{EDM}} - \frac{\Delta h^2}{d_{EDM}^2} \right)^{1/2} \; . \tag{8.4}$$

With a series expansion for the square root and neglecting terms of third and higher order

$$d_{TH} = d_{EDM} + \Delta h \cos z_{TH} - \frac{\Delta h^2}{2 d_{EDM}} , \qquad (8.5)$$

where Δh is defined by Eq. (8.1).

The above reduction procedure is used if, for example, the theodolite observations are carried out independently of the EDM measurements (viz. with different equipment, on different days). On steep and short lines, the heights above the survey mark of theodolite, EDM instrument, target and reflector have to be measured carefully and accurately. The volume of computations may be reduced at least for the reduction of the distance, by using constrained centring. The relationships between h_{TH} and h_{EDM} as well as between h_T and h_R become constant and may be measured with high accuracy, thus reducing the errors of Δh in Eq. (8.5).

8.1.2 Correction for Theodolite-Mounted EDM Instruments

The reduction from the measured distance to the distance along the zenith angle ray path for theodolite (or standard) mounted distance meters follows easily from the derivation given in the previous section as the trunnion axis of the distance meter will always be vertically above the trunnion axis of the theodolite. The correction applies in cases where the slope distance and the zenith angle are measured to the centre of the reflector. (No correction is necessary if the vertical offset of the distance meter is compensated by a special reflector-target assembly with the prism vertically offset from the target by the same amount.)

This necessary correction may be derived from Eqs. (8.1) and (8.5) in Section 8.1.1.

$$d_{TH} = d_{EDM} + \Delta h \cos z_{TH} - \frac{\Delta h^2}{2 d_{EDM}} \qquad (8.5)$$

$$\Delta h = h_{EDM} - h_{TH} + h_T - h_R . \qquad (8.1)$$

Because h_T is equal to h_R and $(h_{EDM} - h_{TH})$ may be renamed as eccentricity e, Eq. (8.5) may be written as

$$d_{TH} = d_{EDM} + e \cos z_{TH} - \frac{e^2}{2 d_{EDM}} , \qquad (8.6)$$

where d_{TH} = distance along zenith angle ray
$\quad\quad\;\; d_{EDM}$ = measured distance (wave path chord)
$\quad\quad\;\; e$ = eccentricity of EDM instrument
$\quad\quad\quad\;\;$ = height of EDM instrument minus height of theodolite
$\quad\quad\;\; z_{TH}$ = measured zenith angle.

The correcting terms in Eq. (8.6) are functions of zenith angle and distance. Their magnitude is worked out in a short example. Assuming an eccentricity $e = 0.2\,m$ the following correction terms are obtained:

z_{TH}	$e \cos z_{TH}$	d_{EDM}	$\dfrac{e^2}{2\,d_{EDM}}$
70°	+ 68 mm	5 m	− 4 mm
80°	+ 35 mm	10 m	− 2 mm
85°	+ 17 mm	20 m	− 1 mm
90°	0 mm		
95°	− 17 mm		
100°	− 35 mm		
110°	− 68 mm		

8.1.3 Correction for Telescope-Mounted EDM Instruments

Telescope-mounted EDM instruments are usually adjusted so as to have an optical axis parallel to the optical axis of the theodolite's telescope. Figure 8.2 indicates that distance and zenith angle measurements require two independent pointings and a reduction from d_{EDM} to d_{TH} as long as both measurements are taken to the centre of the prism.

The problem is explained in Fig. 8.2, which depicts a vertical section through the theodolite (T), EDM instrument (E) and reflector (R). Assuming that the EDM axis (ER', E'R) is always parallel to the optical axis of the theodolite (TR, TR''), the telescope is first pointed at R, taking the zenith angle and the direction, then the EDM instrument is aligned with R. This tilt causes the EDM instrument

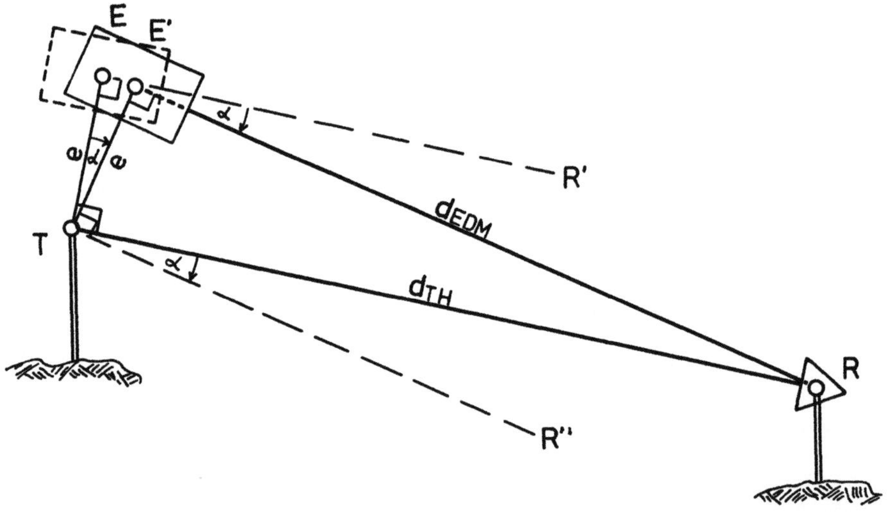

Fig. 8.2. Principle of the reduction of measured distances to the distance along the zenith angle ray for telescope mounted EDM instruments. It is assumed that slope distance as well as zenith angle are measured to the centre of the reflector. The optical axis of the distance meter during the zenith angle measurement is indicated by the line ER'. The optical axis of the theodolite points along the line TR'' during the distance measurement

centre E to move to E'. The distance d_{EDM} is then measured from E'. Defining the eccentricity of the EDM instrument with e, the corrected distance d_{TH} between the horizontal axis of the theodolite T and that of the reflector R may be computed as

$$d_{TH} = (d_{EDM}^2 + e^2)^{1/2} \tag{8.7}$$

$$= d_{EDM} \left(1 + \frac{e^2}{d_{EDM}^2}\right)^{1/2}. \tag{8.8}$$

Expanding the root term yields

$$d_{TH} = d_{EDM} \left(1 + \frac{e^2}{2d_{EDM}^2} - \ldots\right) \tag{8.9}$$

$$= d_{EDM} + \frac{e^2}{2d_{EDM}}. \tag{8.10}$$

The correcting term in Eq. (8.10) depends on the eccentricity of the EDM instrument and the measured distance. A numerical example illustrates the magnitude of the correction. Assuming an eccentricity of 0.1 m, the correction is $+0.5$ mm for $d = 10$ m and $+1$ mm for $d = 5$ m and therefore smaller than the usual precision of such instruments.

8.2 Eye-to-Object Corrections for Zenith Angles and Distances

Occasionally it is of advantage to reduce separately measured zenith angles and slope distances to the corresponding values between the ground marks at the terminals. This approach is commonly used to fulfil the requirement for collinearity between two (reciprocal) zenith angles and a slope distance unless collinearity can be achieved in the field by suitable procedures and equipment.

8.2.1 Eye-to-Object Correction for Zenith Angles

The zenith angle z_G from the survey mark at P_1 to the survey mark at P_2 can be derived from the upper shaded triangle in Fig. 8.3. Using the sine rule in the said triangle yields

$$\sin \Omega = \frac{h_T - h_{TH}}{d_G} \sin (z_{TH} + \delta - \gamma) \tag{8.11}$$

and

$$\Omega = \text{arc sin} \left(\frac{h_T - h_{TH}}{d_G} \sin (z_{TH} + \delta - \gamma)\right) \tag{8.12}$$

or, using a series expansion,

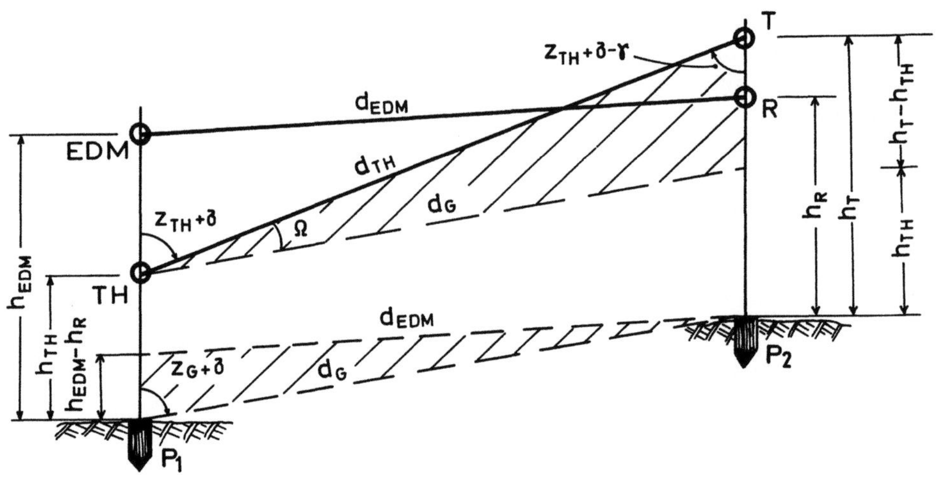

Fig. 8.3. Principle of the eye-to-object correction for zenith angles and EDM distances

$$\Omega = \frac{h_T - h_{TH}}{d_G} \sin(z_{TH} + \delta - \gamma) + \frac{(h_T - h_{TH})^3}{6d_G^3} \sin^3(z_{TH} + \delta - \gamma) + \ldots , \qquad (8.13)$$

where z_{TH} = measured zenith angle (at P_1)
δ = refraction angle [Eq. (7.46)]
γ = angle between spheroid normals throught P_1 and P_2 [see Eqs. (7.50) and (8.25)]
h_T = height of target (at P_2)
h_{TH} = height of theodolite (at P_1).

The second term in Eq. (8.13) amounts to less than one second of arc for distances larger than 16 m, if $(h_T - h_{TH})$ is less than 0.5 m. It can be ignored in most cases. It can also be shown that $(\delta - \gamma)$ in the first term affects Ω by a maximum of 0.02" if $(h_T - h_{TH})$ is less than 0.5 m. For most practical cases, the above equation can be simplified to:

$$\Omega \approx \frac{h_T - h_{TH}}{d_G} \sin z_{TH} . \qquad (8.14)$$

The zenith angle between ground marks computes then as

$$z_G = z_{TH} + \Omega . \qquad (8.15)$$

8.2.2 Eye-to-Object Correction for Distances

The slope distance d_G between the ground marks at P_1 and P_2 can be derived by application of the cosine rule to the lower shaded triangle in Fig. 8.3.

$$d_{EDM}^2 = (h_{EDM} - h_R)^2 + d_G^2 - 2(h_{EDM} - h_R)d_G \cos(z_G + \delta) \qquad (8.16)$$

$$= d_G^2 \left[1 + \frac{(h_{EDM} - h_R)^2}{d_G^2} - \frac{2(h_{EDM} - h_R)}{d_G} \cos(z_G + \delta) \right] . \qquad (8.17)$$

Solution for d_G yields the rigorous equation

$$d_G = d_{EDM} \left[1 + \frac{(h_{EDM} - h_R)^2}{d_G^2} - \frac{2(h_{EDM} - h_R)}{d_G} \cos(z_G + \delta) \right]^{-0.5} . \qquad (8.18)$$

Using a series expansion for the bracket term leads to the following simplied expression

$$d_G = d_{EDM} - \frac{(h_{EDM} - h_R)^2 d_{EDM}}{2 d_G^2} + \frac{(h_{EDM} - h_R) d_{EDM}}{d_G} \cos(z_G + \delta) + \dots , \qquad (8.19)$$

where
d_G = slope distance between ground marks
d_{EDM} = measured slope distance (d_2)
h_{EDM} = height of EDM instrument (at P_1)
h_R = height of reflector (at P_2).

Equations (8.15) and (8.18) include the unknown d_G on the right hand side of the equations. An iterative solution procedure is therefore required. It is suggested to compute an initial value of d_G by using d_{EDM} for d_G and z_{TH} for z_G on the right hand side of Eqs. (8.18) or (8.19). This initial value of d_G is then used to compute z_G through Eqs. (8.15) and (8.12) or (8.14), which, in turn, is then used in the final computation of d_G.

The second computation of d_G can be omitted, if maximum errors of 1 mm and 10 seconds of arc in d_G and z_G are permissible and the following conditions are met:

$$|h_{EDM} - h_R| < 0.5\,\text{m} \;\; ; \;\; |h_T - h_{TH}| < 0.5\,\text{m}$$

$$\text{and} \quad 70° < z < 110° \qquad ; \quad d_{EDM} > 30\,\text{m}$$

$$\text{or} \quad 80° < z < 100° \qquad ; \quad d_{EDM} > 11\,\text{m} .$$

Similarly, maximum errors of 0.1 mm and 0.2 seconds of arc are not exceeded for d_G and z_G, respectively, if the following restrictions apply:

$$|h_{EDM} - h_R| < 0.5\,\text{m} \;\; ; \;\; |h_T - h_{TH}| < 0.5\,\text{m}$$

$$\text{and} \quad 70° < z < 110° \qquad ; \quad d_{EDM} > 200\,\text{m}$$

$$\text{or} \quad 80° < z < 100° \qquad ; \quad d_{EDM} > 60\,\text{m} .$$

8.2.3 Numerical Example

Given are
$d_{EDM} = 400.000\,\text{m}$; $z_{TH} = 81°00'00''$
$h_{EDM} = 2.000\,\text{m}$; $h_R = 0.700\,\text{m}$
$h_{TH} = 1.000\,\text{m}$; $h_T = 1.500\,\text{m}$.

Using $d_G = 400.000\,\text{m}$ and $z_G = 81°00'00''$ in Eq. (8.19) yields

$$d_G^0 = 400.000 - 0.0021 + 0.2034 \text{ m}$$

$$= 400.2013 \text{ m} \ .$$

Substitution in Eq. (8.14) leads to

$$\Omega = 0.0012340 \text{ rad}$$

$$= 4'14.5''$$

$$z_G = 81°04'14.5'' \ .$$

Using $d_G = 400.2013$ m and $z_G = 81°04'14.5''$ in Eq. (8.19) gives finally

$$d_G = 400.000 - 0.0021 + 0.2017 \text{ m}$$

$$= 400.1996 \text{ m} \ .$$

As a check, z_G may be recomputed with the final value for d_G.

8.3 Height Difference from Measured Zenith Angle(s) and Slope Distance

With the introduction of EDM, the determination of height differences by trigonometric levelling has become very convenient. This combination of conventional trigonometric levelling and EDM (and, often, precise electronic theodolites) let the new survey method of EDM-height traversing become a real alternative to spirit levelling, both in precision and in speed. A summary of this new technique may be found in Rüeger and Brunner (1982). More recently, one-kilometre standard deviations of ± 1 mm to ± 2 mm have been reported for simultaneous reciprocal EDM-height traversing (Kasser 1985; Whalen 1985).

In the field, height differences are required for a wide range of setting-out work. Most semi-electronic tacheometers and all fully electronic tacheometers provide a facility for the on-board computation of height differences, with the first type of instrument requiring a manual entry of observed zenith angles. Users of the reduction facilities of such instruments should be fully aware of the equations on which a tacheometer's software is based. This includes the knowledge of the value of the coefficient of refraction used. The basic equations for the computation of height differences are developed below and may be used to critically review the formulae which are implemented in electronic tacheometers.

8.3.1 Single Zenith Angle Measurement

The problem is depicted in Fig. 7.3 (Sect. 7.2.3). The deviations of the vertical are ignored because orthometric heights (= geoidal heights) are sought.

The angle of refraction δ can be computed according to Eq. (7.46) and Fig. 7.2 as follows:

$$\delta = \frac{\beta}{2} = \frac{d_2}{2r} = \frac{d_2 k}{2R} \simeq \frac{k}{2}\gamma \ . \tag{7.46}$$

In triangle $P_1 P_2 P_2'$, the following angles can be derived:

$$v = z_1 + \delta - \gamma = z_1 - \tfrac{\gamma}{2}(2 - k) \tag{8.20}$$

$$\alpha = \pi - (\tfrac{\pi}{2} - \tfrac{\gamma}{2}) - (z_1 + \delta) = \tfrac{\pi}{2} - [z_1 - \tfrac{\gamma}{2}(1 - k)] \ . \tag{8.21}$$

Based on Fig. 7.3 and Eqs. (7.46), (8.20) and (8.21), the sine-rule applied to the triangle $P_1 P_2 P_2'$ gives

$$H_2 - H_1 = d_2 \, \frac{\sin \{\tfrac{\pi}{2} - [z_1 - \tfrac{\gamma}{2}(1 - k)]\}}{\sin (\tfrac{\pi}{2} + \tfrac{\gamma}{2})} = d_2 \, \frac{\cos [z_1 - \tfrac{\gamma}{2}(1 - k)]}{\cos \tfrac{\gamma}{2}} \ . \tag{8.22}$$

Considering that γ is very small, cosine $\tfrac{\gamma}{2}$ is taken as unity.

$$H_2 - H_1 = d_2\{\cos z_1 \cos [\tfrac{\gamma}{2}(1 - k)] + \sin z_1 \sin [\tfrac{\gamma}{2}(1 - k)]\} \ . \tag{8.23}$$

With $\cos [\tfrac{\gamma}{2}(1 - k)] = 1.00$ and $\sin [\tfrac{\gamma}{2}(1 - k)] = \tfrac{\gamma}{2}(1 - k)$ as in Section 7.2.3.2:

$$H_2 - H_1 = d_2 \cos z_1 + \tfrac{\gamma}{2}(1 - k) d_2 \sin z_1 \ . \tag{8.24}$$

The unknown parameter $\tfrac{\gamma}{2}$ may be derived from Eq. (7.44) in Section 7.2.2:

$$\frac{\gamma}{2} = \frac{d_2 \sin z_1}{2(R + H_1) + 2 d_2 \cos z_1} \ . \tag{8.25}$$

Ignoring the second term of the denominator and setting $H_1 = 0$, produces in conjunction with Eq. (8.23) the final form given by Brunner (1973):

$$H_2 - H_1 = d_2 \cos z_1 + \frac{(1 - k)}{2R} \, (d_2 \sin z_1)^2 \ , \tag{8.26}$$

where $H_2 - H_1$ = orthometric height difference between P_1 and P_2
        d_2 = wave path chord
        k = coefficient of refraction of light
        R = radius of curvature of spheroid along the line [see Eqs. (7.1) and (7.6)]
        z_1 = observed zenith angle in P_1.

So far, all data are assumed to be measured or reduced to the points P_1 and P_2. The omission of H from the denominator of Eq. (8.25) causes an error in $(H_2 - H_1)$, computed according to Eq. (8.26) of only 1 mm for a height above sea level of 1000 m and a horizontal distance of 10 km.

A graph depicting the accuracy of Eq. (8.26) is given by Brunner (1973). The maximum error in $(H_2 - H_1)$ is < 0.1 mm for slope distances < 2.5 km and height differences < 1000 m.

To cater for the heights of theodolite (h_{TH}) and target (h_T) at the terminals and considering further the corrections of Section 8.1 (if required), the above equation may be written as

$$H_2 - H_1 = d_{TH} \cos z_{TH} + \left(\frac{1 - k}{2R}\right) (d_{TH} \sin z_{TH})^2 + h_{TH} - h_T \ . \tag{8.27}$$

The coefficient of refraction in Eqs. (8.26) and (8.27) refers to theodolite observations and not to the EDM measurement. It has been discussed in Section 5.7 that the actual coefficient of refraction of "grazing" rays (close to the ground) may vary between (-3.0) and $(+4.0)$. Grazing rays are usually encountered in short range EDM. As the actual coefficient of refraction is usually unknown, the height differences are computed assuming a mean coefficient of refraction of 0.13. (For lines close to the ground, $k = 0.0$ would be as good or bad an assumption!)

The difference $\delta(k)$ between the mean and the actual value of the coefficient of refraction will affect the computed height difference $(H_2 - H_1)$ by an error $\delta(H_2 - H_1)$. The equation for $\delta(H_2 - H_1)$ is obtained by differentiating Eq. (8.26) with respect to k:

$$\delta(H_2 - H_1) = \frac{(d_2 \sin z_1)^2}{2R} \, \delta k \tag{8.28}$$

The table below gives some values for likely errors $\delta(H_2 - H_1)$.

$d_2 \sin z_1 =$	$\delta(H_2 - H_1)$ for $\delta k = 1.0$	$\delta(H_2 - H_1)$ for $\delta k = 2.0$
100 m	0.8 mm	1.6 mm
300 m	7.0 mm	14.1 mm
500 m	19.6 mm	39.2 mm
1000 m	78.5 mm	157.0 mm

The above table should be kept in mind when assessing the reliability of heights obtained by electronic tacheometry. If the above uncertainties are not tolerable, the coefficient of refraction can easily be determined on a particular day and at a particular location by reciprocal zenith angle measurements on one or more lines. The details are discussed in the following section.

8.3.2 Reciprocal Zenith Angle Measurements

The uncertainty in the coefficient of refraction k can be greatly reduced if reciprocal, simultaneous zenith angles are observed. The counterpart of Eq. (8.26) in the case of reciprocal zenith angle observations reads (Brunner 1975):

$$H_2 - H_1 = \frac{d_2}{2} (\cos z_{12} - \cos z_{21}) , \tag{8.29}$$

where $H_2 - H_1$ = orthometric height difference between P_1 and P_2
$\quad\quad\quad d_2$ = wave path chord
$\quad\quad\quad z_{12}$ = zenith angle at P_1 to P_2
$\quad\quad\quad z_{21}$ = zenith angle at P_2 to P_1
$\quad\quad\quad H_1, H_2$ = orthometric (= geoidal) heights.

Equation (8.29) can be easily derived from Eq. (8.26) and assumes colinear measurements of both zenith angles and the distance. If this assumption does not hold, the eye-to-object corrections derived in Section 8.2 should be applied to all three measured quantities prior to the application of the above equation. The corresponding equation for the determination of spheroidal height differences may be found in Brunner (1975).

It was assumed in Eq. (8.29), that the coefficients of refraction are equal for the reciprocal observations. This will rarely eventuate. The uncertainty of the height difference $(H_2 - H_1)$ caused by the uncertainty of the difference of reciprocal coefficients of refraction $(k_1 - k_2)$ may be calculated as follows (Brunner 1975):

$$\sigma_{(H_2 - H_1)} = \frac{(d_2 \sin z_1)^2}{4R} \, \sigma_{(k_1 - k_2)} \, , \tag{8.30}$$

where $\sigma_{(H_2 - H_1)}$ and $\sigma_{(k_1 - k_2)}$ denotes the appropriate standard deviations.

The following values have been estimated for $\sigma_{(k_1 - k_2)}$ (Brunner 1975):

$\sigma_{(k_1 - k_2)} = \pm 0.3$ for simultaneous, reciprocal zenith angle observations, and

$\sigma_{(k_1 - k_2)} = \pm 0.5$ for non-simultaneous, reciprocal zenith angle observations.

A short table may illustrate the effect of the standard deviation of Δk on the standard deviation of the height difference ΔH:

$d_2 \sin z_1$	$\sigma_{\Delta H}$ for $\sigma_{\Delta k} = \pm 0.3$	$\sigma_{\Delta H}$ for $\sigma_{\Delta k} = \pm 0.5$
100 m	± 0.1 mm	± 0.2 mm
300 m	± 1.0 mm	± 1.8 mm
500 m	± 2.9 mm	± 4.9 mm
1 000 m	± 11.8 mm	± 19.6 mm

8.4 Determination of the Coefficient of Refraction from Reciprocal Zenith Angle Measurements

Reciprocal and simultaneous zenith angles are sometimes measured at the same time as EDM observations in order to determine the coefficient of refraction k for the prevailing atmospheric conditions. Sometimes this is because the uncertainty of the coefficient of refraction determines the accuracy of the combined correction for K'', K_1, K_4 [see Eq. (7.35)] and therefore the accuracy of the distance reduction. In other cases, such as EDM tacheometry, the prevailing coefficient of refraction is determined on a few lines to improve the accuracy of computed height differences [see Eq. (8.28)].

8.4.1 Derivation of the Equation for the Coefficient of Refraction

Considering the triangle CP_1P_2 in Fig. 7.2, the following equations may be derived

$$\gamma + (\pi - \zeta_1 - \delta) + (\pi - \zeta_2 - \delta) = \pi \tag{8.3}$$

$$\gamma + \pi = 2\delta + \zeta_1 + \zeta_2 \ . \tag{8.32}$$

All angles are to be taken in radians.

From Fig. 7.2 and Section 7.1.1.1 follows that

$$\delta = \frac{\beta}{2} = \frac{d_1}{2r} = \frac{d_1 k}{2R} \ . \tag{7.46}$$

Substitution of Eq. (7.46) in Eq. (8.32) leads to

$$\gamma + \pi = \zeta_1 + \zeta_2 + \frac{d_1 k}{R} \tag{8.33}$$

$$= z_1 + z_2 + \frac{d_1 k}{R} + (\varepsilon_1 - \varepsilon_2) \ . \tag{8.34}$$

Considering Eq. (7.29), the above equation may be written as

$$\frac{d_1 k}{R} = \frac{d_4}{R} + \pi - (z_1 + z_2) - (\varepsilon_1 - \varepsilon_2) \ , \tag{8.35}$$

thus leading to the final rigorous equation for k

$$k = \frac{d_4}{d_1} + \frac{R}{d_1} \left[\pi - (z_1 + z_2) - (\varepsilon_1 - \varepsilon_2) \right] \ , \tag{8.36}$$

where all angles are in radians and where

k = coefficient of refraction
d_1 = wave path length
d_4 = spheroidal distance
R = radius of curvature of spheroid along the line (or mean radius)
z_1, z_2 = observed zenith angle at P_1 and P_2 respectively
$\varepsilon_1, \varepsilon_2$ = deviation of vertical at P_1 and P_2 respectively (see Fig. 7.2 for sign of ε).

A first simplification of the rigorous Eq. (8.36) may be obtained by assuming $d_4 = d_1 \sin z_1$ and $\varepsilon_1 = \varepsilon_2$ and by expansion with $1/\sin z_1$:

$$\frac{k}{\sin z_1} = \frac{d_1}{d_1} + \frac{R\left[\pi - (z_1 + z_2)\right]}{d_1 \sin z_1} \ . \tag{8.37}$$

Setting $\sin z_1 = 1.0$ for the denominator on the left hand side of Eq. (8.37) leads to the form given by Rüeger and Brunner (1981, 1982):

$$k = 1.0 + \frac{R\left[\pi - (z_1 + z_2)\right]}{d_1 \sin z_1} \ . \tag{8.38}$$

This equation is suitable for most short range determinations of the coefficient of refraction. For the computation of the coefficient of refraction on long EDM lines, Eq. (8.36) may be further simplified, by assuming $d_1 = d_4$ and $\varepsilon_1 = \varepsilon_2$, to the form given by Kahmen and Faig (1988):

$$k = 1 + \frac{R}{d_1} \{\pi - (z_1 + z_2)\} \ . \tag{8.39}$$

It is evident from Eq. (8.36) that an omission of the (usually unknown) deviations of the vertical will not affect the value of k as long as the deviations are equal at both terminals of the line.

8.4.2 Error Analysis

The total differential of Eq. (8.36) yields

$$\delta k = -\left(\frac{R}{d_1}\right)\delta z_1 - \left(\frac{R}{d_1}\right)\delta z_2 - \left(\frac{R}{d_1}\right)\delta\varepsilon_1 + \left(\frac{R}{d_1}\right)\delta\varepsilon_2 \ , \tag{8.40}$$

where δz, $\delta\varepsilon$ are in radians. An error of one second of arc in z_1, z_2, ε_1 or ε_2 produces the following errors δk in k

$$
\begin{aligned}
d_1 &= 10 \ \text{km} & \delta k &= 0.003 \\
&= 30 \ \text{km} & &= 0.001 \\
&= 50 \ \text{km} & &= 0.0006 \\
&= 70 \ \text{km} & &= 0.0004 \ .
\end{aligned}
$$

Assuming that

$$\sigma_{z_1} = \sigma_{z_2} \quad \text{and}$$

$$\sigma_{\varepsilon_1} = \sigma_{\varepsilon_2} = 0 \ ,$$

Eq. (8.40) may be expressed as

$$\sigma_k = \sqrt{2} \left(\frac{R \sin(1'')}{d_1}\right) \sigma_z \ , \tag{8.41}$$

where σ_z is now to be taken in seconds of arc.

To investigate the effect of the accuracy of measured zenith angles on the derived coefficient of refraction and the combined correction $(K'' + K_1 + K_4)$, Eq. (7.35) is differentiated to give [see also Eq. (7.40)]:

$$\sigma_{(K'' + K_1 + K_4)} = \frac{d_1^3}{12 R^2}(1 - k)\sigma_k \ . \tag{8.42}$$

To achieve an accuracy of, say, 0.3 ppm in the combined correction $(K'' + K_1 + K_4)$, the accuracy of the coefficient of refraction should not be less than the values listed below

$$
\begin{aligned}
d_1 &= 10 \ \text{km} & \sigma_k &= \pm 1.46 \\
&= 30 \ \text{km} & &= \pm 0.16 \\
&= 50 \ \text{km} & &= \pm 0.06 \\
&= 70 \ \text{km} & &= \pm 0.03
\end{aligned}
$$

A worst case of $(1 - k) = 1.0$ has been considered in the above table.

On longer EDM lines, the determination of the coefficient of refraction is therefore justified if large deviations of the coefficient of refraction from its mean value of 0.13 are expected. The necessary accuracy of the zenith angle measurements may be computed from

$$\sigma_{(K''+K_1+K_4)} = \frac{\sqrt{2}\,d_1^2\,\sin\,(1'')}{12\,R}(1-k)\,\sigma_z \ . \tag{8.43}$$

This equation can be easily derived from Eqs. (8.41) and (8.42).

To obtain the combined correction $(K''+K_1+K_4)$ to an accuracy of ± 0.3 ppm, the reciprocal and simultaneous zenith angles must be measured with the following accuracies

$$
\begin{aligned}
d_1 &= 10\,\text{km} & \sigma_z &= \pm 5.6' \\
    &= 30\,\text{km} &   &= \pm 1.9' \\
    &= 50\,\text{km} &   &= \pm 1.1' \\
    &= 70\,\text{km} &   &= \pm 0.8'
\end{aligned}
$$

Again, the term $(1-k)$ has been taken as 1.

8.5 Reduction to Centre of Distances

Distances (and angles) are sometimes measured from a satellite station, because the permanent mark is occupied by a beacon or another instrument. Two different possibilities have to be considered. The zenith angles and horizontal directions may be measured at the permanent mark or at the satellite station. A three dimensional solution is given, assuming that the correction for unequal heights has already been applied, where necessary.

8.5.1 Angles and Distances Measured at Satellite Station

The problem is depicted in Fig. 8.4, where

$$
\begin{aligned}
S &= \text{satellite station} \\
C &= \text{permanent station} \\
P_1 &= \text{target (reflector) station number one} \\
(P_1) &= \text{projection of } P_1 \text{ on sphere of radius } r = 1 \\
(C) &= \text{projection of } C \text{ on sphere of radius } r = 1 \\
z_C &= \text{measured zenith angle from S to C} \\
z_1 &= \text{measured zenith angle from S to } P_1 \\
e &= \text{measured slope distance between S and C (eccentric distance)} \\
d^* &= \text{measured slope distance between S and } P_1 \\
d &= \text{centred slope distance between C and } P_1 \\
\alpha &= \text{measured horizontal angle } (C)'S(P_1)', \text{ clockwise from x axis} \\
(C)' &= \text{projection of } (C) \text{ on xy-plane} \\
(P_1)' &= \text{projection of } (P_1) \text{ on xy-plane} \\
\Phi &= \text{angle } (C)S(P_1) = \text{angle } CSP_1 .
\end{aligned}
$$

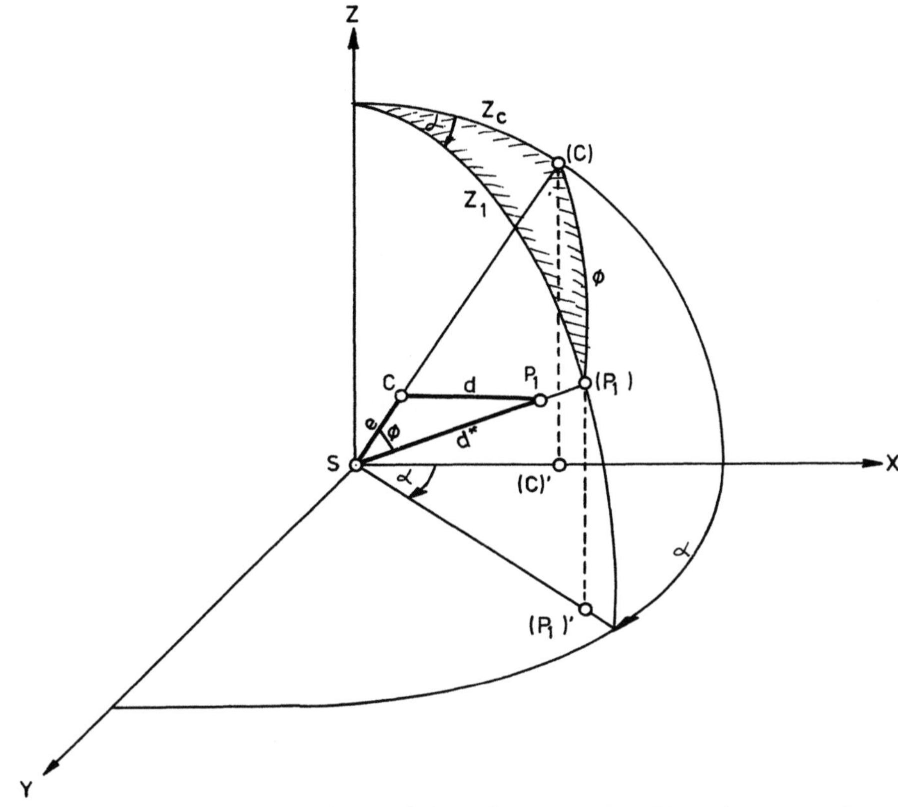

Fig. 8.4. Reduction to the centre station C of slope distances and zenith angles measured at a satellite station S. (Explanation see text)

The slope distance d between C and P_1 is unknown and must be determined.

In Fig. 8.4, the three-dimensional coordinate system has its origin at S and its $x-z$ plane through C; an application of the cosine rule of spherical trigonometry to the hatched triangle yields:

$$\cos \phi = \cos z_C \cos z_1 + \sin z_C \sin z_1 \cos \alpha \ . \tag{8.44}$$

Applying the cosine rule to the plane triangle SCP_1 gives

$$d^2 = d^{*2} + e^2 - 2d^* e \cos \phi \tag{8.45}$$

$$d = d^* \left(1 + \frac{e^2}{d^{*2}} - \frac{2e \cos \phi}{d^*} \right)^{1/2} \ . \tag{8.46}$$

Combining Eqs. (8.46) and (8.44) leads to:

$$d = d^* \left[1 + \frac{e^2}{d^{*2}} - \frac{2e}{d^*} (\cos z_C \cos z_1 + \sin z_C \sin z_1 \cos \alpha) \right]^{1/2} \ . \tag{8.47}$$

Equation (8.47) is a rigorous formula for the slope distance d between the two permanent stations C and P_1.

The zenith angle at S to P_1 has also to be centred. This is achieved by forming the sum of height differences:

$$\Delta H(CP_1) = \Delta H(SP_1) - \Delta H(SC) \tag{8.48}$$

$$d \cos z_{C_1} = d^* \cos z_1 - e \cos z_C . \tag{8.49}$$

The centred zenith angle from C to P_1 (z_{C_1}) now becomes:

$$\cos z_{C_1} = \frac{d^*}{d} \cos z_1 - \frac{e}{d} \cos z_C , \tag{8.50}$$

where d^*, e, z_1, z_C are measured and d is known from Eq. (8.47).

8.5.2 Angles Measured at Centre Station, Distances at Satellite Station

This situation occurs in tacheometry when an EDM instrument cannot be fitted to a theodolite. Distances and angles to radiation points are taken simultaneously at the EDM and theodolite stations.

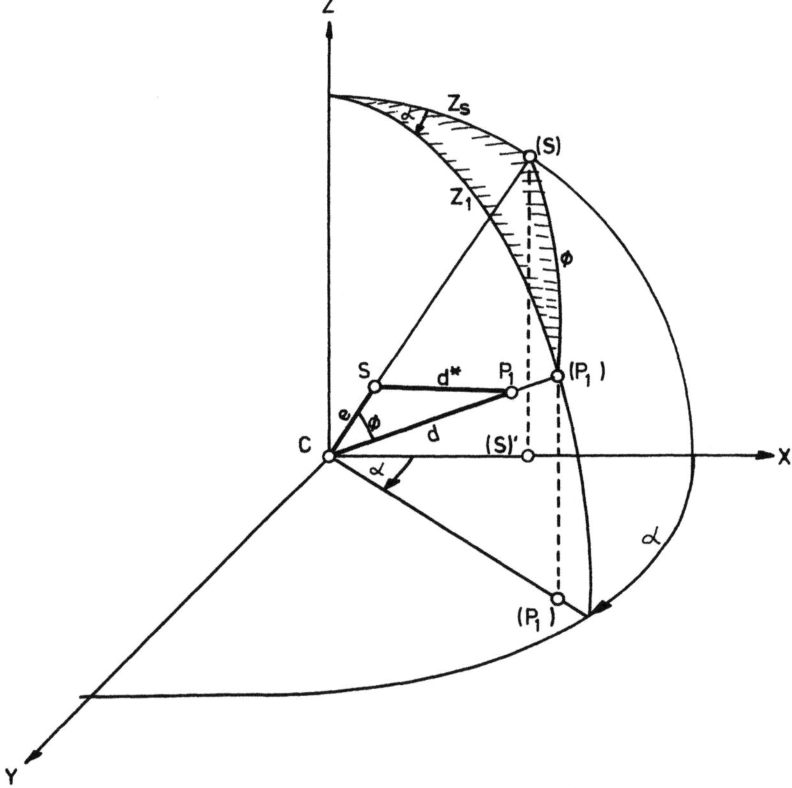

Fig. 8.5. Reduction of slope distances measured at a satellite station S to the centre station C in cases where the horizontal and zenith angles are measured at the centre station

116

The notation in Fig. 8.5 is the same as in Fig. 8.4. As in the previous section, a three-dimensional co-ordinate system is assumed. In this case, the origin is coincident with station C. The unknown slope distance d between centre station C and target station P_1 may be found as

$$d = d^* \left[1 - \frac{e^2}{d^{*2}} + \frac{2\,e\,d}{d^{*2}} \left(\cos z_S \cos z_1 + \sin z_S \sin z_1 \cos \alpha \right) \right]^{1/2} , \qquad (8.51)$$

where d^* = measured slope distance between satellite S and P_1
        e   = measured slope distance between satellite S and centre station C (eccentric distance)
        z_S = measured zenith angle from C to S
        z_1 = measured zenith angle from C to P_1
        α = measured horizontal angle $(S)'C(P_1)'$, clockwise from x-axis.

The Eq. (8.51) is rigorous. Since d appears on both sides of Eq. (8.51) an approximate value d^* is taken for d and improved by iteration, if necessary. The subsequent slope reduction uses the zenith angle z_1. Direct solutions for the computation of the horizontal distance between C and P_1 may be found in Menzies (1972) and LeRoux (1976).

8.6 Numerical Examples

The process of reduction of EDM distances is explained with the aid of two numerical examples. The first example explains the reduction scheme for long distances, the second for short distances.

8.6.1 Reduction of a Long Distance

A distance was measured six times with a Siemens Albis SIAL MD 60 microwave distance meter between the trigonometric stations Perth and Ovens in the vicinity of Bathurst (Australia). Three observations were taken in the master mode and another three in the remote mode of each instrument. The following data set was recorded:

Measured distance $d' = 22\,395.667$ m (mean of six observations, corrected for instrumental constants).

Meteorological data:

Perth   p = 921.3 mb   Ovens   p = 882.6 mb
          t = 6.1 °C dry   t = 5.3 °C dry
          t' = 3.2 °C wet   t' = 3.1 °C wet

All pressures and psychrometer readings, corrected for instrumental constants of barometers and psychrometers, are mean values for the period of distance measurement. The station coordinates and heights are as follows

	E (m)	N (m)	Elevation (m)
Perth pillar	350 579.53	1 295 238.04	879.71 top of pillar
Ovens pillar	372 099.22	1 301 416.35	1 273.90 top of pillar

where the coordinates refer to the (Australian) New South Wales Integrated Survey Grid (I.S.G.) and the elevations to the Australian Height Datum (A.H.D.). Instrument heights above the top of the pillar may be ignored for this case as they are equal on both sides.

8.6.1.1 First Step: First Velocity Correction

Applying Eq. (6.21)

$$K' = \left\{ 320.0 - \frac{77.624(p-e)}{(273.15+t)} - \left[1 + \frac{5748}{(273.15+t)} \right] \frac{64.70e}{273.15+t} \right\} 10^{-6}d' \quad (6.21)$$

and Eq. (5.23)

$$e = E'_W - 0.000662\,p\,(t - t') \qquad (5.23)$$

allows the computation of the first velocity correction K' on both stations:

	Perth		Ovens	
t	6.1	°C	5.3	°C
t − t'	+2.9	°C	+2.2	°C
p	921.3	mb	882.6	mb
E'_w	7.68	mb	7.63	mb
e	5.91	mb	6.34	mb
p − e	915.39	mb	876.26	mb
K'	+0.8061 m		+0.9818 m	

Mean of first velocity corrections $K' = +0.8940$ m.

According to Eq. (6.8) the wave path length d may be computed as

$$d = d' + K' = 22\,395.667 + 0.8940 \text{ m}$$

$$= 22\,396.5610 \text{ m}$$

8.6.1.2 Second Step: Slope Correction

$$K_2 = -\frac{(H_2 - H_1)^2}{2\,d_2} - \frac{1}{8}\frac{(H_2 - H_1)^4}{d_2^3}\ . \qquad (7.22)$$

$$H_2 - H_1 = 394.19 \text{ m}$$

$$d_2 \simeq d = 22\,396.5610 \text{ m}$$

$$K_2 = -3.4690 \text{ m} - 0.0003 \text{ m}$$

$$K_2 = -3.4693 \text{ m}\ .$$

118

8.6.1.3 Third Step: Sea Level Correction

According to Eq. (7.28) and adopting $R = 6\,370\,100$ m yields

$$K_3 = -\frac{H_1 + H_2}{2R}\, d_2 + \frac{(H_1 + H_2)}{4R} \frac{(H_2 - H_1)^2}{d_2} \; . \tag{7.28}$$

$$K_3 = -\frac{(1\,273.900 + 879.71)}{2\,(6\,370\,100)}\,(22\,396.5610) + \frac{(2\,153.6100)\,(394.19)^2}{4\,(6\,370\,100)\,(22\,396.5610)}$$

$$K_3 = -3.7859 + 0.0006 \text{ m}$$

$$K_3 = -3.7853 \text{ m}$$

8.6.1.4 Fourth Step: Combined Second Velocity and Arc-to-Chord Corrections

The combined correction is defined by Eq. (7.35):

$$K'' + K_1 + K_4 = (1 - k)^2 \frac{(d_1)^3}{24\,R^2}.$$

Assuming $k = 0.25$ (mean value for microwaves) and $R_M = 6\,370\,100$ m (N.S.W., I.S.G.) yields:

$$K'' + K_1 + K_4 = (0.75)^2 \frac{(22\,396.561)^3}{24\,(6\,370\,100)^2} = +0.0065 \text{ m}$$

8.6.1.5 Fifth Step: Spheroidal Distance

Combining all former steps:

$$
\begin{aligned}
d &= 22\,396.5610 \text{ m}\\
K_2 &= -\quad 3.4693 \text{ m}\\
K_3 &= -\quad 3.7853 \text{ m}\\
K'' + K_1 + K_4 &= +\quad 0.0065 \text{ m}\\
\hline
d_4 &= 22\,389.3129 \text{ m}
\end{aligned}
$$

8.6.1.6 Sixth Step: Reduction to Map Grid

In most cases, geodetic computations are not carried out on the reference spheroid but rather in a (map) projection. The point or line scale factors which apply to specific national or state-wide projection systems may be taken from relevant publications.

In the context of this numerical example, the equation and parameters applicable to the Integrated Survey Grid of the State of New South Wales (Australia) have been selected. The equation below for the line scale factor is valid for all Transverse Mercator Projections, the constants and false origin definition, however, not.

$$K = k_0 \left(1 + \frac{y_1^2 + y_1 y_2 + y_2^2}{6 r_m^2} \right) , \tag{8.52}$$

where K = the line scale factor,
k_0 = the central scale factor (0.999 940),
r_m = $k_0 R_m$ = 6 369 700 m,
R_m = 6 370 100 m, the mean radius of curvature of spheroid (N.S.W. I.S.G.) and
y = Easting minus 300 000 m.

Thus: $K = 0.999\,940\,(1 + 0.000\,046\,839)$

$$= 0.999\,986\,836$$

The plane distance on grid (grid distance) S may be written as:

$$S = K d_4 \tag{8.53}$$

$$= 0.999\,986\,836\,(22\,389.3129)\ \text{m}$$

$$= 22\,389.018\ \text{m} \ .$$

This is the grid distance on the I S G Grid, Zone 55/3, equivalent to the measured EDM distance between Ovens and Perth.

8.6.2 Reduction of a Short Distance

A distance was measured from station A to station B with a HP 3805 A Distance Meter. The following observations were recorded:

Measured distance d'_{AB} = 587.134 m

Height of EDM instr. h_{EDM} = 1.652 m

Height of reflector h_R = 1.724 m

Measured zenith angle z_{TH} = 85°41′53″

Height of theodolite h_{TH} = 1.673 m

Height of target h_T = 1.534 m

Temperature at A t = 23.8 °C

Pressure at A p = 1008.3 mb

The following coordinates of station A are given for the NSW I.S.G. System, Zone 56/1:

	E	N
Station A:	236 637.897 m	1 271 028.524 m

The elevation H_A of A is 1058.21 m (A.H.D.). Both distance and zenith angle were measured at station A.

8.6.2.1 First Step: First Velocity Correction

The manufacturer states the following first velocity correction for the HP 3805 A:

$$K' = \left(278.7 - \frac{79.148\,p}{(273.15 + t)}\right) 10^{-6} d'$$ (8.54)

$$= \left(278.7 - \frac{(79.148)(1008.3)}{(273.15 + 23.8)}\right) 10^{-6}\,(587.134)$$

$$= +0.0058\ \text{m}\ .$$

Substituting K' in Eq. (6.8) leads to:

$$d = d' + K' = 587.134 + 0.0058 = 587.1398\ \text{m}\ .$$

The fourth decimal place is carried through the computation in order to obtain reductions correct to three decimal places. For the reduction of distances accurate only to several mm, the fourth decimal place may be omitted.

8.6.2.2 Second Step: Correction for Unequal Heights

Equation (8.1) in Section 8.1.1 reads:

$$\Delta h = h_{EDM} - h_{TH} + h_T - h_R = -0.211\ \text{m}$$ (8.1)

and Eq. (8.5)

$$d_{TH} = d_{EDM} + \Delta h \cos z_{TH} - \frac{\Delta h^2}{2\,d_{EDM}}\ .$$ (8.5)

Substituting $d_{EDM} = 587.1398$ m and $z_{TH} = 85°41'53''$ yields:

$$d_{TH} = 587.1398 - 0.0158 - 0.00004 = 587.1240\ \text{m}\ .$$

8.6.2.3 Third Step: Reduction to Sea Level

Repeating Eq. (7.48) of Section 7.2.2:

$$d_4 = R \arctan \left[\frac{d_1 \sin\,[z_1 + \varepsilon_1 + (d_1 k/2R)]}{R + H_1 + d_1 \cos\,[z_1 + \varepsilon_1 + (d_1 k/2R)]}\right]\ .$$ (7.48)

Assuming a coefficient of refraction of light $k = 0.13$, a mean radius $R_m = 6\,370\,100$ m and a deviation of the vertical $\varepsilon_1 = 0$ (and with $d_1 = d_{TH}$ and $H_1 = H_A + h_{TH}$) the above equation becomes:

$$d_4 = 6\,370\,100 \arctan \left(\frac{585.470\,09}{6\,371\,203.9}\right) = 585.3687\ \text{m}\ .$$

The second velocity correction only amounts to 1.4×10^{-13} m; it may be ignored.

8.6.2.4 Fourth Step: Reduction to Map Grid

It is sufficient to replace the *line scale factor* by the *point scale factor* for relatively short distances. In the case of the Integrated Survey Grid of the State of New South Wales (Australia) the point scale factor is given by

$$K = 0.99994 + 1.23\,(10^{-7}\,y)^2 \tag{8.55}$$

for $y = $ Easting $- 300\,000$ m. Considering Eq. (8.53) in Section 8.6.1 leads to:

$$S = K\,d_4$$
$$= [0.99994 + (1.23)\,(0.0063)^2]\,585.3687$$
$$= 585.362 \text{ m} \ ,$$

where S is the grid distance on ISG, Zone 56/1.

9 Electro-Optical Distance Meters

Distance meters featuring visible light or near infrared (NIR) radiation as carrier waves are called electro-optical EDM instruments. These carrier waves follow the laws of geometrical optics; normal telescopes are used for transmitting and receiving the signals.

9.1 Classification of Electro-Optical Distance Meters

Several criteria may be used to classify electro-optical distance meters. A few are mentioned here:

1. range
2. accuracy
3. degree of integration with a theodolite.

Other criteria for classification have been discussed in Sections 4.1.3 (phase measurement), 4.1.1 (light source) and 4.1.2 (modulation technique).

9.1.1 Classification According to Range

Three classes of instruments may be distinguished with respect to range:

1. short range EDM instruments
2. medium range EDM instruments
3. long range EDM instruments.

Short range instruments have a range from around 0.1 m to between 1000 m and 2000 m using a single prism. They are usually equipped with infrared (IR) emitting or lasing diodes.

Long range instruments may have a larger minimum distance (for example, for the Geodimeter 8, 15 m) and are usually equipped with a HeNe laser. The maximum range of such instruments depends very much on visibility (see Sect. 5.1) and the number of prisms used. With a large number of prisms (20 − 40) distances up to 70 km may be measured on clear days. Night time ranges of 120 km have been reported for the Keuffel & Esser/Cubic Precision Rangemaster III.

A class of medium range instruments with a maximum range of 5 − 10 km is also available; manufacturers usually classify these as long range instruments. Medium range instruments may eventually disappear, because the maximum range of short range instruments is steadily increasing towards the medium range.

9.1.2 Classification According to Accuracy

Five categories of instruments may be distinguished with respect to their accuracy. The accuracy specifications presently used by manufacturers are based on the assumptions that the prescribed observation procedures are followed and that at least the additive constant and the scale are calibrated by the user. It should be noted, for example, that the uncertainty of the measurement of the atmospheric parameters is not included in the ppm values listed below. (More details on the definition of accuracy specifications may be found in Section 13.5.)

1. Instruments with a standard deviation of one observation greater than $\pm(5\,\text{mm}+5\,\text{ppm})$
2. Instruments with a standard deviation of one observation equal or smaller than $\pm(5\,\text{mm}+5\,\text{ppm})$
3. Instruments with a standard deviation of one observation of about $\pm(5\,\text{mm}+1\,\text{ppm})$
4. Instruments with a standard deviation of one observation equal or smaller than $\pm(1\,\text{mm}+2\,\text{ppm})$
5. Instruments with a standard deviation of one observation equal or smaller than $\pm(0.2\,\text{mm}+0.2\,\text{ppm})$

Group (1) includes a number of pulse distance meters used for measurements to passive targets or to moving targets. The second group of instruments includes most short range distance meters found in general surveying practice. Group (3) refers mainly to specialized long range distance meters such as the AGA Geodimeter Model 8 or the Keuffel & Esser/Cubic Precision Rangemaster III. The instruments of Group 4 are often referred to as precision distance meters and include some infrared instruments. The instruments of Group 5 may be called high precision distance meters and include the laser instruments Kern Mekometer ME 5000 and Terra Technology Terrameter LDM-2. A more comprehensive list of instruments is given in Appendices E, F and G.

9.1.3 Classification According to the Degree of Integration with Theodolites

As the combination of angle and distance measurements is typical in general survey practice, the integration of EDM instruments with theodolites is very important from an operational point of view. Because of the large variety of integration concepts, any classification of the degree of integration between EDM instruments and theodolites will necessarily remain incomplete. The major concepts of EDM-theodolite integration are outlined below and their relative merits discussed.

9.1.3.1 Full Electronic Tacheometers

Electronic tacheometers are fully integrated EDM instruments and electronic theodolites featuring coaxial optics and digital output of all measured data. They are best used in conjunction with electronic data processing and plotting in large detail or contour surveys and also for large setting-out surveys. Electronic tacheometers have become smaller, lighter and more versatile during the last few years.

The development of electronic tacheometers was pioneered by Zeiss (Reg ELTA 14 in 1970) and AGA (Geodimeter 700 in 1971). Second generation instruments by Hewlett-Packard (HP 3820 A), Wild (Tachymat TC 1), Zeiss (ELTA 2 and ELTA 4) became available around 1978. By 1985, electronic tacheometers were available from most surveying instrument manufacturers, with the Topcon ET-1 (1983) being the first Japanese instrument of this type.

The reduction of distances causes no problem because of the coaxial design which integrates the EDM instrument and theodolite. The centre of the reflector is used as a target for the measurement of horizontal directions and zenith angles. For more precise angular work the precision of pointing the reflector to the electronic tacheometer becomes critical. The virtual displacement of the reflector's glass cube corner due to refraction at the reflector's front surface depends on its size, construction and pointing facilities. It will not exceed a few millimeters, as shown in Section 10.2.5.3.

9.1.3.2 Modular Electronic Tacheometers

A combination of a telescope-mounted EDM instrument with an electronic theodolite may be called a modular electronic tacheometer. Modular electronic tacheometers feature on-line data transfer between EDM instrument and electronic theodolite as well as on-board computation of horizontal distance and height difference. Such modular electronic tacheometers should require only single pointing for angular and distance measurements and no cables for the connection of the EDM instrument to theodolite and battery ("hot shoe" connection between theodolite and distance meter).

Some further aspects relating to telescope mounted EDM instruments are discussed below in Section 9.1.3.4.

9.1.3.3 Semi-Electronic Tacheometers

Such instruments may be fully integrated as far as the optics are concerned but may feature optical (rather than electronic) circles in the theodolite part of the instrument. Alternatively, the integration of the optical paths may be lacking but at least the zenith angle may be read electronically. Two groups of semi-electronic tacheometers are selected for further discussion:

1. Fully integrated EDM instrument and theodolite with visually read glass circles featuring coaxial EDM and theodolite telescopes. Angles and distances are measured simultaneously to the prism, thus requiring only one pointing in FL (as long as the angles need not be repeated in FR!). Typically, the instrument permits on-board computation of horizontal distance and height difference after manual input of the measured zenith angle.

This type of semi-electronic tacheometer was pioneered by Zeiss with the SM 11 (1970) and SM 4 (1977). It became very popular after the first Japanese instruments became available in the early 1980's (e.g. Topcon GTS-2 in 1981).

2. EDM instrument with built-in zenith angle sensor featuring on-line computation of horizontal distance and height difference without external input. Such instruments are either mounted on the telescope of an optical theodolite or are fully integrated instruments of the type described above. In both cases, the built-in level sensor can easily be checked and calibrated by comparison with the zenith angle measurements obtained with the theodolite.

9.1.3.4 Telescope-Mounted EDM Instruments

This group includes all EDM instruments which are mounted on the telescope of a theodolite. The advantage of this system is the combined pointing for EDM, zenith angles and directions, and the virtual absence of any additional corrections. The weight of such instruments is kept low because of the strain on the vertical and horizontal axis of the theodolite. Three designs may be distinguished:

1. Only one pointing for angles and distances is necessary. The offset of the EDM instrument axis relative to the theodolite's telescope axis is taken into account by a special reflector-target system with the same offset between reflector and theodolite target. The reflector-target has to be tiltable about a horizontal axis through the target to compensate for the tilt of the EDM instrument about the trunnion axis of the theodolite. It should be noted that poor vertical pointing of the reflector leads to relatively large errors in measured distances.

2. EDM instruments with the transmitter and receiver on opposite sides of the theodolite telescope need only one pointing to the centre of the reflector for direction, zenith angle and distance, as described in Section 9.1.3.1 (e.g. Kern DM 500/501/502/503/504/550).

3. EDM instrument-theodolite combinations, using a reflector as target, normally require two pointings, one for zenith angle and direction and one for distance. The zenith angle measured to the reflector centre is subsequently used for distance reduction. Theoretically, a small additional correction should be applied to the measured distance on very close range. However, this correction is so small that it may be omitted in most cases. This (small) distance correction for telescope-mounted EDM instruments has been discussed previously in Section 8.1.3.

Most telescope-mounted EDM instruments permit theodolite observations in one face only as long as the EDM instrument is attached. However, some EDM instrument-theodolite combinations allow a change of face with mounted EDM instrument. For example, all Kern EDM instruments of the DM 500 Series on Kern theodolites permit full 360 degree rotation of the telescope and most Wild distance meters on Wild theodolites permit theodolite measurements in two faces.

9.1.3.5 Theodolite-Mounted EDM Instruments

EDM instruments which are fixed on the standards of a theodolite are called
theodolite-mounted instruments. They feature a horizontal axis and a sighting
telescope of their own, because the EDM pointing has to be independent of the
theodolite pointing. Transitting of the theodolite telescope is possible. The advantage of such a design is that no strain is put onto the transit axis. The vertical axis
system of the theodolite has to carry the additional weight. Because of the fact
that the EDM instrument has a large eccentricity relative to the horizontal axis
of the theodolite and does not tilt with the telescope of the theodolite, relatively
large (a few centimetres) distance corrections apply if zenith angles and slope
distances are observed to one and the same prism. The relevant equations as well
as some numerical values have been given in Section 8.1.2.

In order to avoid the additional distance corrections, special reflector-target
combinations may be used. These reflector-target assemblies must have a vertical
separation of target and reflector equal to the vertical separation of theodolite
and EDM instrument and separate tilting axes of prism and target (although the
latter may not be necessary in some cases).

9.1.3.6 Separate EDM Instruments

Most instruments in this category are either long range or high precision instruments. At long distances, simultaneous angle measurements are usually not
required because the distances are reduced with the aid of given station heights.
Any direction measurements would be carried out with first-order theodolites
which should not carry the additional weight of an EDM instrument. (Eccentric
theodolite set-ups are used if reciprocal zenith angles are to be measured for the
determination of the coefficient of refraction.)

Some telescope- or theodolite-mounted short range EDM instruments are also
available as stand-alone instruments with appropriate yoke mounts, and thus also
fall into this category. In short range applications with such instruments, as well
as with high precision or precision distance meters, the measurement of zenith
angles is often required. In these cases, the geometrical correlation between zenith
angle and distance measurements may be achieved by

either 1. replacing the EDM instrument by a theodolite which fits the same con-
           strained centring system and measuring the zenith angle to the centre
           of the reflector
or   2. replacing the EDM instrument by a theodolite, and the reflector by a
           traversing target in a common constrained centring system and then
           measuring the zenith angle to the target.

The first method requires distance corrections according to Eq. (8.6) in Section
8.1.2. The second method needs no corrections for unequal instrument and target
heights, if the target has the same "height" as the theodolite and the reflector the
same "height" as the EDM instrument; "height" relates to the common constrained centring system. Such a system was adopted for the precision distance
meter Mekometer ME 3000 in conjunction with Kern theodolites (DKM 2A,

DKM 3) and a combined reflector-target. The method has been further improved by the use of "fixed height" two-foot-screw instrument bases. Such fixed height levelling bases are a feature of the Kern Mekometer ME 5000, its reflector, the Kern electronic theodolites E 1 and E 2 and the Com-Rad Geomensor 204 DME. If the second method is used in conjunction with equipment which does not fulfil the conditions of corresponding "heights" specified above, corrections according to Eq. (8.5) have to be applied (see Sect. 8.1.1).

9.1.4 Special Features of Modern Short Range Distance Meters

Most short range EDM instruments provide a wide range of features which make the operation of the instruments more efficient and convenient. A selection of such special features is explained below.

9.1.4.1 On-Board Application of First Velocity Correction

Most modern distance meters feature a facility for the on-line application of the first velocity correction, as discussed previously in Section 6.3. Use of the facility leads to first velocity corrected display distances. A manual input is required to enable this facility. Typical input quantities are:

- ppm correction value
- temperature, pressure (humidity)
- refractivity.

Traditionally, the ppm correction is derived from measured temperature and pressure (or, alternatively, mean sea level height) with the aid of correction charts or slide rulers supplied by manufacturers and then entered through a (environmental correction) dial, switches or a numerical keyboard. The least count of the input facility should be such that the specified accuracy of the instrument is not downgraded by the use the ppm input. Early instruments such as the Wild Distomat DI 3 and the Hewlett-Packard HP 3805 A had dials graduated in steps of 30 ppm and 10 ppm, respectively, leading to maximum errors in the display distances of 15 ppm and 5 ppm. More recent instruments feature a least count of the ppm input of 2 ppm, 1 ppm or even 0.1 ppm. To eliminate any round-off errors of the ppm input or to achieve the most accurate first velocity correction, the ppm input may be set to zero and the first velocity correction applied by computation in the office using the equations derived in Section 6.2.

As an alternative to the ppm input, the direct input of temperature, pressure, and possibly humidity is becoming more popular with manufacturers. No ancilliary charts or slide rulers are required as directly measured (or estimated) quantities are entered through a keyboard. The alternative of using heights above mean sea level rather than pressure is usually not available unless the values in Appendix C are considered. Based on the input values, the EDM instrument's computer calculates the first velocity correction according to formulae similar to those shown in Section 6.2. Humidity is typically ignored. The accuracy of the on-board computation is limited by the resolution of the temperature and pressure input, the inaccuracies of the equation used and the numerical accuracy of the computer. To eliminate any residual errors (or the omission of the humidity term) of the on-line correction, reference temperature and pressure (see Section 6.1) may be entered in the field and the normal correction applied in the office following the equations given in Section 6.2.

As a third option, refractivity may be used as an input variable. This approach has rarely been used. This method has no advantages over the ppm method because the value of refractivity must be computed by using charts or slide rulers. For example, the Precision International/Cubic Precision Citation CI 450 and the Pentax PM-81 (see Section 6.2.2) distance meters feature a first velocity correction input based on refractivity.

On-line first velocity corrections are appropriate for short distances, where temperature and pressure are typically recorded at the instrument station only (for further details, refer to Sects. 5.8, 5.9.1 and 5.9.2). On longer distances, atmospheric data are recorded at both terminals. Although a mean correction could be entered into (or computed by) the instrument, it is usually preferred to set the distance meter to zero ppm (or equivalent reference temperature and pressure) and to apply the first velocity correction subsequently by office computations.

The environmental (or first velocity) correction system of a distance meter may be easily checked by measuring a 1-km distance with differing dial settings; a change of 1 mm in distance then corresponds to 1 ppm and the internal accuracy of the instrument (e.g. ± 5 mm) affects the test result by an insignificant amount (e.g. ± 5 ppm). Test measurements may be carried out at $+50$ ppm, $+25$ ppm, ± 0 ppm, -25 ppm and -50 ppm, for example.

EDM instruments equipped with a ppm correction facility may be used to apply a combined scale correction consisting of:

1. the first velocity correction
2. the sea level correction
3. the point scale factor of the map projection.

Such a combined correction is useful for setting out work based on map grid coordinates. However, it is important to realize that such a combined correction would also be applied automatically to any height differences computed by the distance meter. Combined scale corrections should therefore be restricted to cases where solely horizontal distances are required, unless the errors caused in the height differences are negligible.

The full advantage of an on-board first velocity correction is only gained in setting out and if other computations (e.g. slope correction) are also carried out by the distance meter. Distance reductions which are executed on office computers may well include all parameters and corrections and thus render separate field reductions less efficient and accurate. It is important to note that all parameters required for the correction are booked in the field, as well as the applied correction itself, to allow a checking in the office.

9.1.4.2 Computation of Horizontal Distance and Height Difference

A large number of short range instruments provide the facility of computing the horizontal distance and the height difference. Electronic tacheometers provide

these data automatically; with other instruments, the zenith angle has to be di-
alled in manually. Automation is very convenient for setting out work in which
horizontal distances are provided. Users of such instruments should know exactly
how the internal programs determine the horizontal distance and height difference and the accuracy of the determination. For longer distances and higher
accuracies it may be better to reduce the measured slope distances with formulae
given in Section 7.2.

9.1.4.3 Tracking Mode

The tracking mode of an EDM instrument is that measuring routine which
measures the distance to a moving reflector continuously or at constant time intervals (e.g. every 2 s), if continuous pointing is provided. This facility is therefore
very useful for setting out. Distances obtained in the tracking mode are usually
less accurate than those in the normal mode because the distances are based on
smaller observation samples. The full advantage of the tracking mode is obtained
only if the instrument reduces slope distances automatically to horizontal
distances, thus providing the tracking of horizontal distances and possibly the
tracking of height differences.

9.1.4.4 Audio Signal

An audio signal may supplement or replace the traditional galvanometer for the
display of the return signal strength. The pitch of a tone becomes a measure for
the strength of the return signal. The volume of an audio signal should be adjustable. The audio signal may be transmitted by radio to the reflector-man, thus
enabling him to align the reflector. This possibility should be very helpful in setting out.

9.1.4.5 Automatic Data Recording

All full and modular electronic tacheometers, as well as some semi-electronic
tacheometers and a few other EDM instruments, provide on-line recording
facilities. The actual recording media may consist of

- solid state memory
- magnetic bubble memory
- cassette tape
- diskettes (floppy disks),

and may reside inside the EDM instrument (or tacheometer) or inside an external
unit, such as

- on-board recording module (instrument-specific)
- external field data recorder (instrument-specific)
- pocket calculator (through interface, e.g. HP 41 C)
- field computer (through interface)
- office computer (through interface).

Please note that manufacturers typically support only a limited number of the above options for specific instruments. Make-specific field data recorders are common and may support between one and all EDM instruments and electronic tacheometers of the same make. The provision of interfaces for calculators and computers is increasing.

In the case of recording modules, no additional cables are required. In all other cases, the cable between the EDM instrument (or tacheometer) and the external recording unit should not interfere with the operation of the EDM instrument and its host theodolite. Ideally, the EDM data are transmitted through the host theodolite to a cable port in the (non-rotating) base of the theodolite. Often, this is achieved by use of the internal power supply cabling as communication link. In these cases the power supply port doubles as communication port.

Additional equipment is needed to read data from recording modules residing in or on EDM instruments and to transfer such data to a computer for processing. External recording units typically allow data transfer to computers, possibly requiring additional interfaces.

It is important that only one command is required to record a full data set comprising, for example, point number, horizontal direction, zenith angle, slope distance, height of instrument and target. The sequence and redundancy of storage is very important. Some manufacturers provide processing programs to bring stored data into a usable form. Automatic data recording is of limited use unless a full chain from field measurements to computations and plan is secured.

9.1.4.6 Computer-Assisted Surveying

The first electronic tacheometer to provide any integrated computing facilities in excess of slope reduction, height difference, standard deviation and possibly bearing and distance computations was the Zeiss ELTA 2. Already the first electronic tacheometer REG ELTA 14 had an attachment for the same purpose in later years.

The Zeiss ELTA 2 allows the storage of control points and detail points to be set out in the solid state recording unit prior to the fieldwork. In the field, an additional computer unit (slide-in) will determine, for example, the station coordinates of a free stationing point (from radiations to 4 stations) and will assist in setting out by indicating left-right and forward-backward movements to get from the reflector station to final point to be set out.

Ten years after the introduction of the Zeiss ELTA 2 in 1978, most fully integrated and modular electronic tacheometers provide, in connection with their respective data recorders, some form of computing assistance. Some manufacturers put the computing power into the tacheometers, some others into the corresponding data recorders. This leads to combinations of "dumb" data recorders with "intelligent" tacheometers and "intelligent" data recorders with "dumb" tacheometers.

As an example, some of the EDM-related computing tasks provided by the Geodimeter Geodat 126 data recorder (in connection with the Geodimeter System 400) are listed below:

- Horizontal distance and height difference (see Sect. 9.1.4.2).
- Station coordinates (x, y, z) from resection with directions and distances.
- Traversing: Carrying forward of coordinates and traverse adjustment after closure of traverse.
- Bearing and distance between instrument station and reference object (point of known coordinates used for the orientation of the horizontal circle), including computation of orientation unknown and its subsequent application for the conversion of measured directions into measured bearings.
- Detail point coordinates (x, y, z) from radiations.
- Setting out data from given coordinates and comparison with actual measurement; display of necessary move of reflector forward/backward and left/right.
- Setting out of straight lines defined by two coordinated points: the distance from the starting point of the line as well as the offset from the line to be set out is computed.
- Height difference between reflector (or ground level) and inaccessible point directly above (or below), e.g. height of buildings, ground clearance of power lines. The same routine can also be used for the setting out of heights.
- Distance and elevation difference between two radiated points.

Similar (and additional) computing routines are available for some other makes of tacheometers and data recorders and might include routines for the setting out of circular curves and EDM height traversing. Some "intelligent" data recorders as well as all interfaced calculators and computers can be programmed by the user. This allows the implementation of any other required field computing tasks.

9.1.4.7 Computer-Controlled Distance Measurement

A number of electronic tacheometers and other distance meters permit two-way communication with computers through standard RS-232C interfaces, either directly or through their respective data recorders. Computers may be linked to the instrument or data recorder through cables, telephone lines (using modems) or even through radio communication links.

If a single distance needs to be measured repeatedly, the computer can be programmed to start distance measurements at predetermined times and to analyze and store these observations. Such an application for the remote monitoring of unstable slopes has been reported by Kern (1987). A similar arrangement can be used for the real-time control of linear motions of machines and cranes. In such cases, the computer becomes part of a feed-back loop which monitors and controls the movements of the machines or cranes. A special distance meter (HP 3850A) was developed at one time for this very purpose by Hewlett-Packard and used in-house for the control of the manufacturing of printed-circuit boards (Brown et al. 1981). AGA offers the Total Control System (TCS) for similar purposes.

As soon as computer control for the measurement of a number of different distances is required, the distance meter must be mounted on or be part of a motor-driven electronic theodolite or tacheometer, so that the pointing of the distance meter can be automated. A complete slope monitoring system has been developed by Geodimeter (Model 140 SMS) based on the Geodimeter 140 and a IBM-PC compatible computer (Geotronics 1987). The initial pointing to all reflectors is done manually. Afterwards, the instrument is pointed automatically to the initially recorded reflector position, after which an electronic fine pointing is carried out and the horizontal direction, the zenith angle and the slope distance measured. The first velocity correction is applied using electronically measured temperatures and pressures at the instrument site. Prototypes of a similar instrument based on the motorized version of the electronic theodolite Wild T2000 and a Wild distance meter were reported by Kahmen and Steudel (1988) and Schneuwly and Celio (1988). The automatic survey station Wild TM3000D was released in 1989.

A related application is the automatic tracking of moving ships or other targets. Such positioning systems are required in hydrographic surveying, for example. Typically, only distances and directions (bearing) are required, with the third dimension given by the water surface. The instrument manufactured by Krupp Atlas Elektronik (Krupp 1983) features a remote controlled measuring station on shore and the reflector, control unit and computer on the moving target. Two-way data communication is by a SHF telemetry link. The resolution of the pulse distance measurement (laser diode) is given as ± 0.1 m. Similar instruments are available from other manufacturers.

9.1.4.8 Setting-Out Aids

The use of EDM tacheometry for setting-out purposes requires frequent communications between the observer at the instrument and the assistant at the reflector in order to bring the reflector to its design position. Handheld two-way radios may be used for the purpose. However, less radio communications lead to a faster completion of the job.

Some Geodimeter instruments feature a one-way voice link between the instrument and the reflector, with a microphone at the instrument and a small infrared detector and a loudspeaker at the reflector end. The voice signal is transmitted with an additional modulation of the EDM beam. Some Geodimeter instruments can be equipped further with a visible flash light which projects three coloured sectors. The assistant at the reflector sees a white light when on line and a red or green light when outside the centre of the beam. The flash rate doubles as soon as the EDM instruments receives a return signal. When using both facilities, the initial alignment into the measuring line is done by the assistant without help and the backward/forward information is communicated over the infrared voice link.

For the same purpose, Kern supplies a reflector mounted infrared receiver which is able to receive and display the "distances" measured by the distance meter. Depending on the host theodolite, the following displays can be selected: slope distance, horizontal distance or height difference. The latter two options are only available in conjunction with electronic Kern theodolites. Infrared detectors serving as an aid for the positioning of the reflector in the EDM beam are supplied by a number of manufacturers.

9.1.4.9 Pointing Aids

Pointing aids are designed to facilitate the pointing of EDM instruments to reflectors in special cases. The aids can be based at the instrument or at the reflector station. No problems are usually encountered during close range distance measurements, unless the reflector/target is hidden behind leaves, for example. Instruments using visible light sources, such as HeNe lasers, can make use of the returned laser beam for pointing purposes. In the case of infrared distance meters, the pointing can be facilitated by mounting a flash light near the reflector or by mounting an aligned flash light on the distance meter. Flash lights for reflector mounts are available from Wild, for example. The second option is implemented in some Geodimeter instruments with the three-colour track light (see above).

When using infrared distance meters on medium or long ranges, it is often impossible to see the reflector or its tripod through the telescope. A flash light, centrally mounted in a reflector array (such as the Wild target lamp GEB 72), can greatly assist with the pointing of the distance meter and can double as a target for angular observations, which might be impossible otherwise. On sunny days, an efficient alternative is available: The assistant at the reflector can assist the EDM pointing by flashing the sun light towards the instrument with a small mirror. At night, a torch may be pointed towards the reflector from the EDM instrument side. The light reflected by the reflector may then be used for pointing purposes.

The pointing of HeNe laser instruments on very long lines can be similarly difficult, due to their narrow beams and the small amount of light returned by the prisms. Aids similar to the above may be used.

9.1.4.10 Other Features of Short Range Distance Meters

A few other distinguishing properties of short range distance meters are mentioned because they may be important in selecting an instrument.

In 1988, most instruments feature fully automatic attenuation ("balancing") of the return signal strength. This is typically achieved in two ways, namely optically, through internal variable neutral density filters, and electronically, through automatic gain control (AGC) circuits. Older instruments may require manual "balancing" or "tuning" of the return signal, manual setting of an aperture diaphragm or manual attachment of a so-called attenuator to the transmitter or receiver optics. The latter may be in the form of a neutral density or infrared filter or a perforated disk.

Most short range distance meters permit user access to one or more preset instrument corrections or operating parameters. For competent instrument operators, this is advantageous. For incompetent operators, it may be disastrous. Typically, the user has access to the prism constant and/or the instrument constant. Most (but, unfortunately, not all) manufacturers assign the sign of a correction to these values. The resolution of the input of these quantities should be in agreement with the stated accuracy of the respective instrument. A record of all preset constants should be kept in a log book for future reference and should be (automatically) transferred to the data recorder at each set-up.

Since the 1980's, distance meters are usually able to detect beam breaks and to stop and re-start measurements automatically. The beam break routines differ from instrument type to instrument type, with some instruments coping better with repeated beam breaks than others. Intermittent loss of signal occurs typically when measuring through traffic, but also in severe heat shimmer conditions.

For a limited number of applications, it is essential that the distance meter or the tacheometer can measure in plumb line direction (upwards). Only a limited number of instruments permit such vertical measurements. Diagonal eyepieces are required for the main telescope and, possibly, for the circle reading telescope and the distance reading should remain visible for the upward-pointing instrument.

Changing face of the theodolite is possible with some telescope-mounted EDM instruments but not with others. For lower accuracy work, this criterion is not really important because the measurement of directions and zenith angles in two faces may not be required.

The advantage of having no obstructing cables between the rotary part of the EDM instrument or the EDM instrument-theodolite combination and the tripod or ground may be another factor affecting the ease of EDM and angular measurements. Cables cause additional unbalanced strain on the instrumental axes. Of the two types of interfering cables, namely power supply and data transfer (electronic tacheometers), the former may be easier to replace by better design.

9.2 Design of Some Electro-Optical Distance Meters

The working principle and some components of electro-optical EDM instruments were discussed in Section 4.1. The design of some instruments is now explained in more detail to demonstrate the different design principles and the interaction of different components. A wider range of instruments is covered in a number of books (e.g. Saastamoinen 1969; Burnside 1991; Laurila 1976; Kahmen 1978; Zetsche 1979; Bolsakov et al. 1985), patent applications and numerous technical papers.

9.2.1 Kern DM 500

The Kern DM 500 is a representative of the large group of electro-optical short range distance meters employing techniques such as direct modulation, direct demodulation and digital phase measurement. This telescope-mounted distance meter was first released in 1974. More recent models of the same basic design interface with electronic theodolites and thus become part of modular electronic tacheometers. The block diagram of the original DM 500 is depicted in Fig. 9.1 (Münch 1973, 1974).

The transmitter includes a master oscillator, a digital divider (1 : 100), a modulator and two infrared emitting diodes. The master oscillator (RTXO) produces the fine measurement frequency of 14.985 400 MHz (~ 15 MHz) corresponding to a unit length of 10 m. The coarse measurement frequency of

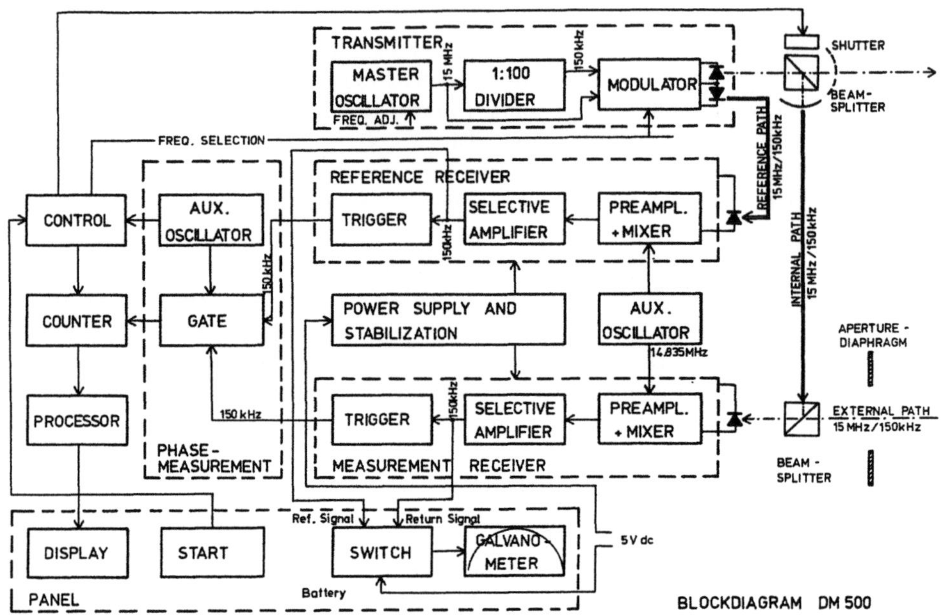

Fig. 9.1. Block diagram of the Kern DM 500 distance meter. The transmitting and receiving optics are not shown

149.854 kHz ($\sim$150 kHz) is then derived by the divider. The modulator has an input for both frequencies and selects one of them according to the commands coming from the program control through a glass fibre. The power supply of the transmitter block is not shown in Fig. 9.1. The modulator drives two infrared emitting diodes. The radiation of the first travels either through the external or the internal light path, the second one through a glass fibre to the reference signal receiver, thus providing an optical coupling between the transmitter and receiver. Optical coupling has the advantage of full electrical isolation.

In front of the first emitting diode, a turning shutter opens either the interval or external light path according to commands from the program control. The measuring as well as the reference beam are amplitude-modulated by the 15 MHz or 150 kHz signal.

Two almost identical receiver blocks process the measuring and reference signal. Avalanche photodiodes transform the incident IR beams into modulated electrical currents which pass first through preamplifiers and then through mixers. In the mixers a signal from an auxiliary oscillator, having a frequency of 150 kHz below the master oscillator's frequency (14.835 MHz), is multiplied with the 15 MHz or 150 kHz signal coming from the preamplifier thus generating signals of different frequencies. In the following selective amplifier, only the low frequency (LF) signal of 150 kHz is amplified. The output of the selective amplifier has therefore always a LF of 150 kHz irrespective of the modulation frequency (either 15 MHz or 150 kHz) received at the photodiode. It can be shown that the phase difference between the signals in the reference and the measurement channel, either LF or HF, does not change through the frequency transformation in the mixers (Münch 1973). Low frequencies are easier to handle in a digital phase measurement and are therefore preferred for that purpose. This principle of down-converting HF signals into LF signals was first implemented in the AGA Geodimeter 6A in 1967 after a proposal by Bjerhammar (1971).

The reference and measurement signals pass a trigger before they enter the block of the digital phase measurement as square waves. This type of phase measurement is described in Section 4.1.3.3 and depicted in Figs. 4.9 and 4.10. The auxilliary oscillator of the phase measurement section is also used as a time base for the program control of the distance meter.

The program control organizes the entire distance measurement once the start button has been pressed. It controls the selection of the frequencies, the selection of the measurement signal path (turning shutter) and the counter.

The processor subtracts the number of counts of the internal path from those of the external path for the fine measurement. This differencing between internal and external path is essential as the phase drifts with time and temperature. Without it, the additive constant of the distance meter would be strongly time and temperature dependent. The processor then converts counts into length units, adds the additive constant (built-in) and brings all three digits after the decimal point onto the display. The number of full metres within the unit length of 10 m is stored, until the internal and external path of the coarse measurement (150 kHz) is also measured and processed. The value of full metres in the zero to 10-m interval is then compared between the coarse and the fine measurement and the former adjusted. The digits of the coarse measurement then go onto the display.

The adjustment of the two values for the first digit on the left of the decimal point has to be made in such a way that the fine measurement takes precedence over the coarse measurement. In order to always obtain a positive correction to the 0 to 10 m value of the coarse measurement, 5 m is subtracted from the coarse measurement first by setting the start value of the display at 995 m. The reduced coarse value is then compared with the fine value and the necessary additional and integer terms computed. This system corrects for errors of up to 5 m in the coarse measurement. An example may illustrate the procedure:

Calculated (measured) data:		Displayed data:
Coarse measurement (m)	Fine measurement (m)	(m)
995.0	0.068	995.068 (start constant)
	(+4.260)	
995.0	4.328	004.328
(+268.6)		
263.6	4.328	
+1.0		
264.328		264.328

The start constant of 995.068 m includes the -5 m offset ($= +995$ m) as well as the preset (built-in) additive constant of $+0.068$ mm. The results of the actual phase measurements at the fine and coarse unit lengths are shown in brackets. Before merging the coarse and fine measurements, 1 m is added to the coarse measurement to match the metre value of the fine measurement.

Like all distance meters with digital phase measurement, the DM 500 has to check whether the measured phase differences of the external path are close to zero. Phase difference measurements around zero could lead to errors because of the erroneous averaging of, say, 0.01 m fine measurements with 9.99 m fine measurements. The problem can easily be overcome by inverting (shift by 180 degrees or 5 m) the signal in the reference path and by elimination of the introduced 5-m bias later, prior to the display of the result.

The galvanometer has differing functions which may be selected by turning a switch located on the instrument panel. It measures the battery voltage, the strength of the reference or return signal, as may be seen on the different inputs on the block diagram in Fig. 9.1. The aperture diaphragm is used to adjust the strength of the return signal to that of the reference signal by changing the aperture of the receiver optics.

The two beam splitters and the connecting glass fibre are part of the internal light path. The measuring sequence of the instrument is as follows:

1. 2 seconds of time: fine measurement (15 MHz), external path
2. 2 seconds of time: fine measurement (15 MHz), internal path
3. 2 seconds of time: coarse measurement (150 kHz), external path
4. 2 seconds of time: coarse measurement (150 kHz), internal path.

The equal time intervals for all four measurements result from the use of the same phase measurement routine for all four steps. About 300 000 phase measurements can be made in 2 seconds of time at a frequency of 150 kHz.

The original DM 500 had a range of 300 m to a single prism and a specified accuracy of $\pm(5\text{ mm} + 5\text{ ppm})$. The subsequent development of this instrument series may illustrate the rapid progress made in terms of built-in features, range and accuracy. The first (DM 501) of the later models of the DM 500 featured automatic attenuation (with liquid crystal filters), tracking mode and extended range (1.2 km). The DM 502 featured LCD displays for better visibility and reduced power consumption, reduced measuring time (8 s) and an improved accuracy of $\pm(3\text{ mm} + 5\text{ ppm})$. The introduction of the DM 503 extended the range further (to 2.5 km), provided a ppm correction facility, a user-accessible instrument correction and a display range to 16 km for the first time. The latter was achieved by adding a second LF modulation frequency, differing from the first one by 1.5 kHz. This provides a coarse measuring unit length of 100 km when using the techniques discussed in Section 3.2.1.3. The accuracy improved to $\pm(3\text{ mm} + 2\text{ ppm})$, most likely through the use of a temperature-compensated crystal oscillator (TCXO). The measuring time was reduced eventually to 6.5 s in 1987 with the DM 504 model, which also includes a distance averaging mode, a flexible $5-15$ V supply voltage, a number of remote control options, a reduced power consumption and, in the case of the DM 550, a zenith angle sensor.

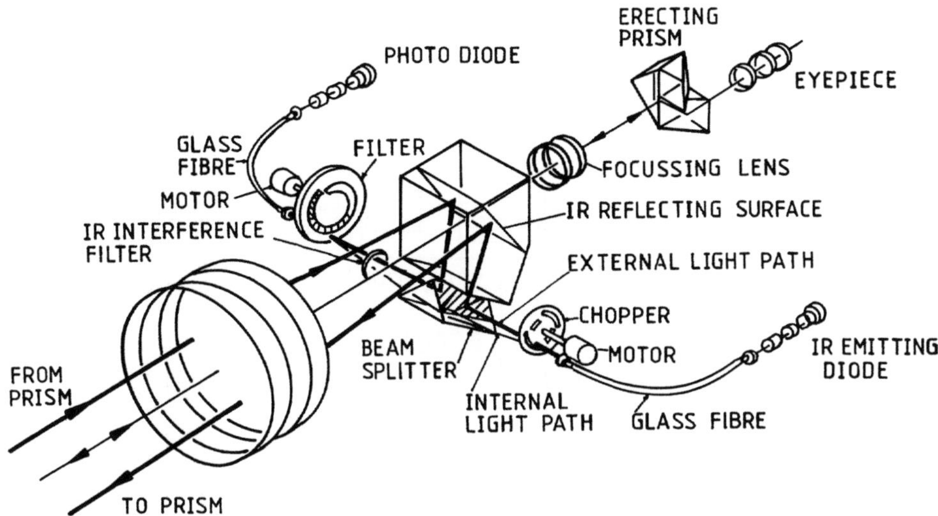

Fig. 9.2. EDM and visual ray paths through the telescope of the Topcon ET-1 electronic tacheometer. The diagram shows the horizontal separation of the two EDM beams. The transmitted rays are emitted through the right hand side of the telescope and the reflected rays are collected through the left hand side

9.2.2 Topcon ET-1

The distance meter section of the Topcon ET-1 electronic tacheometer (released in 1984) may be discussed as a second representative of the group of electro-optical short range distance meters. Here, the distance meter is fully integrated into an electronic theodolite, using the same telescope for angular and distance measurements. Distances are measured by digital phase measurement at three different modulation frequencies. The optical lay-out and integration is shown in Fig. 9.2. The electronic circle reading and level sensor systems do not make use of the distance meter's phase measurement capability. They are not discussed in this context [see Kondo (1985) for example]. The principle of the design is discussed below. Further details may be obtained from Hamada and Ohtomo (1981).

Figure 9.2 explains the optical parts of the distance meter section of the ET-1. The carrier wave is produced by a GaAlAs diode of 820 nm wavelength. The bandwidth (to half-power points) is about 50 nm. The GaAlAs diode is mounted on a printed circuit board and connected to the telescope by a glass fibre. The use of glass fibres reduces the effects of phase inhomogeneities of transmitting and receiving diodes. Through a chopper, one part of the transmitted beam reaches the reflecting part of a prism, which deflects the transmitted beam onto another infrared reflecting surface in the optical axis of the telescope. From there the transmitted beam reaches the objective lens and the reflector. The latter returns the beam through the objective onto the same infrared reflecting surface in the telescope axis. The return beam then reaches via the partially reflecting prism the end of a glass fibre after passing through an interference filter (820 nm) and a variable neutral density filter.

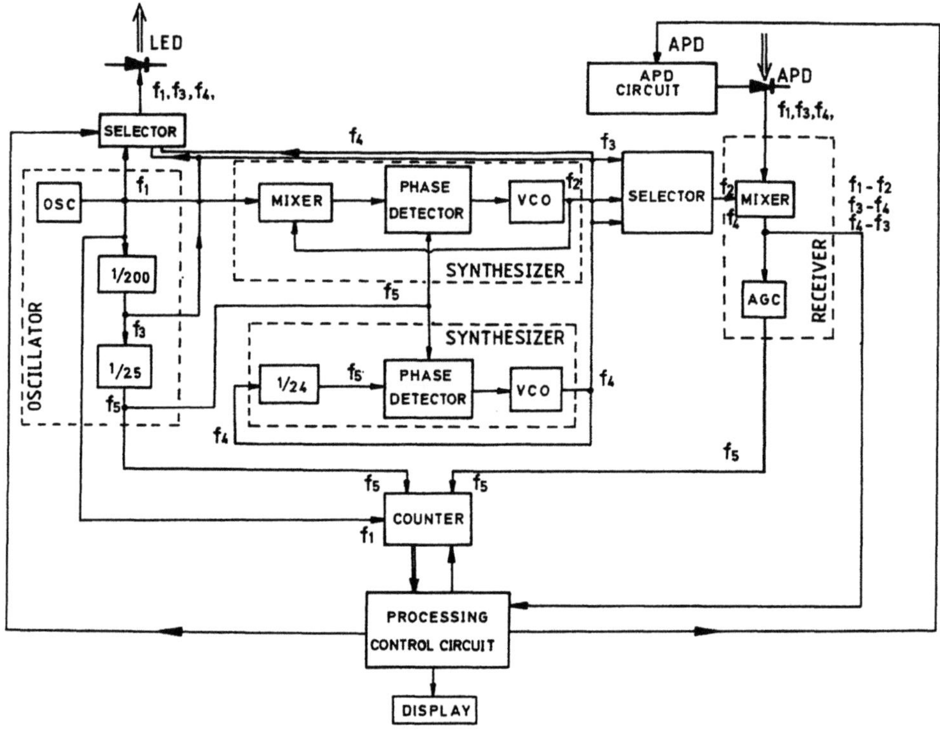

Fig. 9.3. Block diagram of the electronic circuitry of the distance meter section of the Topcon ET-1 electronic tacheometer

Another part of the transmitted beam (reference beam) travels through an uncoated part of the deviating prism directly into the receiver. Two variable neutral density filters maintain a constant signal strength of the return and reference beams, respectively. The chopper operates in a sector of $\pm 60°$ with a frequency of 4 Hz (FINE MODE). The transmitted beam has a width of 7 minutes of arc and exits through the bottom half of the ET 1 (in face left). The top half of the telescope (focal length: 115 mm, fixed focus) receives the return beam. This construction permits the use of small prisms, even on close range.

The interference filter is an optical band pass filter with a centre wavelength of 820 nm. It improves the signal-to-noise ratio of the receiver by preventing all other wavelengths (of ambient day light, scattered sun light) from reaching the photodiode. The neutral density filters (in the chopper windows as well as in the rotary filter) are necessary to reduce the large return signal strengths obtained on close range to levels experienced on long range, as the dynamic range of the electronic gain control (AGC in Fig. 9.3) is limited.

A temperature-compensated crystal oscillator (TCXO) serves as master oscillator. It runs at 14 985 437 Hz and features specifications of better than ± 5 ppm between $-20°C$ and $+50°C$. The first frequency produces a unit length of 10 m, the f_3 frequency one of 2000 m. The f_5 frequency is used for the phase measurement. The f_1 frequency is used as the clock signal for the digital phase measurement, leading to $14\,985\,437/2997 = 200 \times 25 = 5000$ pulses per LF period. Two additional frequencies are required to mix the received frequencies down to about 3 KHz. One synthesizer (with a phase locked loop) produces the f_2 frequency being $f_2 = f_1 - f_5$. A second synthesizer (with phase locked loop) produces the f_4 frequency where $f_4 = f_3 - f_5$. The f_4 signal is also transmitted to resolve the multiples of 2000 m, by using the $(f_3 - f_4)$ unit length of 5000 m.

The signal received by the (temperature stabilized) avalanche photodiode (APD) is mixed according to the received frequency, so that a LF of 2997 always results for phase measurement purposes:

$$f_1 - f_2 = 2997 \text{ Hz}$$

$$f_3 - f_4 = 2997 \text{ Hz}$$

$$f_4 - f_3 = (-)2997 \text{ Hz} \ .$$

The automatic gain control (AGC) stabilizes the signal level for the phase measurement. Additional filters will remove all frequencies other than the required f_5.

The measuring sequence in the FINE measuring mode is as follows:

$$
\begin{array}{ll}
\left.\begin{array}{ll} f_4 \text{ EXT } 400\times \\ f_4 \text{ INT } 400\times \end{array}\right| & 2\times \quad (1.0\text{ s}) \\[1.2em]
\left.\begin{array}{ll} f_3 \text{ EXT } 400\times \\ f_3 \text{ INT } 400\times \end{array}\right| & 2\times \quad (1.0\text{ s}) \\[1.2em]
\left.\begin{array}{ll} f_1 \text{ EXT } 400\times \\ f_1 \text{ INT } 400\times \end{array}\right| & 4\times \quad (2.0\text{ s})
\end{array}
$$

A total of 1600 fine measurements are carried out.

The COARSE measurement sequence comprises, after an initial full FINE measurement:

$$
\begin{array}{ll}
\left.\begin{array}{ll} f_3 \text{ EXT } 200\times \\ f_3 \text{ INT } 200\times \end{array}\right| & 1\times \quad (0.25\text{ s}) \\[1.2em]
\left.\begin{array}{ll} f_1 \text{ EXT } 200\times \\ f_1 \text{ INT } 200\times \end{array}\right| & 1\times \quad (0.25\text{ s})
\end{array}
$$

The on-board microprocessor checks the spread of the measurments and extends the measuring time if required.

9.2.3 Distomat Wild DI3000

The Distomat Wild DI3000 is a representative of the growing number of pulsed infrared distance meters. The principle of the pulse method has already been discussed in Section 3.1.1. In the case of the DI3000, the total flight time is obtained

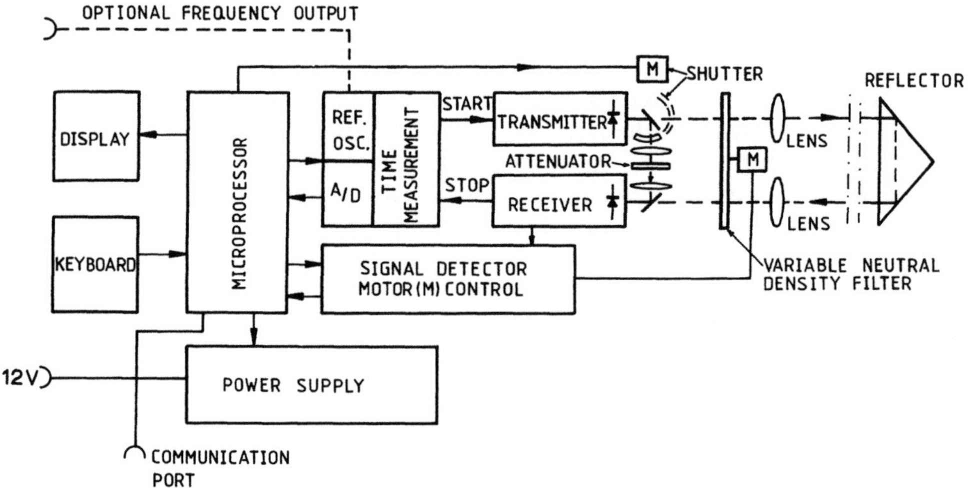

Fig. 9.4. Simplified block diagram of the Distomat Wild DI3000. (Grimm et al. 1986a, b)

as the sum of an integer number of periods of the time base and fractions of the time base period. This coarse and fine measurement of the flight time resembles to a certain degree the procedures used in phase measuring distance meters.

As Grimm et al. (1986a, b) point out, pulse distance measurements offer a number of advantages when compared to phase distance measurements. Firstly, a single internal and external measurement is sufficient to derive a distance. No switching of modulation frequencies is required. Secondly, the range is markedly increased by the larger output power of the short laser pulses. Thirdly, some systematic errors of phase measuring distance meters (cross-talk between transmitter and receiver) are directly eliminated and some others (multipath) by careful monitoring of (multiple) return pulses.

A simple block diagram of the DI3000 is given in Fig. 9.4. The switching between the internal and external ray paths is carried out by the motor-driven shutter. The internal and external paths are attenuated by a constant and a variable neutral density filter, respectively. Following a start command from the microprocessor, the time measurement is commenced and an infrared pulse [of about 12 ns (or 3.5 m) width] is emitted by the GaAs laser diode. The drive current of the laser diode of several 10 amperes is varied with temperature to ensure a constant power output over the specified temperature range (see Sect. 4.1.2.1).

After travelling over the internal or external light path, the pulse is focussed onto the avalanche photodiode (APD). The resulting current surge through the diode is not used directly to stop the time measurement. Instead, the power surge is used to generate a $(\sin x)/x$ wave in a parallel resonance circuit (Chaborski 1978). The second crossover of this sine wave is largely unaffected by changes in the power, shape and width of the return pulse and, in consequence, used as a stop signal for the time measurement. The flight time measuring procedure is explained in more detail in Fig. 9.5.

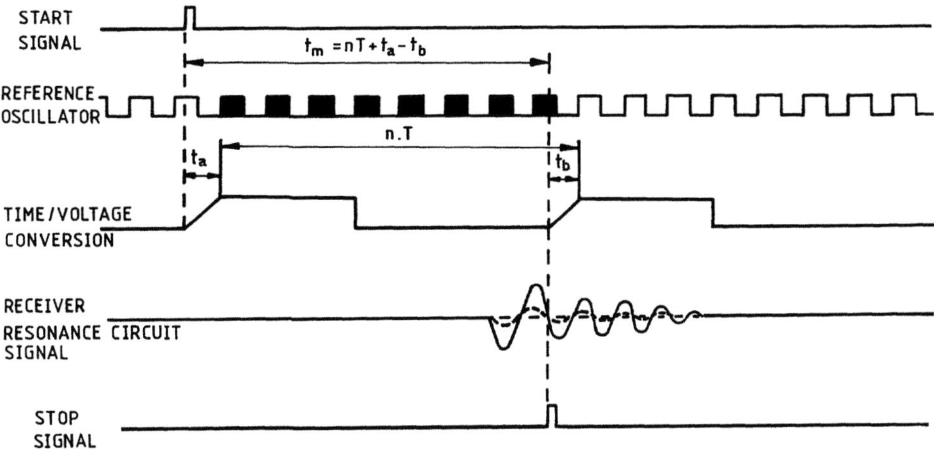

Fig. 9.5. Principle of flight time measurement used by the Distomat Wild DI 3000. (Grimm et al. 1986a, b)

Before the display distance can be computed according to Eq. (3.1), and with consideration of a reference refractive index of 1.000 281 5, the flight time t_m must be measured for the external as well as for the internal path. As indicated in Fig. 9.5, the total flight time is derived from three separate measurements. The term nT is easily obtained by counting the number of full cycles of the time base frequency of nominally 15 MHz between the start and the stop marks. The two fractions t_a and t_b of the time base period (of 66.66 ns) are determined by a fast analogue time interpolation procedure. The start pulse activates a constant current charge cycle of a capacitor, leading to the voltage-versus-time ramp shown in Fig. 9.5. The first leading edge of the 15 MHz time base signal stops the charging of the capacitor. The voltage across the capacitor is then amplified and fed through an analogue to digital converter into the microprocessor. Use of the full voltage range of 10 V of the 12 bit A/D converter would lead to a resolution of 2.4 mm of the time interpolation process. In order to minimize the effects of component tolerances and temperature drifts, the voltage range of the A/D converter is not fully used. The effective resolution of the time-to-amplitude (TAC) converter is about 4 mm. After completion of the first timing sweep, the capacitor is discharged and is then ready for the measurement of t_b. In this case, the timing sweep is started by the stop pulse and terminated by the leading edge of the next time base signal.

Some of the finer details of the design are depicted in Fig. 9.6. The reference oscillator is a quartz crystal oscillator (most likely a TCXO), the frequency versus temperature characteristic of which is further improved (to better than 1 ppm) by factory calibration between -30 and $+60\,°$C. During (frequency calibration and) distance measurements, the temperature of the oscillator is monitored by a thermally coupled temperature sensor and used to correct the displayed distances. Two key components are explained in detail before discussing the operation of the distance meter as shown in Fig. 9.6.

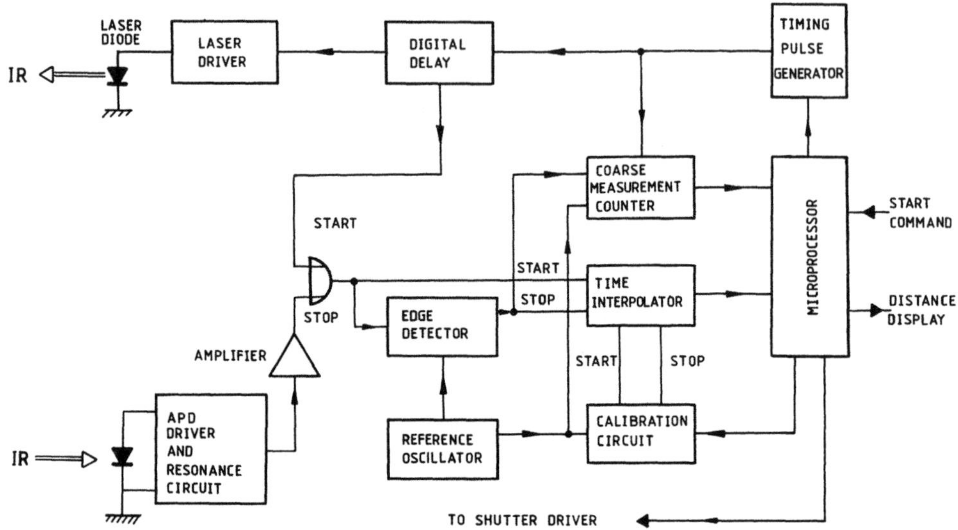

Fig. 9.6. Diagram of the components of the flight time measurement circuit and their interactions. (Giger 1983)

The digital delay is required to permit the execution of the timing sweep for t_a before the return pulse arrives. An auxilliary oscillator, a counter and two gates form the delay circuit (Giger 1983). Upon receipt of a timing (or synchronization) pulse, the counter starts accumulation of pulses of the auxilliary oscillator. After reaching a first predetermined number of pulses, a first gate (leading to the time interpolator) is activated. The next pulse of the auxilliary oscillator then starts the t_a measurement. After reaching a second predetermined number of pulses of the auxilliary oscillator, a second gate (leading to the laser diode driver) is activated. The next pulse of the auxilliary oscillator then activates the laser driver and the emission of the IR pulse. The delay between the start of the time measurement and the emission of the pulse will be as constant as the auxilliary oscillator. Depending on the latter, it will be the same for the measurements of the internal and external paths.

The calibration circuit is used to calibrate the slope of the timing ramp of the time interpolator. Prior to a distance measurement, the signal of the reference oscillator is used to generate well-defined start and stop pulses. The resulting time interpolation values are sent to the microprocessor for the purpose of the correction of subsequent time interpolation measurements. The effect of the small non-linearities of the timing ramp (10−20 mm) are minimized by multiple time interpolation measurements with variation of the t_a and thus t_b, over the full period T of the reference oscillator. The variation of t_a is achieved by synchronizing the timing pulse generator with the auxilliary oscillator of the digital delay (rather than with the reference oscillator) and by control of the auxilliary oscillator by the microprocessor.

Following Giger (1983) and Fig. 9.6, the measurement of a single distance may now be discussed as follows. With the shutter set appropriately, the internal path

144

is measured first. After requesting a distance measurement, the calibration circuit is activated and the resulting calibration values of the timing ramp slope are stored in the microprocessor. The timing pulse generator is activated next. It has a pulse repetition rate of about 2 kHz. Each timing pulse initiates a distance measurement. The timing pulse resets the coarse measurement counter to zero and triggers the digital delay circuit. This in turn starts the time interpolation circuit, and, later, the emission of the laser pulse. The first time interpolation measurement (t_a) is stopped and the coarse measurement counter started by the next leading edge of the reference signal. The return pulse triggers the resonance circuit of the photodiode. The stop signal is derived from the second crossover of the damped oscillation and used to start the second time interpolation. The edge detector is also triggered by the stop signal and enables the leading edge of the next reference oscillator signal to stop the coarse measurement counter as well as the second time interpolation measurement (t_b). The coarse and both fine measurements are stored in the processor, and averaged after the required number of repeat measurements.

The external light path is then opened by the shutter and the procedure repeated for the external path. Again, repeat measurements are taken and averaged by the processor. Finally, the internal light path is again enabled by the shutter for a second set of internal measurements. [This second internal measurement permits the elimination of linear drifts (with time) of the electronic circuits.] The two sets of internal measurements are meaned and then subtracted from the mean of the external measurements. The display distance is derived by using Eq. (3.1) and the reference refractive index given above.

9.2.4 Kern Mekometer ME 5000

The Kern Mekometer ME 5000 differs markedly from the previously discussed instruments. It uses a HeNe laser as light source and, in consequence, the classical principle of indirect modulation. The short unit length of about 0.3 m is the same as with the earlier Kern Mekometer ME 3000 and the Com-Rad Geomensor 204 DME and follows from the high modulation frequency of 500 MHz. Rather than measure the phase difference between transmitted and received signals (as with the ME 3000, Sect. 4.1.3.1), the modulation frequency is changed until a zero phase difference results (similar to the Geomensor 204 DME, but fully automatic). The distance can then be calculated according to the equations given in Section 3.2.1.4.

Following Meier and Loser (1986), the operating principle of the Mekometer ME 5000 may be briefly described. Linearly polarized light at a wavelength of 632.8 nm is emitted continuously by the HeNe laser and travels through an attenuator and (unaffected) through the beam splitter of parallel polarization into the (unspecified) modulating crystal. In the modulator, the polarization of the laser beam is modulated at the frequency (about 500 MHz) provided by the frequency synthesizer. The latter is based on a quartz crystal oscillator with an accuracy specification of better than ± 0.3 ppm between -10 and $+40\,°C$. The emerging beam then travels through a quarter wavelength path, a beam expander

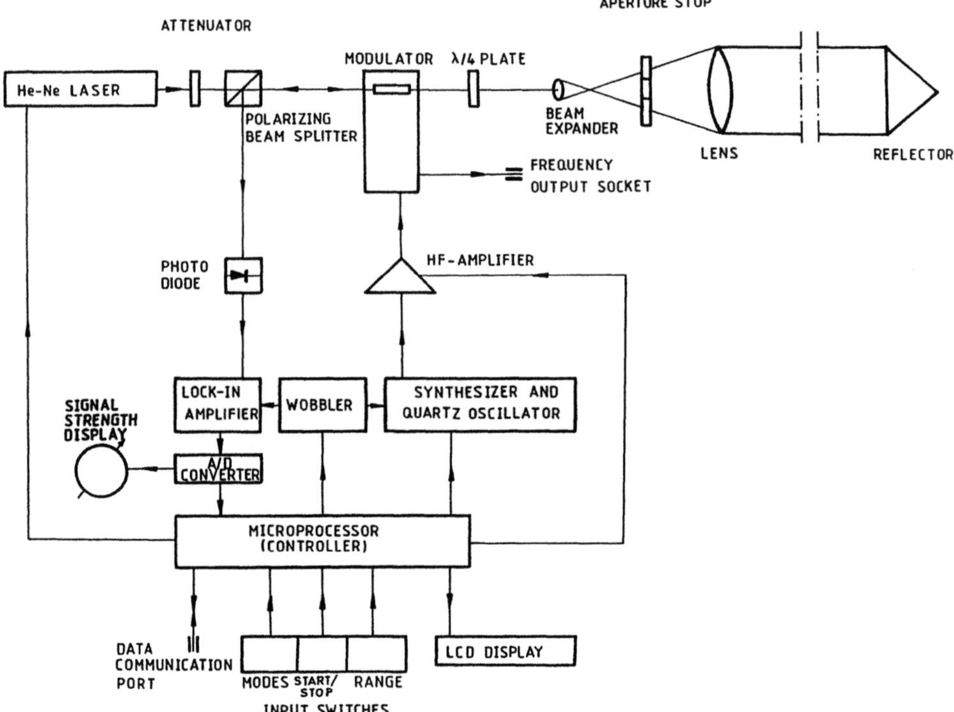

Fig. 9.7. Block diagram of the optical and electronic components of the Kern Mekometer ME 5000. (After Meier and Loser 1986)

and a diaphragm to the reflector. The quarter wavelength plate is used to remove the effects of the temperature dependency of the polarization modulation of the modulating crystal (Meier and Loser 1988). The return beam travels on the same axis through the now collimating beam expander, through the quarter wavelength plate into the now demodulating crystal. No light will fall onto the photodiode whenever the transmitted and received signals are in phase or, in other words, whenever the measured distance is an exact multiple of the unit length of about 0.3 m.

A technique similar to the earlier Mekometer ME 3000 is used to improve the resolution of the zero phase which would be given otherwise by the minimum of the photodiode current. The modulation signal is frequency modulated by the wobbler at a rate of 2 kHz and with frequency offsets of ±5 kHz (for distances larger than 500 m) or ±25 kHz (for distances shorter than 500 m), depending on the range knob setting. In the case of the transmitted and received signals being in phase, the photodiode will receive light pulses of equal intensity at a rate of 4 kHz. If the two signals are not in phase, the 4 kHz pattern recorded by the photodiode will exhibit a saw-tooth pattern, with the odd and even pulses having different amplitudes. This amplitude difference is measured by the lock-in amplifier and passed on to the controller through the analogue-to-digital (A/D)

146

converter. In turn, the controller will shift the synthesized frequency in steps of 160 Hz or 0.3 ppm until the transmitted and received signals are in phase and the amplitude of the 4 kHz signal on the photodiode constant.

The A/D converter also feeds the signal strength display of the ME 5000. Measurements are interrupted whenever the amplitude of the 4 kHz signal on the photodiode falls below a critical level. The attenuator and the diaphragm are activated during the measurement of short distances so that the photodiode is not overloaded. The instrument can be operated in two ways, either with the on-board controls or with an external computer through the RS-232 compatible communication port. The distances are derived according to the routine discussed in Section 3.2.1.4 and displayed to 0.1 mm. The displayed distance values are based on a reference refractive index of 1.000 284 514 844.

The built-in frequency synthesizer has a range of 30 MHz, 15 Mhz of which are utilized by the on-board measuring routine. Considering the differences of the modulation frequency for adjacent phase minima at different distances [see Eq. (3.24)], the following number of phase minima are available in the 15 MHz range:

Distance (m)	20	100	300	3000	8000
Frequency difference (MHz)	7.5	1.5	0.5	0.05	0.02
Number of phase minima	2	10	30	300	800

The requirement for a minimum of two independent distance measurements restricts the minimum range of the ME 5000 usually to 20 m. When operating the ME 5000 from an external computer and with additional input of approximate distances, the minimum range can be reduced to about 5 m (Maurer et al. 1988). Similarly, the upper range of normally 8 km can be extended by operation from an external computer and using least-squares techniques for the resolution of the "ambiguities". A successful distance measurement of 17 km was reported by Maurer et al. (1988). The same authors quote the resolution of the ME 5000 as ±0.03 mm.

10 Reflectors

10.1 Introduction

Apart from military laser rangers, electro-optical EDM instruments need a device at the target station which reflects the light (or infrared) beam back to the instrument. Reflecting devices should have the following properties:

1. good reflectivity
2. complete illumination of the receiver optics of the instrument
3. no change direction of emerging rays through small movements of the reflecting device, thus rendering a continuous alignment unnecessary.

Such devices are:

1. plane front surface mirror
2. spherical reflector
3. solid glass prism reflector (corner cube reflector)
4. hollow corner cube reflector
5. acrylic retroreflector
6. reflective sheeting.

Plane front surface mirrors were used in conjunction with the Geodimeter NASM2. Although they had the best reflective properties, their use was soon discontinued because of the requirements for a very stable and precise alignment. Spherical reflectors allow for a misalignment of ± 30 minutes of arc and consist basically of a spherical mirror and a small plane mirror in its focal point. Their reflectivity (65%) is much smaller than that of the plane mirrors (90%) due to the greater number of reflections involved. Spherical reflectors were used in conjunction with early Geodimeter models. Glass prism reflectors have proved to be the most suitable reflector system. They were first employed by the U.S. Army Map Service in 1953 (Rinner and Benz 1966) and are now almost exclusively used in terrestrial EDM as well as in satellite and lunar laser ranging (see Sect. 3.1.2). They are discussed in more detail in Section 10.2.

Hollow corner cubes operate on the same principle as the solid corner cubes discussed in Section 10.2. Three plane front surface mirrors are assembled at right angles to each other. Hollow corner cube reflectors are lighter than ordinary solid prism and adapt more quickly to changes in temperature. Coating may be aluminium or gold (for IR reflection). Typical diameters are 25 mm and 63 mm. The reflecting faces may become dirty as they are exposed to the elements. The coating of the plane mirrors may be subject to corrosion, particularly in humid climates, and may be scratched in the process of cleaning. Hollow corner cube reflectors are not necessarily cheaper than solid corner cube reflectors, if a high

Table 10.1. Experimentally determined distance range of a selection of infrared distance meters with acrylic and reflective tape retroreflectors

EDM instrument	Distance range applicable with reflector type:				
	A (m)	B (m)	C (m)	D (m)	E (m)
AGA Geodimeter 14	$-$N/A$-$	2.7 $-$ 77	2.0 $-$ 108	1.8 $-$ 203	1.6 $-$ 312 +
Nikon NTD-3	1.4 $-$ 43	1.4 $-$ 79	1.5 $-$ 105	1.4 $-$ 197	1.5 $-$ 312 +
Pentax PX-06D	1.4 $-$ 30	3.1 $-$ 44	1.4 $-$ 60	1.4 $-$ 109	1.5 $-$ 327 +
Sokkisha SDM-3ER	4.1 $-$ 27	4* $-$ 51	2.8* $-$ 66	3* $-$ 120	1.3 $-$ 311 +
Topcon GTS-2	12 $-$ 29	7 $-$ 57	2.2 $-$ 77	2.0 $-$ 129	1.3 $-$ 312 +
Topcon DM-C2	5 $-$ 52	2.3 $-$ 100	1.5 $-$ 135	1.5 $-$ 250	1.5 $-$ 323 +
Wild TC 1600	$-$N/A$-$	$-$N/A$-$	13 $-$ 67	11 $-$ 269	1.5 $-$ 320 +
Wild DI 3000	2.3 $-$ 145	2.3 $-$ 267	2.3 $-$ 319 +	2.6 $-$ 327 +	2.4 $-$ 319 +

Where no measurements were possible, this is indicated by N/A. Plus signs indicate that longer distances could have been measured, and asterics that some 10 m errors were recorded on close range. Reflector A: Scotchlite (3 M) enclosed lens sheeting (flat-top wide angle reflective sheeting), red, 33 × 33 mm; Reflector B: Scotchlite (3 M) encapsulated lens sheeting (high intensity grade reflecting sheeting) with honeycomb pattern, yellow, 60 × 60 mm; Reflector C: Simsonite acrylic retroreflector, clear, 28 mm diameter; Reflector D: red road delineator acrylic retroreflector, 77 mm diameter; Reflector E (for comparison): small glass prism reflector, 25 mm diameter of aperture

accuracy of the 90 degree angle is required. The reflectivity depends on the type and the cleanliness of the reflective coating. The reflectivity of clean open corner cube reflectors may be estimated with 94% and 65% for gold and aluminium coatings, respectively.

Acrylic retroreflectors consist of arrays of small corner cubes, with each cube having typically a diameter of 2 mm. Such retroreflectors are commonly used for road traffic control purposes such as road delineators (red, yellow or white, 85 mm diameter), sign buttons (white, diameters between 13 and 41 mm) and vehicle reflectors. Although they have a very much smaller reflectivity than glass prism reflectors, they may permit distance measurements on close range. They may be used as disposable and/or permanent reflectors for setting out and for displacement monitoring applications. Because of the small aperture of the individual corner cubes, the usable range will not depend solely on the output power of the distance meter but also on the separation between the transmitter and receiver. The first criterion affects the upper limit of the range, the second the lower limit. Maximum distances in excess of 1 km have been reported for HeNe laser instruments (Geodimeter 600, 1500 m). Table 10.1 gives the working ranges for some infrared distance meters. Red retroreflectors perform best in connection with infrared distance meters. Further details on the use of acrylic reflectors may be found in Johnson (1981), Kennie (1983) and Rüeger (1989).

Reflective sheeting (or tape) is an even cheaper alternative to glass prisms. It is offered in basically three versions by the Minnesota Mining and Manufacturing (3M) Company under the tradename of Scotchlite. The "flat-top wide angle reflective sheeting" has a layer of glass beads embedded in a durable transparent plastic and a metallic reflector coating underneath. The high intensity grade (encapsulated lens) sheeting is normally used for traffic signs and features a layer of glass beads, which is partially embedded in a plastic resin, and an enclosed air space above the beads. The transparent top film is supported by walls in a honeycomb pattern. The diamond grade cube corner sheeting replaces the glass beads with a plastic corner cube pattern sealed in a hexagonal pattern. Reflective sheeting is self-adhesive and may be attached to walls as temporary reference marks for setting out or displacement monitoring surveys. The usable range is governed by the reduced reflectivity as well as the small size of the glass beads of, say, 0.25 mm. Table 10.1 gives an indication of the practical working range of infrared distance meters with reflecting tape. Maximum ranges of about 200 m have been reported for HeNe laser instruments. A detailed description of reflective sheeting was given by Kennie (1983).

10.2 Glass Prism Reflectors

These common reflectors consist of a corner of a glass cube which was cut in such a way that the plane of the cut is perpendicular to the cube diagonal (equilateral cube corner). It can be shown that if the angles between the cube planes are exactly equal to 90 degrees, all incident rays are reflected in such a way that incident and emerging rays are exactly parallel and symmetrical to the centre ray (ray to the prism corner), independent of the alignment of the reflector. The ray path through a prism is depicted in Fig. 10.1. The figure shows clearly that all rays are refracted two times at the front surface of the prism and reflected three times internally on the perpendicular prism faces.

The alignment of the prism has an effect on the effective aperture of the reflector and therefore on the retro-reflected radiant power. Figure 10.2 shows how the effective aperture A_{eff} expressed as a percentage of the aperture A varies with the angle of incidence α for a prism of circular aperture and a refractive index of 1.5.

The reflecting faces of the glass prisms are either uncoated or coated. In the former case, the rays are reflected by total internal reflection as long as their angles of incidence (on the reflecting faces) exceed the critical angle of 41.1 degrees. It can be shown that the rays must enter the prism within a Y-shaped area around the centre of the aperture in order the meet the critical angle on the reflecting surfaces (Arnold 1979). As long as the reflecting faces are perfectly clean, no power is lost by the three total internal reflections. (It should be noted that fingerprints on the surfaces can cause loss of the total internal reflection.) Some power is, however, lost due to reflection on the front face and absorption inside the glass. Under the assumptions that the acceptance angle limitations are fulfiled (all input rays enter through the Y-shaped area defined above), the reflectivity of uncoated prisms may be estimated as 90−95%. Unpolarized input beams become partially polarized by the total internal reflection. Polarized incident beams have their polarization changed, depending on which of the six sectors (defined by the three real edges and their three reflected images) the beam enters and exits. Chang et al. (1971) demonstrated that the far-field diffraction pattern of the exiting

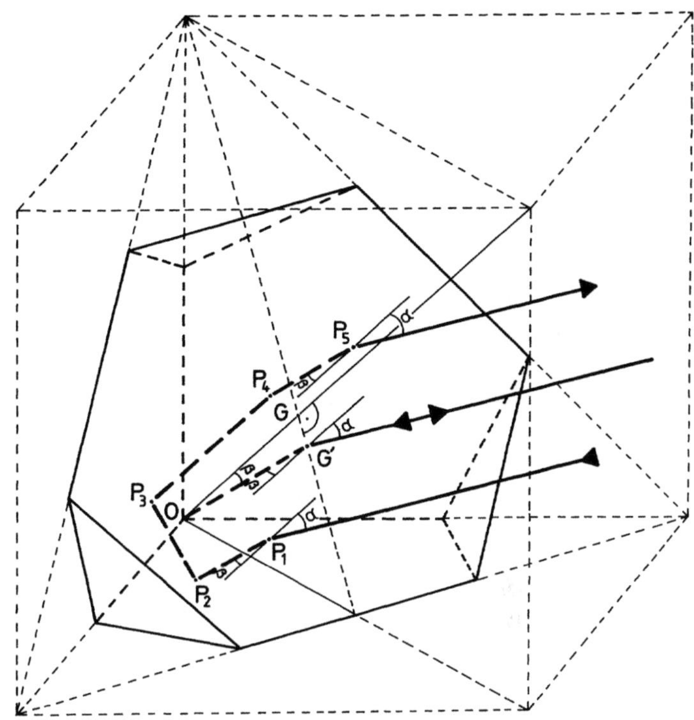

Fig. 10.1. Ray path through a solid corner cube reflector. (After Pauli 1969, 1973). The apex of the prism is denoted by O and the intersection between the cube diagonal and the front face by G. The line OG is perpendicular to the prism's front face. The angle of incidence and refraction are denoted by α and β, respectively. The point G' is the point of incidence and emergence of the centre ray. Points P_1 and P_5 indicate the same two points for an eccentric ray. The eccentric ray is reflected at the points P_2 to P_4

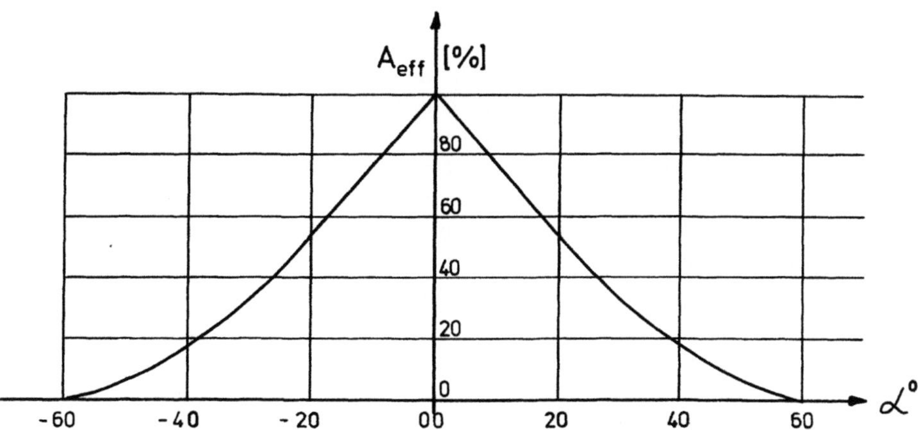

Fig. 10.2. Effective prism area of a circular prism for different angles if incidence. (After RCA 1974)

beam of original polarization (same polarization as incident beam) differs markedly from that of the orthogonal polarization. This aspect would be of particular relevance to distance meters using polarization modulation.

Most EDM prisms have coated back faces and carry an anti-reflex coating on the front face. No acceptance angle limitations apply as no critical angles must be met on the reflecting surfaces. The coating of the reflecting faces may be silver, copper or gold, which all perform very well at infrared wavelengths. The quoted reflectivities of silver-coated prisms vary considerably from about 70% (Rinner and Benz 1966) to < 92% (Chang et al. 1971). The reflectivity depends heavily on the actual reflective and anti-reflex coatings as well as on the absorption losses in the glass. Coated prisms do not introduce additional polarizations. However, polarized light will be changed in amplitude and phase depending on the coating, the angle of incidence and the azimuth of incidence. Considering the real edges and their images in a prism, six sectors can be distinguished. Changes in polarization depend on the sector, onto which the incident EDM beam falls (Chang et al. 1971). The effects must be considered for prisms used in connection with distance meter using polarization modulation. The orientation of prisms may have to be specified in such cases. (In the case of the Terrameter LDM-2, the problem is avoided altogether by using spherical reflectors.)

At the front face of the prism, about 4% of the beam is reflected (rather than refracted) each time it enters or exits the glass corner cube. The beam, which is reflected at the time of exit, travels again through the prism before exiting. The strength of this multiple reflection beam is small (0.16%) and unlikely to cause significant distance measurement errors. The beam reflected at the time of entering the prism is returned to the distance meter at normal incidence. This case of the front face of the prism being perfectly normal to the incident EDM beam will apply on rare occasions, and thus affect the distance measurements. The problem can be reduced by antireflex coating of the front face of the prism or by mounting the prism at a slight tilt against the prism case.

Two principles are followed by manufacturers (more or less rigorously) when mounting the glass prisms inside the reflector housing. Minimizing the reflector constant leads to so-called *zero error prisms*. Alternatively, the distance and angular errors caused by poorly pointed prisms can be minimized by mounting the prisms in such a way that their "optical centres" coincide with the mechanical vertical and horizontal reflector axes (Peck 1948). The absolute reflector constants of such prisms amount typically to a few centimetres (e.g. *30 mm prisms*). In connection with electronic tacheometers, the second design is to be preferred, because angular pointings are usually carried out to the refracted image of the apex of the glass prism. The errors caused by misaligned prisms will be discussed in detail in Sections 10.2.5.2 and 10.2.5.3.

10.2.1 Accuracy of Reflectors

Glass prism reflectors may be classified according to their accuracy and their shape. The accuracy of a prism may be defined in terms of the maximum deviation of the angle between its faces from the ideal theoretical angle of exactly 90°

and the maximum deviation of its faces from an ideal plane. The old AGA long range prisms were ground to 2 seconds of arc maximum deviation from a true right angle and the old Aga short range prisms only to 20 seconds (AGA 1975). The facing of Keuffel and Esser prisms is true within ± 546 nm and is at right angles within 4 seconds of arc (Keuffel and Esser 1976).

Prisms are costly. For example, the cost of single prism reflectors is between 6 and 14% of the price of the EDM instrument. It is therefore wise to be perfectly clear about the required range, accuracy and number before ordering them. Their performance is checked by manufacturers through inspecting interference patterns in an interferometer (Hewlett-Packard 1973).

10.2.2 Shape and Size of Reflectors

The shape of the reflectors is determined in advance by the relative position of the transmitter and receiver optics of the corresponding EDM instrument because it is required that the return beam always fully illuminates the receiver optics. EDM instruments with coaxial transmitter and receiver optics require reflectors with a circular aperture; instruments with parallel but separated transmitter and receiver optics need, in principle, prisms with a rectangular aperture to displace the rays at the reflector by an offset equal to that of the two instrument optics. Reflectors of rectangular aperture are used in conjunction with the following instruments: Wild DI3, DI3S, Kern DM500, DM501, and Keuffel & Esser Autoranger. It may be mentioned that differing solutions were adopted for the prisms of earlier Geodimeters (models 2, 3, 4). The prisms were either ground to an angle larger than 90° or were fitted with a special wedge in front of the prism to spread the beam in order to reach the eccentric receiver optics.

The reflector's size is determined by the diameter of transmitter and receiver optics and by the requirement that the receiver optics must always be fully illuminated by the reflected beam. For an instrument with coaxial optics it can be easily proved that

$$\phi_B = \phi_T + 2\phi_P \ , \tag{10.1}$$

where ϕ_B is the diameter of the return beam at the receiver, ϕ_T the diameter of the transmitter optics and ϕ_P the diameter of the reflector's aperture. Denoting the diameter of the receiver optics by ϕ_R the above condition may be written in the form:

$$\phi_B \geq \phi_R \ . \tag{10.2}$$

The minimal reflector diameter may then be computed as

$$\phi_P = \tfrac{1}{2}(|\phi_R - \phi_T|) \ . \tag{10.3}$$

The actual diameter of reflectors is usually larger than the minimal value given in Eq. (10.3), thus increasing the radiant power of the return signal.

The more common case of parallel transmitter and receiver axis is depicted in Fig. 10.3. This case applies also to all fully integrated tacheometers, where the

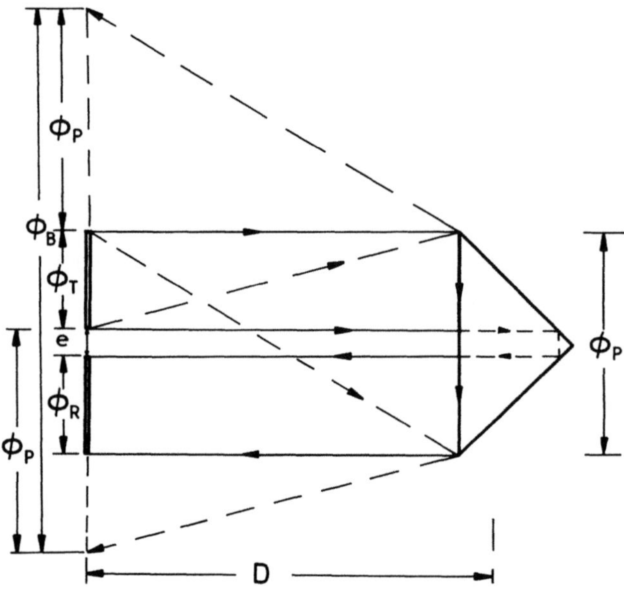

Fig. 10.3. Footprint of returned EDM beam for distance meters with parallel transmitter and receiver telescopes. The *solid ray paths* are critical for short range measurements. The *dashed ray paths* are exaggerated. They apply during long range measurements, where the beam divergence exceeds $2\phi_P/D$. The indices of the diameters (ϕ) B, P, R and T refer to the return *B*eam, *P*rism, *R*eceiver and *T*ransmitter, respectively

telescope is shared (two 180 degrees sectors) by the transmited and received beams and where the eccentricity, e, does not exceed a few millimetres. On close range, the ideal diameter of the prism is given by

$$\phi_P = \phi_T + \phi_R + e \; , \tag{10.4}$$

where all the parameters are defined in Fig. 10.3. At longer distances, where the beam divergence exceeds $2\phi_P/D$, the minimum prism diameter computes as

$$\phi_P = \phi_R + e \; . \tag{10.5}$$

It further follows from the solid rays in Fig. 10.3, that the diameter of the smallest possible prism to return any part of the transmitted beam must exceed the eccentricity e of the optics. This consideration explains why some distance meters cannot measure to reflective tapes and acrylic reflectors on close range. The problem is reduced to a certain degree by the large divergence of these inexpensive reflectors. Table 10.1 gives experimental values for these cases.

EDM instruments with separate transmitter and receiver optics and relatively large eccentricities e may use reflectors with rectangular apertures. Assuming equal diameters of transmitter and receiver, such a prism would have widths of ϕ_T and lengths of between $\phi_T + e$ and $2\phi_T + e$.

10.2.3 Phase and Group Refractive Index in Glass

Prisms used in EDM are usually made of BK7 optical glass. The phase refractive index of BK7 may be computed from the dispersion formula (Melles Griot 1975):

$$n_{G_{ph}}^2 = A_0 + A_1\lambda^2 + A_2\lambda^{-2} + A_3\lambda^{-4} + A_4\lambda^{-6} + A_5\lambda^{-8} \qquad (10.6)$$

with the constants
$$
\begin{aligned}
A_0 &= 2.2718929 \\
A_1 &= -1.0108077\times10^{-2} \\
A_2 &= +1.0592509\times10^{-2} \\
A_3 &= +2.0816965\times10^{-4} \\
A_4 &= -7.6472538\times10^{-6} \\
A_5 &= +4.9240991\times10^{-7} ,
\end{aligned}
$$

where $n_{G_{ph}}$ = phase refractive index in optical glass BK7
λ = wavelength of carrier wave of EDM instrument (μm).

The dispersion formula (10.6) is accurate to ± 5 ppm in the spectral region between 365 nm and 1014 nm.

Because of the EDM ray being modulated, the group refractive index according to Eq. (5.11) has to be considered:

$$n_G = n_{G_{ph}} - \frac{dn_{G_{ph}}}{d\lambda}\lambda . \qquad (10.7)$$

Differentiation of Eq. (10.6) provides the differential in Eq. (10.7). The group refractive index in BK7 optical glass therefore yields:

$$n_G = n_{G_{ph}} - \frac{1}{n_{G_{ph}}}(A_1\lambda^2 - A_2\lambda^{-2} - 2A_3\lambda^{-4} - 3A_4\lambda^{-6} - 4A_5\lambda^{-8}) , \qquad (10.8)$$

where n_G = group refractive index in BK7 optical glass
$n_{G_{ph}}$ = phase refractive index of BK7 optical glass
λ = wavelength of carrier wave of EDM instrument (μm)
A_i = constants of dispersion formula.

Table 10.2. Group and phase refractive indices in optical glass (BK7) at some wavelengths used in EDM

λ (nm)	n_G (BK7)	$n_{G_{ph}}$ (BK7)
550	1.5462	1.5185
633	1.5367	1.5151
840	1.5252	1.5100
860	1.5246	1.5097
880	1.5240	1.5093
900	1.5235	1.5090
920	1.5230	1.5087

The phase and group refractive indices of glass are listed in Table 10.2 for some wavelengths commonly used in EDM. Only group refractive indices should be considered in EDM applications.

10.2.4 Reflector Constant

If only one particular type of reflector is used in conjunction with a particular EDM instrument, the reflector constant is combined with the additive constant for the EDM instrument. The determination of this combined constant is described in Sections 13.2.4.3 and 13.2.4.4.

Various type of reflectors may be employed if their relative or absolute constants are determined. The relative reflector constant may be determined by measuring a large number of different distances with all reflectors. The mean differences of distances using two reflectors is a measure of the *relative reflector constant*. In practice, relative constants are usually computed in relation to one specified reflector (reference reflector). The *absolute reflector constant*, which refers to the vertical axis of a reflector, may be worked out from the known or measured dimensions of a reflector. Technical data of a selection of reflectors is given in Appendix H. Figure 10.4 illustrates the situation.

The first velocity correction of a distance measurement assumes that the EDM beam travels in air between instrument, reflector and instrument. In consequence, the glass path (a) needs to be reduced to the equivalent air path.

Using Eq. (6.1)

$$d = \frac{c_0}{n} \frac{\Delta t'}{2}$$

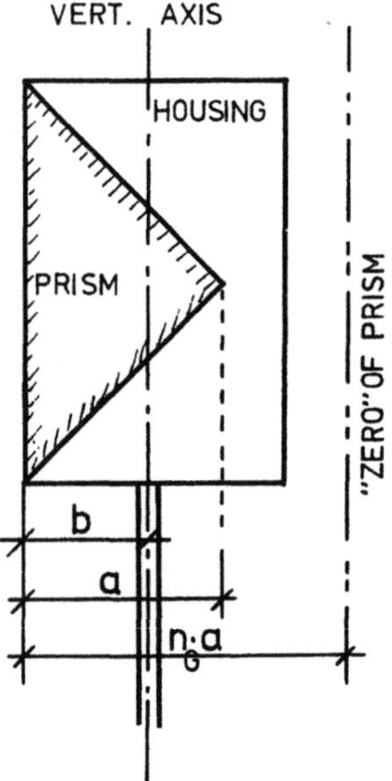

Fig. 10.4. Critical dimensions of a reflector for the computation of the absolute reflector constant. The height of the apex of the prism above the front surface is denoted by a and the offset of the vertical and horizontal axes from the front face by b. The *zero of prism* indicates the plane to which the distance measurement refers to (air equivalent path). Errors due to reflector misalignment can be minimized by selecting $b = a/n_G$

the vacuum equivalent flight time Δt^* of a signal travelling a distance "a" in glass is

$$a = c_0 \frac{\Delta t^*}{2 n_G} \tag{10.9}$$

$$\Delta t^* = \frac{2 a n_G}{c_0} \; . \tag{10.10}$$

For the same flight time in air, a signal would travel the air path a*:

$$a^* = \frac{c_0 \Delta t^*}{2 n_A} \; . \tag{10.11}$$

Inserting Eq. (10.10) leads to

$$a^* = \frac{c_0}{2 n_A} \frac{2 a n_G}{c_0}$$

$$a^* = \frac{n_G}{n_A} a \; , \tag{10.12}$$

where a* is the air path equivalent to the glass path a.

To correct any distance to the vertical axis of the prism assembly subtract

$$a^* = \frac{n_G}{n_A} a \quad \text{and} \quad \text{add b} \; .$$

This leads to the

$$\text{Absolute Reflector Constant} = - \left\{ \frac{n_G}{n_A} a - b \right\} \; , \tag{10.13}$$

where a is the height of the cube corner above the front surface, n_G the group refractive index of light in glass, n_A the group refractive index of light in air and b the distance between the front surface of reflector and the vertical axis of reflector assembly. The refractive index of glass prisms depends on the type of glass and the carrier wave (dispersion). The group refractive index of glass has been discussed in the previous section. A table of reflector-specific a and b is given in Appendix H.

The reflector constant should be added to the distance measurement to obtain the corrected distance. It follows from Fig. 10.4 that it is possible to build so-called zero constant prisms which can be fixed on ranging rods. The front surface of the reflector is then at a distance of $(n_G/n_A)a$ from the centre line of the ranging rod (zero of prism).

The method of measuring absolute reflector constants can also be used to derive relative constants. The reflector constants of several reflectors of the same design and brand can be expected to be within a range of ± 1 mm (Hewlett-Packard 1973), although worse results were reported for old AGA prisms and one make of newer prisms (McLean 1984).

Example: Computation of Relative Reflector Constant.

An instrument Zeiss SM4 was calibrated using a prism Zeiss TR2. What relative reflector constant would have to be applied, if the SM4 was used in conjunction with a Kern DM501 reflector?

From Appendix H:
   Zeiss Reflector TR2 $a = 40$ mm , $b = 27$ mm
   Kern DM501 Reflector $a = 81$ mm , $b = 34.5$ mm

From Appendix D:
   λ_{carr} of SM4 $= 910$ nm $n_{REF} = 1.0003 = n_A$

With the group refractive index of glass taken as 1.523, the Absolute Reflector Constants (A.R.C.) for the two reflectors compute as:

$$\text{Zeiss Reflector:} \quad \text{A.R.C.} = -\left\{\frac{1.523}{1.000}\ 40 - 27\right\} = -33.9 \text{ mm}$$

$$\text{Kern Reflector:} \quad \text{A.R.C.} = -\left\{\frac{1.523}{1.000}\ 81 - 34.5\right\} = -88.9 \text{ mm}$$

The ARC of -33.9 mm is already taken into account by the additive constant of the SM4, as the Zeiss prism was employed during calibration. The additional additive constant to be applied if the Zeiss SM4 is used with the Kern DM501 reflector amounts to

$$\text{Relative Reflector Constant (R.R.C.)} = -88.9 - (-33.9) = -55.0 \text{ mm}$$

Another way of solving the problem would be to recalibrate the EDM instrument with the Kern prism on the baseline!

10.2.5 Effects of Errors of Reflector Alignment

10.2.5.1 Tiltable Reflector-Target with Eccentric Prism(s)

It was mentioned in Section 9.1.3.4 that several telescope-mounted instruments employ reflector-target combinations which compensate for the offset between telescope and distance meter. The horizontal axis of such reflector-targets is usually located through the centre of the target. Any erroneous vertical pointing of the reflector-target leads to a small error in the measured distance. The effect on distance of an error in vertical pointing α together with an eccentricity e of the reflector centre from the tilting axis may be written as follows:

$$\Delta d = e \sin \alpha \ , \tag{10.14}$$

where Δd is the error in distance and α is given in degrees. An angle α of $5°$ and an eccentricity e of 45 mm (Wild DI3) would lead to a distance error of 4 mm. It can be seen that reflectors need to be aligned carefully in cases where the axis of the prism has an eccentricity relative to the tilting axis of the reflector-target.

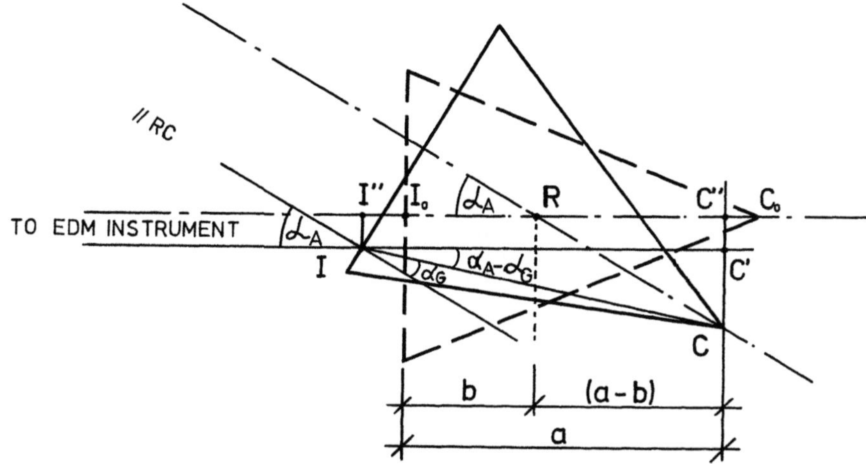

Fig. 10.5. Change of point of incidence of the central ray between a properly aligned and a tilted solid corner cube reflector. The apex of the prism is denoted by C and the point of rotation (vertical and/or horizontal axis) by R. The distance II'' is a measure of the angular errors incurred when measuring angles to the refracted image of the prism corner. The distance errors can be derived from a comparison of the distances I'' to C_0 and I to C

10.2.5.2 Effect of the Reflector's Alignment on Measured Distance

Another effect of inaccurate alignment, which affects all prisms whether they are tiltable or not, is that the EDM rays travel a longer distance within the glass of the prism if the prism's front surface is not exactly perpendicular to the EDM rays. Because of the different refractive indices of glass and air a different distance is then measured. The situation is depicted in Fig. 10.5, where the broken lines depict a properly aligned prism and the solid lines a misaligned prism.

The corner of the aligned prism is denoted by C_0 and that of the misaligned one by C. The point of incidence of the EDM ray which travels to the prism corner C_0 and back is termed I_0 and that of the ray which proceeds to the prism corner C (and back) is denoted by I. Only the corner rays are considered because they allow a two-dimensional solution of the problem. The final statement, however, holds for all rays in the three-dimensional case.

The vertical or horizontal axis is denoted by R and the dimension of the reflector and its relationship to the axis are given by the parameters a and b. The *angle of incidence* α_A and the *angle of refraction* α_G refer to the maladjusted prism (A = air, G = glass). The corresponding angles in the case of the correctly aligned prism are zero.

The *law of refraction* is defined by the formula

$$\frac{n_A}{n_G} = \frac{\sin \alpha_G}{\sin \alpha_A} \, . \tag{10.15}$$

It permits the computation of the unknown angle of refraction α_G for a given angle of incidence α_A, equal to the error angle in alignment, as

$$\alpha_G = \arcsin\left(\frac{n_A}{n_G}\sin\alpha_A\right) . \tag{10.16}$$

In order to compare the distance measurements in the cases of an aligned and a deflected prism, the measured lengths $I''C_0$ and IC are compared. Considering the appropriate refractive indices the difference Δd between the measured distances to the aligned and the deflected prisms amounts to

$$\Delta d = I''I_0 + (I_0C_0)\frac{n_G}{n_A} - (IC)\frac{n_G}{n_A} . \tag{10.17}$$

If Δd is added to the distance measured to the deflected prism, the correct distance, equal to the distance measured to the aligned prism, is obtained. In Eq. (10.17), the refractive indices of glass and air are denoted by n_G and n_A, respectively.

The line elements in Eq. (10.17) may be derived from Fig. 10.5 as

$$IC \quad = a\sec\alpha_G \tag{10.18}$$

$$\begin{aligned}
IC' \quad &= IC\cos(\alpha_A - \alpha_G) \\
&= a\sec\alpha_G\cos(\alpha_A - \alpha_G) \tag{10.19}
\end{aligned}$$

$$RC'' = (a-b)\cos\alpha_A \tag{10.20}$$

$$\begin{aligned}
I''I_0 \quad &= IC' - RC'' - b \\
&= a\sec\alpha_G\cos(\alpha_A - \alpha_G) - (a-b)\cos\alpha_A - b . \tag{10.21}
\end{aligned}$$

Substitution of Eqs. (10.18) and (10.21) in Eq. (10.17) yields

$$\begin{aligned}
\Delta d &= a\sec\alpha_G\cos(\alpha_A - \alpha_G) - (a-b)\cos\alpha_A - b + a\frac{n_G}{n_A} - a\frac{n_G}{n_A}\sec\alpha_G \\
&= a\left(\cos(\alpha_A - \alpha_G) - \frac{n_G}{n_A}\right)\sec\alpha_G - (a-b)\cos\alpha_A + a\frac{n_G}{n_A} - b , \tag{10.22}
\end{aligned}$$

where α_G is defined by Eq. (10.16).

Using the substitutions

$$\alpha_A = \alpha \quad \alpha_G = \beta$$

$$\frac{n_G}{n_A} = n \tag{10.23}$$

$$\text{and} \quad \sin\beta = \frac{\sin\alpha}{n} ,$$

a simplified form of Eq. (10.22) can be given

$$\begin{aligned}
\Delta d &= a[\cos(\alpha - \beta) - n]\frac{1}{\cos\beta} - (a-b)\cos\alpha + na - b \\
&= a\left[\frac{\cos(\alpha - \beta)}{\cos\beta} - \frac{n}{\cos\beta} - \cos\alpha + n\right] - b(1 - \cos\alpha) . \tag{10.24}
\end{aligned}$$

Using $\cos(\alpha - \beta) = \cos\alpha\cos\beta + \sin\alpha\sin\beta$ leads to

$$\Delta d = \left(\frac{\sin\alpha\sin\beta}{\cos\beta} - \frac{n}{\cos\beta} + n\right) - b(1-\cos\alpha)$$

$$= a\left(n - \frac{1}{\sqrt{1-\sin^2\beta}}(n - \sin\alpha\sin\beta)\right) - b(1-\cos\alpha)$$

$$= a\left(n - \frac{1}{n\sqrt{1-\sin^2\beta}}(n^2 - \sin^2\alpha)\right) - b(1-\cos\alpha)$$

$$= a\left(n - \frac{(n^2 - \sin^2\alpha)}{\sqrt{n^2 - \sin^2\alpha}}\right) - b(1-\cos\alpha) \tag{10.25}$$

$$\Delta d = a(n - \sqrt{n^2 - \sin^2\alpha}) - b(1-\cos\alpha) \ , \tag{10.26}$$

where n is defined as the quotient (n_G/n_A) of the group refractive indices of glass and air and α as the angle of incidence. The distance correction Δd for misaligned prisms given in Eq. (10.26) is equivalent to the formulae given by Pauli (1969, 1973) and Heister (1988). It can be seen that the possible errors in distance depend largely on the size of the prism and on the position of the vertical or horizontal axis (R) in relation to the prism.

According to Peck (1948), the distance correction can be minimized by the reflector manufacturer for small angles of incidence by selecting

$$b = a\,n_A/n_G \tag{10.27}$$

or, considering a previous substitution [Eq. (10.23)],

$$b = a/n \ . \tag{10.28}$$

It can be shown that substitution of the above equation in Eq. (10.26) leads to the following distance correction equation (Heister 1988)

$$\Delta d = a\,n(1-\cos\beta) - a(1-\cos\alpha)/n \ , \tag{10.29}$$

where β is the angle of refraction α_G. It should be noted that Eq. (10.29) applies only to prisms which have their tilting and rotation axis in the "optical centre" as defined by Eq. (10.27). Normally, either Eq. (10.22) or Eq. (10.26) has to be used.

A numerical example may illustrate the magnitude of the error in distance Δd caused by the misalignment of the reflector. The following parameters are assumed (Wild GDR 11 reflector):

$a = 60$ mm	$n_A = 1.00$
$b = 22$ mm	$n_G = 1.52$ ($\lambda = 900$ nm)
$\alpha_A = \ \ 1°$	$\Delta d = 0.00$ mm
$= \ \ 5°$	$= 0.07$ mm
$= 10°$	$= 0.26$ mm
$= 15°$	$= 0.58$ mm
$= 20°$	$= 1.01$ mm
$= 25°$	$= 1.53$ mm
$= 30°$	$= 2.13$ mm

An alignment of the reflector to $\pm 15°$ will lead to a maximum error in distance of only 0.5 mm (Wild reflector GDR 11). The alignment of a centrally supported reflector is therefore not critical at all with regard to the distance measurement. See Section 10.2.5.1 for additional errors of eccentrically supported prisms and Appendix H for technical data of other reflectors. Numerical examples for other reflector types may be found in Rüeger (1978a) and Heister (1988).

10.2.5.3 Effect of the Reflector's Alignment on Angular Measurement

The corner of a prism is sometimes used as a target for the measurement of zenith angles and horizontal directions. Because the prism corner is observed through the glass, the ray is refracted at the incident plane of the prism, if the reflector is not exactly aligned. As can be seen from Fig. 10.5, the image of C appears in I, which has an offset of II'' from the true direction to the axis of the reflector R.

Considering Eqs. (10.16) and (10.18) of the previous section and inspection of Fig. 10.5 lead to

$$II'' = CC'' - CC' \tag{10.30}$$

$$CC' = IC \sin(\alpha_A - \alpha_G)$$
$$= a \sec \alpha_G \sin(\alpha_A - \alpha_G) \tag{10.31}$$

$$CC'' = (a-b) \sin \alpha_A . \tag{10.32}$$

The displacement of the ray is then given by:

$$II'' = (a-b)\sin\alpha_A - a \sec\alpha_G \sin(\alpha_A - \alpha_G) , \tag{10.33}$$

where the angle of refraction α_G is computed according to Eq. (10.16):

$$\alpha_G = \arcsin\left(\frac{n_A}{n_G}\sin\alpha_A\right) .$$

The displacement II'' produces the following error Δz in a zenith angle measured at the EDM instrument station or $\Delta\alpha$ in a measured direction:

$$\Delta z'' (\text{or } \Delta\alpha'') = \frac{II''}{d \sin(1'')}$$
$$= \frac{(a-b)\sin\alpha_A - a \sec\alpha_G \sin(\alpha_A - \alpha_G)}{d \sin(1'')} , \tag{10.34}$$

where α_A is the vertical or horizontal misalignment of the reflector and d the distance between EDM instrument and reflector. The effects of the misalignment of the reflector on angular measurements taken to the refracted image of the prism's apex can again be minimized by the manufacturer of the reflector by selecting b according to Eqs. (10.27) or (10.28).

Table 10.3 gives a numerical example of the magnitude of angular errors incurred when pointing to a misaligned prism. The selected reflector type does almost fulfil the relationship of Eqs. (10.27) and (10.28), namely that the reflector rotates

162

Table 10.3. Angular errors ($\Delta\alpha$, Δz) incurred when pointing to a misaligned old type AGA reflector (non-tiltable, with metal housing, part Nos. 571.125.001 and 570.590.383)

$\alpha_A =$	$\alpha_G =$	$II'' =$	$\Delta z(\Delta\alpha)$ for d = 50 m	$\Delta z(\Delta\alpha)$ for d = 100 m	$\Delta z(\Delta\alpha)$ for d = 500 m
1°	0.66°	0.14 mm	0.6"	0.3"	0.1"
3°	1.97°	0.42 mm	1.7"	0.9"	0.2"
5°	3.29°	0.70 mm	2.9"	1.5"	0.3"
10°	6.56°	1.43 mm	5.9"	3.0"	0.6"
15°	9.80°	2.21 mm	9.1"	4.6"	0.9"
20°	13.00°	3.07 mm	12.7"	6.3"	1.3"
25°	16.14°	4.04 mm	16.7"	8.3"	1.7"
30°	19.20°	5.13 mm	21.1"	10.6"	2.1"

The prism parameters are taken from Appendix H as a = 41 mm, b = 35 mm, $n_A = 1.00$ and $n_G = 1.52$ (from Table 10.2, for $\lambda = 900$ nm). The angles of incidence (for misalignment) and refraction are denoted by α_A and α_G respectively. The length II'' indicates the shift of the apex image caused by the rotation or tilt of the prism

about its "optical centre" (Peck 1948). Experiments have shown that the reflector in question can be pointed in the horizontal direction with an accuracy of about ±40 minutes of arc (Rüeger 1978a). The resulting errors in measured directions are therefore minimal. As the particular type of reflector is not tiltable, zenith angle measurements (z) are heavily affected as $\alpha_A = 90° - z$.

Table 10.3 indicates clearly that, for example, the cube corner of a non-tiltable AGA reflector (old type) should not be used as theodolite target if accurate zenith angle observations are required over short ranges. Angular errors caused by the misalignment of other types of reflectors may be computed using Eq. (10.34). The technical data of some reflectors are given in Appendix H. Further numerical examples may be found in Rüeger (1978a), for example.

In order to avoid the problems involved in theodolite pointings to cube corners of reflectors, many reflectors are equipped with additional targets for angular measurements. However, additional eccentricities may affect the angular measurements unless such targets are strictly in the tilting axis and in the vertical axis for zenith angle and horizontal direction measurements, respectively.

10.2.6 Temperature Effects

The reflectivity of a reflector is only optimal if its entire glass mass has the same temperature. Unequal temperature distribution in the prism causes deformations of the faces and thus divergence of the emerging beam. Prisms need 1 h or more to adapt to a new temperature after a change in ambient temperature (Dalcher 1975). Naturally, the larger the prism the longer it will take to settle down in new temperature conditions. If distances at the upper limit of the range of an EDM instrument and its reflectors are attempted, it may be advisable to shade not only the instrument but also the reflector and to expose the reflector to (shade) field temperatures well before the measurement.

10.2.7 Care of Reflectors

Basically, reflectors should be treated with the same care as theodolites. Wet or dusty surfaces should be cleaned with a soft cloth, avoiding any scratching of the glass surface. Wet prisms should be allowed to dry before they are stored. Fingerprints on the glass should be avoided, but may be cleaned with a soft cloth dampened with ether or alcohol.

During measurements in rain or snow, the reflector should be protected to avoid water drops on the front surface of the prism. Water drops on the plane of incidence of the prism decrease the reflectivity and thus the range of the EDM instrument. In wet conditions the front surface of the reflector may also be treated with an "antimist" towel or covered with a thin soap layer to produce a thin layer of water rather than water drops (AGA 1969).

11 Batteries and Other Power Sources

11.1 Review of Power Sources

Power sources are very important in electronic distance measurement as no power means no distance measurements. The most common types of power sources are:

1. batteries
2. solar cells
3. generators
4. mains-operated DC power supplies.

Batteries are by far the most common power sources in EDM. They are discussed in the next section in more detail. Usually, batteries are also employed in connection with solar cells and generators, where the latter are used to charge the batteries or to operate the EDM instrument through an intermediate battery.

Commercial panels of *photovoltaic* (solar) *cells* are usually made of semi-crystalline silicon (Si). The voltage of each cell is about 0.5 V, irrespective of light intensity or size. The current output of solar cells is proportional to light intensity and size. The voltage is increased by placing cells in series. The output current can be increased by connecting cells in parallel. A single solar cell panel of 1.0 by 0.5 m may, for example, produce 2.6 A at 16.2 V at peak power (42 W). This would be sufficient to operate most infrared short range distance meters. Typically, solar cells would not supply EDM instruments directly but rather charge their batteries, either concurrently with distance measurements or off-line, while a second battery powers the distance meter. Figure 11.1 shows the wiring and components required for the former case. To prevent the battery from discharging through the solar cells (at night, for example), a blocking diode must be put in series with the positive line. The charge controller is required to protect the battery from extended overcharging. Charge controllers or regulators are usually supplied by the manufacturers of solar panels. Some suitable circuits may be found in Byers (1984).

In EDM, solar panels are of interest whenever overnight charging from mains supply is not possible. Such cases might arise when surveying in remote locations. Solar panel back-up might also be of interest in certain monitoring surveys where continuous measurements are carried out over days and weeks. The required power output may be computed as follows (Solarex 1986):

$$\text{Solar system peak output} > (\text{Ah/day} + \text{system loss})/\text{peak sun hours} , \quad (11.1)$$

where the peak output is taken in amperes (A), the system loss as 20% and 40% for lead-acid and nickel-cadmium batteries, respectively, and the peak sun hours from diagrams published by solar cell manufacturers. Solarex (1986) suggests the

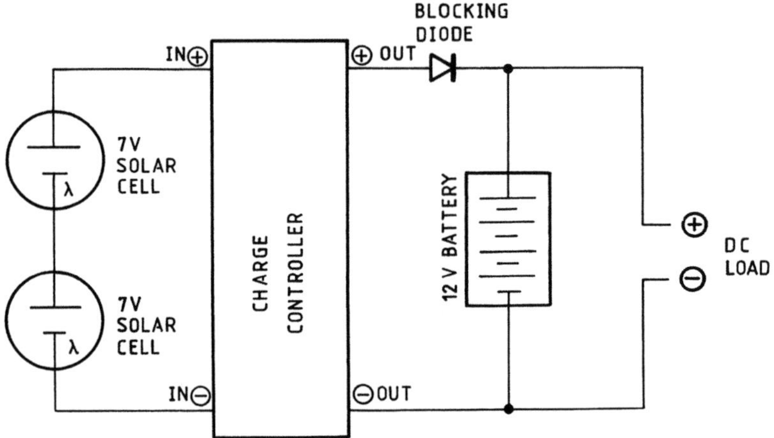

Fig. 11.1. Wiring diagram of a solar panel used to charge the EDM instrument's battery during EDM operation. (After Byers 1984)

following peak sun hours per day (yearly average): Sydney (Australia) 5.5, Alice Springs (Australia) 6.5, Central Europe 3.65; Alaska 2.75. Considering the solar panel mentioned above (2.6 A), a 40% system loss of NiCd batteries and Sydney (Australia), the 12 V nickel-cadmium battery of an EDM instrument could be charged with 10 Ah/day (yearly mean value).

Small portable petrol *generators* can be used in similar ways to charge the batteries of EDM instruments. However, it is not advisable to connect an EDM instrument directly to the battery being charged because of temporary voltage peaks and static discharges. Portable generators usually provide AC and DC (12 V). Typical units weigh about 20 kg and may provide 1 A at 240 V AC and 8 A at 12 V DC. The DC output may be used for the direct charging of 12 V (car) batteries. Where required, the charge current through the batteries can be reduced by charging two batteries in parallel.

Mains-operated *DC power supplies* are of advantage in stationary (often indoors) applications. The power supply should deliver rectified (and stabilized) 12 V direct current, with a current rating which exceeds the current requirements of the distance meter. It should be noted that mains supplies often suffer from interference, such as voltage peaks at the time of turn on/off of large machines. Such interference may reach the distance meter and cause malfunctioning. Wild (1988) suggests the use of a commercially available mains filter between the power point and the power supply as well as a special cable between power supply and distance meter. The suggested cable features a basic interference suppressor and a protection against wrong polarity. However, full protection from conducted electromagnetic interference can only be provided by batteries!

Most manufacturers of EDM instruments supply mains operated battery chargers for the make specific battery packs. Apart of a charge regulator, these battery chargers also include a AC-DC converter. Usually, these battery chargers cannot be used to operate distance meters directly.

11.2 Batteries Used in EDM

The batteries presently used as power sources for EDM instruments may be grouped as follows:

1. Primary batteries
   a) Alkaline cells
   b) Lithium cells
2. Secondary batteries (rechargeable storage batteries)
   a) Automotive lead-acid batteries
   b) Sealed lead-acid cells and batteries
   c) Sealed nickel-cadmium (NiCd) cells.

11.2.1 Primary Batteries

The use of primary batteries in EDM is not common, as the cost would be prohibitive. However, some distance meters permit alternative operation with primary batteries. This option may be helpful in cases were the normal power source is exhausted. When using primary batteries, only alkaline manganese oxide batteries should be used (e.g. Duracell). Alkaline batteries perform much better than the older carbon-zinc (dry cell, Leclanché cell) batteries. The former exhibit better low temperature operation, a much improved discharge characteristic, larger capacity and smaller volume in comparison with the latter. The negative electrode (anode) is formed by jelled zinc powder. Electrolytic manganese dioxide (EMD, MnO_2) and graphite is used as positive electrode (cathode). An aqueous solution of potassium hydroxide (KOH) forms the electrolyte. Mantell (1983) gives the following overall equation of the chemical reaction:

$$2\,Zn + 2\,KOH + 3\,MnO_2 \rightarrow 2\,ZnO + 2\,KOH + Mn_3O_4 \ .$$

Lithium cells and batteries are a relatively new type of primary battery. The capacity of lithium cells is four times that of alkaline cells of the same weight and two times that of alkaline cells of the same volume. Common types of lithium cells (size AA to D) are: lithium thionyl chloride (Li-SOCl$_2$), lithium sulphur dioxide (Li-SO$_2$) and lithium copper oxide (Li-CuO). They have very good discharge characteristics and perform very well at very low as well as very high temperatures. Lithium cells and batteries are expensive. Present usage includes (military) communication equipment, CMOS and other electronic circuit back-up applications and life-support devices (e.g. heart pacemakers). EDM instruments and electronic theodolites may make use of the second type of application of lithium cells.

11.2.2 Secondary Batteries

Automotive lead-acid batteries (car batteries) normally supply power for starting, lighting and ignition (SLI) of road vehicles. Accordingly, the specifications state cranking current in amperes (CCA) and reserve capacity, rather than the more familiar capacity in Ah. The International Electrotechnical Commission (IEC) defines CCA as the current (in amperes), delivered at a temperature of $-18\,°C$ for not less than 1 min to a voltage of not less than 8.4 V for a 12 V battery. IEC defines the reserve capacity as the number of minutes a fully charged battery can maintain a current of 25 A to a voltage of not less than 10.5 V for a 12 V battery at a temperature of 25 °C (Barak 1980). Although automotive batteries are not optimized for use with EDM instruments, they usually serve as a stand-by power source. All manufacturers of EDM instruments supply the necessary cables (and DC/DC converters, if required) to connect their distance meters to a 12 V car battery. Wild (1988) suggests that the car engine and car lights be turned off before the EDM instrument is connected. Distance meters should be protected from wrong polarity of the connection to external batteries.

The cathode (positive electrode) of lead-acid batteries is made from lead oxide (PbO_2) and the anode (negative electrode) from spongy lead (Pb). Diluted sulphuric acid (H_2SO_4) is used as electrolyte. A full description of the chemical reaction as well as performance data may be found in Barak (1980). It should be noted that car batteries generate highly flammable hydrogen and oxygen during charging. Considering that, in Australia, 250 people per year suffer eye injuries due to exploding lead-acid batteries, great care should be used when charging and handling batteries.

Since about 1965, non-spill lead-acid batteries were developed for so-called cordless appliances. EDM instruments fall into this category. The development started with gel batteries and led later to sealed lead-acid cells (e.g. Gates) and batteries. The sealed lead-acid (or lead-calcium) batteries approach the energy density of the nickel-cadmium batteries discussed below. Gas re-combination techniques are employed to control the generation of gases. Sealed lead-acid (SLA) cells have similar performance characteristics as nickel-cadmium alkaline cells, although the capacity of the former is slightly less in terms of weight and volume. In comparison to NiCd cells, the SLA cells have a better discharge characteristic at elevated temperatures, feature a much smaller self-discharge rate and are cheaper than NiCd cells. However, NiCd cells can sustain more charge/discharge cycles during their service life. This is most likely the reason why most EDM instrument manufacturers supply NiCd batteries. The principle and characteristics of these batteries are discussed in the next section in more detail.

11.3 Sealed Nickel-Cadmium Batteries

11.3.1 Construction and Principle

Most EDM instruments employ batteries which are series combinations of individual cylindrical cells assembled in packs. The individual cells are usually rated at 1.2 V DC. They consist of a nickel-plated steel case as the negative terminal and a cell cover as the positive terminal. The positive and negative electrode plates are wound to form a compact roll and are isolated from each other by a porous separator. An electrolyte furnishes the ions for conduction between the positive and negative electrodes. As a safety measure, a high-pressure vent is usually incor-

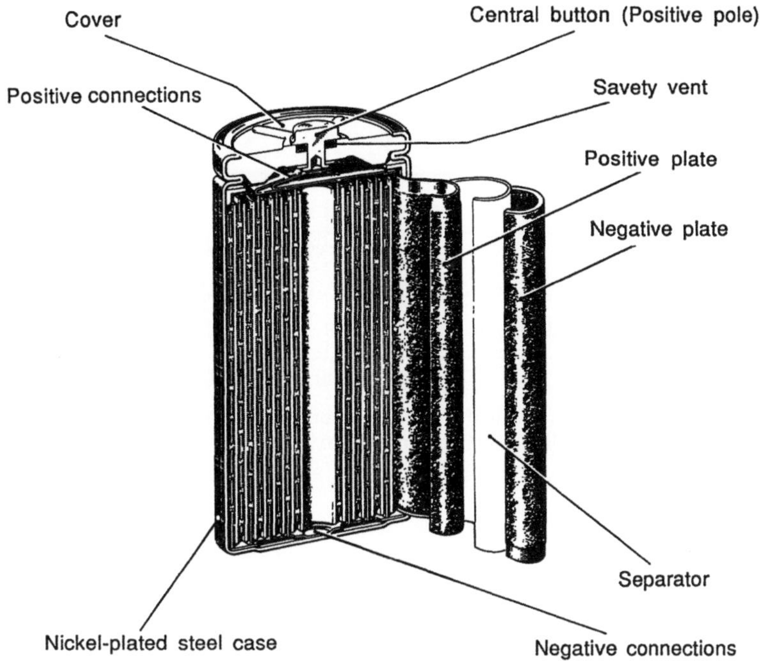

Fig. 11.2. View of a cut nickel-cadmium (NiCd) cell. (Courtesy of Plessey)

porated in the button of the cell cover. The construction of a sealed Ni-Cd cell is depicted in Fig. 11.2.

The electrodes in Ni-Cd cells undergo changes in oxidation state without any change in physical state and, because of this, have a long life. Nickel hydroxide is the active material in the positive plate. The charged nickel hydroxide (NiOOH) is transformed to a lower valence state, namely $Ni(OH)_2$, during discharge. In the negative plate, cadmium (Cd) metal is the active material and is oxidized to cadmium hydroxide $[Cd(OH)_2]$ during a discharge of the cell. The reactions in the electrolyte (KOH, potassium hydroxide), during charge and discharge may be summarized according to General Electric (1975) as

$$Cd + 2\,H_2O + 2\,NiOOH \underset{\text{Charge}}{\overset{\text{Discharge}}{\rightleftharpoons}} 2\,Ni(OH)_2 + Cd(OH)_2 \ .$$

The pressure relief safety vent opens if the internal pressure of the cell, caused by hydrogen and oxygen gases, exceeds 1 to 2 MPa (General Electric 1975).

Note of Caution. Ni-Cd batteries should not be incinerated or multilated because they may burst or release toxic materials. They should not be short circuited either because this may cause burns. Even a small amount of the electrolyte causes serious problems if it gets in the eye. Immediate flushing with water for 15 min and subsequent medical attention is absolutely necessary in this case.

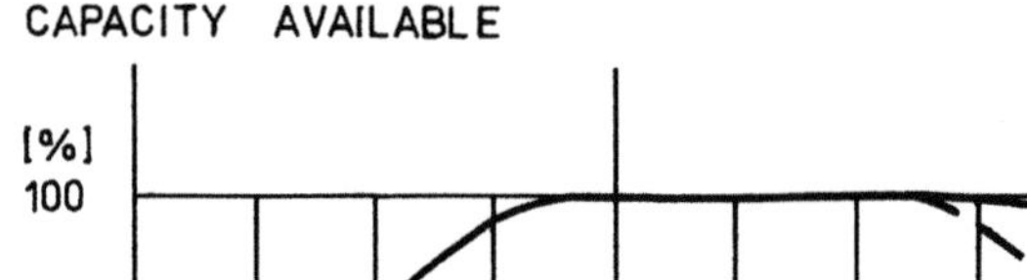
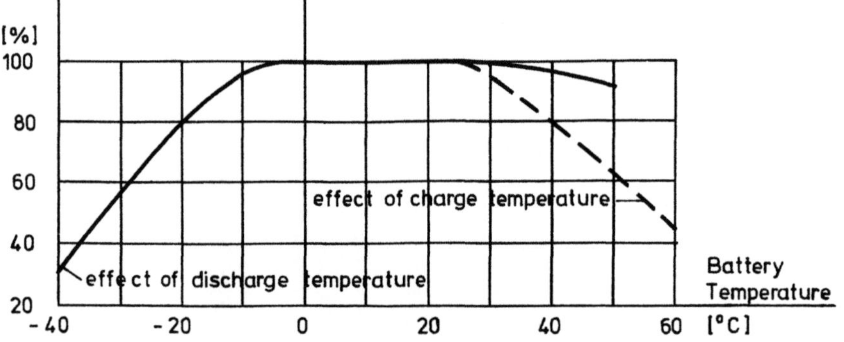

Fig. 11.3. Capacity of sealed nickel-cadmium cells versus ambient temperature during charge and discharge of cell. (After General Electric 1975)

11.3.2 Discharge Characteristics

The discharge characteristics depend on various operating variables such as discharge rate, discharge duration, temperature during the discharge and the previous charge, previous charge rate and time and the previous cycling history.

The capacity of Ni-Cd cells of 1.2 V nominal voltage is given in Ah and is based upon a discharge down to 1.0 or 1.1 V at a specified discharge rate. At lower drain rates, the actual capacity of a cell will be greater than the rated capacity; at higher drain rates the available capacity will be smaller. The temperature during the discharge as well as during the previous charge also has an effect on the available capacity. Both effects are depicted in Fig. 11.3. The drop in available capacity at low temperatures is considerable. However, the low temperature performance of Ni-Cd batteries is better than that of most other battery types. A decrease in available capacity also occurs at temperatures above 40° to 50 °C.

The voltage of Ni-Cd cells remains relatively flat throughout most of the discharge period, with only a 100 mV drop for the rated discharge current. Higher drain rates produce curves of steeper slopes and at lower voltage levels. The discharge voltage is also affected by previous long term overcharging. The effect called *voltage depression* may lead to a 0.1 V smaller discharge voltage during the entire discharge. One or more discharge/charge cycles will remove the effects of this voltage depression.

A multi-cell battery should never be discharged down to zero voltage because it may cause reverse charging of the lowest capacity cells within the battery, leading to a build-up of pressure in these cells. It is recommended that the discharge should be terminated when the battery voltage reaches a value equivalent to the product (cell voltage)$\times$(n$-$1) where n is the number of cells in a battery. A ten-cell battery (12 V) should therefore not be discharged below 10.8 V. However, deeper discharges of single cells are possible and sometimes required for reconditioning purposes.

170

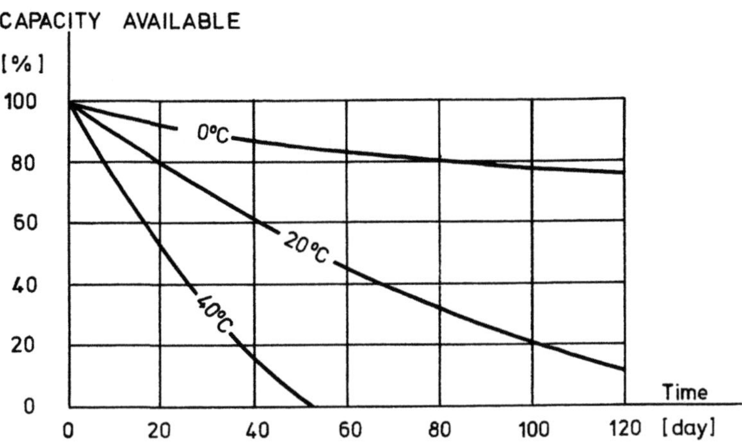

Fig. 11.4. Charge retension of nickel-cadmium cells at storage temperatures of 0°, 20° and 40 °C. (After General Electric 1975)

11.3.3 Charge Characteristics

Four types of charge may be distinguished:

1. slow charge (14−16 h, "overnight")   Charge current = 0.1 C
2. quick charge (3−7 h)   Charge current = 0.3 C
3. rapid charge (15−60 min)   Charge current = 1.0 C
4. ultra-rapid charge (1−3 min).

The parameter C is the current in amperes equal to the numerical value of the nominal ampere-hour capacity of the cell. In EDM the technique of overnight charge is usually adopted because overcharging of batteries is not critical at low charge rates of 0.1 C and because charging during field work is usually not possible. A few high rate cells allow overcharging at quick charge rates. Quick, rapid and ultra-rapid charging may be possible with normal Ni-Cd cells, but requires a sophisticated charge control to avoid overcharging, which would cause excessive temperatures and pressures in the cells.

Nickel-cadmium batteries should not be charged at temperatures below 5 °C unless the charge rate is reduced according to specifications. The batteries can be fully charged at temperatures between 5° and 25 °C and partially charged (≤ 70%) at higher temperatures (≤ 45 °C). Long periods of overcharging should be avoided because they produce the effect of voltage depression, discussed in the previous section, during the next discharge. If batteries are overcharged beyond their design capability, a permanent loss in capacity will result.

The *charge retention* is the ability of a stored battery to retain its energy once it has been charged and this is both time and temperature dependent as shown in Fig. 11.4. The *self-discharge* may be reduced and the charge retention increased by storing batteries below 20 °C. Charged Ni-Cd batteries will eventually lose all of their charge through this chemical self-discharge, which is, however, in no way harmful either to their life or to their storage characteristics.

The *memory effect* is an almost constant reduction of the discharge voltage level (about 0.1 V) during the discharge period and subsequently a reduction in the stated capacity to a predetermined discharge voltage cut-off point. It is developed in the battery from repetitive usage patterns. If repeated partial charge and discharge cycles of exact magnitudes are applied, the cell may become so conditioned that it will deliver only slightly more capacity than in the previous repetitive cycles. The memory effect does not occur when the battery is discharged to random end-voltages or overcharged for random amounts of time as is usually the case in EDM applications.

11.3.4 Capacity and Life of Battery

The capacity rating of a cell is generally stated in terms of ampere-hours and is referred to as an operating variable, indicating the integral of the discharge current-time relationship under certain qualifying conditions, rather than as a rating of total electrical charge capability. The qualifying conditions define the charge current and time, the discharge current and the end-of-discharge voltage after a discharge.

The only way to measure the capacity of a battery accurately is to discharge it while measuring and integrating time and current of discharge. This should be done under the qualifying conditions specified for the battery. The battery is usually discharged at a constant current rate (specified by qualifying conditions) to the cut-off voltage (end-of-discharge voltage) laid down for the battery. The calculation of the ampere-hour capacity is simply obtained by multiplication of time and current.

Two types of loss of capacity may be distinguished:
1. temporary (reversible) loss
2. permanent loss.

Temporary loss results from continuous overcharging at high temperatures, programmed cyclic charges and discharges (memory effect) of repetitive nature and other long-term constants in the manner of operation. The full capacity of the battery may be restored by simply discharging the battery completely at a low rate of discharge and recharging it at the 0.1 C rate for 20 h at 25 °C. Only one or two of these cycles are generally needed. It is possible to extend the useful battery life by such periodic *reconditioning*.

Permanent loss of capacity may result from an internal short circuit (physical contact of two plates of opposite polarity after decomposition of the separator material) or from an open circuit caused by loss of water in the electrolyte (dry-outs). The life span of a battery to its permanent failure is affected by a large number of factors. The battery temperature is one such factor. Temperatures above room temperature (20−25 °C) accelerate the degradation of the separator and seal of a cell. Frequent exposure to high temperatures will therefore shorten the life expectancy, as will charging at very low temperatures. The time element is also very important, both on its own (electrolyte loss) and in conjunction with temperature (separator decomposition). High pressure may cause venting and thus loss of electrolyte. However, only frequent venting will have an effect on the life and performance of a battery. The depth of discharge also has an impact on the life expectancy because the probability of an internal short circuit is greater during as low state of charge than during a partial or full state of charge.

The end of the useful life of a battery is reached when the remaining capacity reaches 50 to 60% of the rated value (Barak 1980). It should be noted that some suppliers use 80% of original capacity to define the life-time of batteries. It is usually given in time and/or number of charge/discharge cycles. Plessey (1975) state a cycling life for SAFT/VR cells of approximately 500 cycles with 100% depth of discharge and 2000 cycles with 50% depth of discharge. General Electric (1975) state a range of life expectancy from 500 cycles to more than 30000 cycles depending on the conditions of use. The expected life may amount to more than 10 years with regard to the loss of electrolyte through gas diffusion. To achieve the above values the following operating conditions have to exist:

storage -40 to $+45°$ $(+65°)$
discharge (-40) $-20°C$ to $+45°C$ $(+65°)$
charge $+5°C$ to $+45°C$ $(+65°)$ (at $0.1C$ rate)

Data in brackets refer to specific cells only.

12 Errors of Electro-Optical Distance Meters

All electro-optical distance meters suffer from a large number of usually very small instrumental errors, irrespective of the use of the pulse measurement or phase measurement principle. The errors may be inherent to the electrical and optical design and/or caused by manufacturing and component tolerances. The magnitude of these errors is kept small by the manufacturers and usually accounted for in the accuracy specifications of instruments. In view of the fact that a small number of errors must be calibrated by the user of EDM instruments and that errors occasionally exceed the specified accuracy and may change with time, the user must be aware of the main error pattern of instruments.

Before giving a review of most known error effects, the traditional components of the overall instrument correction (I.C.) are discussed. These are: the additive constant, the scale correction, the non-linear distance-dependent correction and the short periodic error correction.

12.1 Additive Constant

Because the virtual electro-optical origin or zero of an EDM instrument is usually not located on the vertical axis of the instrument, a small correction has to be added to all distance measurements to refer the distance to the instrument's vertical axis. This correction is usually called the additive constant and compensates for electrical cable and component delays as well as optical path lengths. The terms "zero error" or "instrument error" are sometimes used to describe the equivalent error (negative value of the additive constant). A correction which combines the additive constant of the distance meter and the constant of the corresponding reflector (see Sect. 10.2.4) is usually determined by the manufacturer and incorporated in the instrument. This built-in correction is usually correct to 1 or 2 mm for instruments with an internal accuracy of ± 5 mm at the time of the factory adjustment and when using the specified reflector. The built-in constant can amount to a few centimetres and is accessible to the user in some instruments.

The residual additive constant can change with time and in the course of service or repair work. In consequence, it should be determined periodically and applied to all measured distances by either changing the built-in correction or by computation.

Test measurements have shown that the additive constant may also be affected by a number of other parameters. For example, the additive constants of most distance meters are temperature-dependent. However, the temperature coefficient is usually very small. An exceptionally large temperature coefficient of $+0.19\,\text{mm}/°\text{C}$ was reported by Rüeger (1987). Because of the existence of phase inhomogeneities in the emitting and photodiodes, the additive constant may change due to changes in attenuator or diaphragm settings or when changing the number of prisms. The introduction of glass or plastic attenuators obviously delays the EDM signal in function of the refractive index and thickness of the material (similar to the reflector delays).

Some instruments were found to exhibit voltage-dependent or signal strength-dependent additive constants. The return signal strength depends on ambient atmospheric conditions as well as on attenuator and diaphragm and internal neutral density filter settings. Some older instruments also exhibited day-night changes of the additive constant due to changes in the operating point of the photodiode. In the case of telescope-mounted EDM instruments, the additive constant may change between mountings on different host theodolites.

12.2 Short Periodic Errors

EDM instruments based on the phase measuring principle may exhibit periodic errors with wavelengths equivalent to the fine measuring unit length U (often 10 m) or its harmonics. The periodic errors, which repeat at multiples of the unit length U, may be due to electrical or optical crosstalk (feedthrough) or due to a systematic error in the phase measuring system. The former are typically sinusoidal. The latter may be any pattern repeating at intervals of U. Periodic errors of wavelength U are called first-order short periodic (or cyclic) errors. Periodic errors with wavelengths equivalent to U/2, U/3, U/4 etc. are termed higher-order short periodic (or cyclic) errors. Mathematically, these errors are expressed as forced period Fourier series unless the errors follow clearly another pattern (e.g. linear trend).

Pulse distance meters are free of the crosstalk problem, as the transmitted and received pulses are separated in time. (Time interpolation procedures may, however, exhibit systematic errors repeating at the distance equivalent of the time interpolation interval.)

12.2.1 Electrical or Optical Crosstalk Errors

Short periodic errors may be caused by electrical coupling between the reference signal and the measurement signal and by optical crosstalk between the transmitter and receiver optics in electro-optical distance meters. It occurs in all distance meters to a greater or lesser degree, and is independent of the type of phase measuring system. The amplitude of cyclic errors in short range instruments with ± 5 mm internal accuracy is usually smaller than 5 mm.

The effect of electrical or optical crosstalk may be explained by a mathematical model. Considering Eqs. (2.5a) and (2.7a, b) the transmitted signal y_1 and the return signal y_2 may be written as

$$y_1 = A_1 \sin(\omega t) \tag{12.1}$$

$$y_2 = A_2 \sin(\omega t + \Delta\Phi) . \tag{12.2}$$

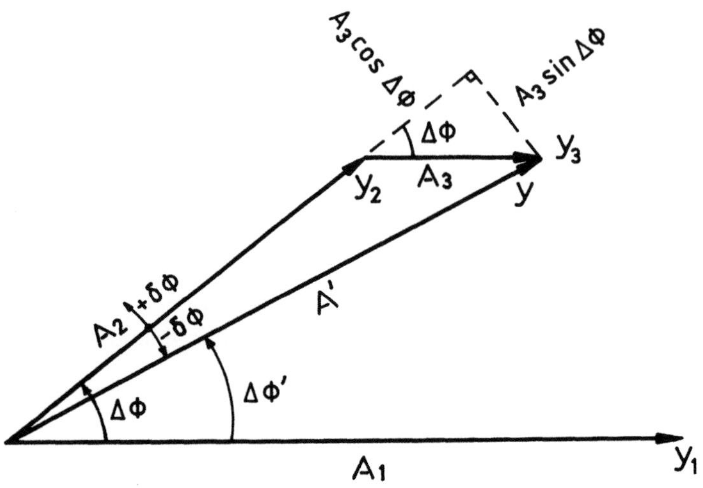

Fig. 12.1. Periodic error ($\delta\Phi$) caused by vector addition of a crosstalk signal (y_3) to the received signal (y_2). The distance meter uses the combined and erroneous signal (y) rather than the correct return signal (y_2) and thus measures the erroneous phase ($\Delta\Phi'$) rather than the correct one ($\Delta\Phi$)

A component y_3 of the transmitted signal y_1 is superimposed on the return signal y_2. The contaminating signal y_3 has the same phase and frequency as the transmitted signal, but a much smaller amplitude:

$$y_3 = A_3 \sin(\omega t) \, , \tag{12.3}$$

where $A_3 \ll A_1$. The superimposed return signal in the receiver channel then becomes

$$\begin{aligned} y &= y_2 + y_3 \\ &= A_3 \sin(\omega t) + A_2 \sin(\omega t + \Delta\Phi) \, . \end{aligned} \tag{12.4}$$

Because two waves of equal frequency are superimposed, the resulting wave has the same frequency but different phase and amplitude

$$y = A \sin(\omega t + \psi) \, . \tag{12.5}$$

Under the condition of $A_3 \ll A_2$, Eq. (12.5) may be written as

$$y = (A_2 + A_3 \cos \Delta\Phi) \sin \left\{ \omega t + \left(\Delta\Phi - \frac{A_3}{A_2} \sin \Delta\Phi \right) \right\} \, . \tag{12.6}$$

Proof of Eq. (12.6) follows easily from Fig. 12.1. In the figure, the amplitude of the periodic function in Eq. (12.6) is denoted by A' and the periodic error by $\delta\Phi$.

Because electro-optical EDM instruments are based on the measurement of phase differences, only the last term of Eq. (12.6) is important:

$$\Delta\Phi' = \Delta\Phi + \delta\Phi \tag{12.7}$$

$$\Delta\Phi' = \Delta\Phi - \frac{A_3}{A_2} \sin \Delta\Phi \, , \tag{12.8}$$

where $\Delta\Phi'$ is the time lead of the contaminated return signal. Using the terms unit length U and fraction of unit length L, Eq. (3.8) of Section 3.2.1 yields

$$L' = L - \frac{A_3}{A_2} \left(\frac{U}{2\pi}\right) \sin\left(\frac{2\pi L}{U}\right) , \qquad (12.9)$$

where
- L' = measured fraction of unit length of the contaminated return signal
- L = fraction of unit length of return signal
- U = unit lenght of EDM instrument
- A_2 = amplitude of the return signal
- A_3 = amplitude of the contaminating part of the reference signal.

The second term in Eq. (12.8) is called the first-order short periodic error because it is mainly a function of the fraction of the unit length and thus repeats itself every U metres. The error is, however, also a function of the return signal strength A_2 which decreases with increasing distance. Depending on the design of the distance meter and the point where the crosstalk(s) occur, the first-order short periodic may or may not increase with distance. Assuming a unit length U of 10 m and an amplitude of the contaminating signal $A_3 = 0.003\,A_2$ the amplitude of the cyclic error will be

$$\frac{A_3}{A_2} \left(\frac{U}{2\pi}\right) = 0.003 \left(\frac{10\,\text{m}}{2\pi}\right) = 4.8\,\text{mm} . \qquad (12.10)$$

For example, the first-order short periodic error may be distance-dependent if the crosstalk occurs between the optical attenuators (typically neutral density filters) and the reflector, that means before the return signal is attenuated to the appropriate level. In such cases, the constant crosstalk signal adds to a return signal which becomes weaker with increased distance: The effect of the crosstalk increases with distance. On the other hand, if the crosstalk occurs after the return signal has been attenuated to the appropriate level, then the ratio between the two signal as well as the periodic error is constant. It should be noted that often both types of crosstalks (or feedthrough) occur, although one or the other type may be more dominant.

12.2.2 Analogue Phase Measurement Errors

Some analogue phase measuring systems may display non-sinusoidal periodic errors over one unit length U of the instrument. The first Hewlett-Packard distance meter (HP 3800 B) belonged to this category. The Kern Mekometer ME 3000 was an instrument with another type of "analogue" phase measurement system. The optical-mechanical phase measurement system of the ME 3000 (see Sect. 4.1.3.1) was sometimes found to exhibit a small linear error within the 0.3 m unit length.

Early EDM instruments generally used phase shifting resolvers for the analogue phase measurement (see Sect. 4.1.3.2). These devices often displayed sinusoidal errors with wavelengths equivalent to one half of the distance meter's unit length, or second-order short periodic errors (Kahmen 1978). The amplitudes of all periodic errors caused by analogue phase measurement systems are constant, and thus independent of distance.

12.2.3 Multipath Errors

Short periodic errors of first and higher order can also be caused by signals travelling more than two times over the distance from the instrument to the reflector. The return signal may be reflected by the distance meter's optics or highly reflective GaAlAs and/or Si diodes (or by the optics of the host theodolite) and travel once more to the reflector and back. This process can repeat itsself a few times, although the signal strength is reduced greatly by each reflection. Uncoated glass surfaces may reflect a maximum of 4% of the incident ray and GaAlAs, with its refractive index of 3.6 (see Sect. 4.1.2.1), a considerable 32%. Normal incidence is assumed in both cases. Considering Eq. (12.7), the periodic errors caused by multipathing may be described by the following equation (Covell and Rüeger 1982):

$$\delta\Phi = A_5/A_2 \sin(\Delta\Phi) + A_6/A_2 \sin(2\,\Delta\Phi) + A_7/A_2 \sin(3\,\Delta\Phi) \ , \qquad (12.11)$$

where the amplitudes of the double, triple and quadruple path signals are denoted by A_5, A_6 and A_7, respectively. Some numerical examples may be found in Covell (1979) and Covell and Rüeger (1982). As all amplitudes A decrease with distance, the short periodic errors caused by multipathing become smaller with increased distance. It should be noted that the first term in Eq. (12.11) has opposite sign of the last term in Eq. (12.8) and that these two first-order short periodic errors may cancel each other.

12.2.4 Experimental Results

Experimental results indicate that first-order short periodic errors may change their amplitudes (and phase) with distance, changes in signal strength, time, ambient light, and in the course of service and repair work. Amplitudes may exceed 10 mm in some cases. Please note that fingerprints on the front lens of an electronic tacheometer may generate crosstalk and, thus, a short periodic error.

Multipath errors are usually small and restricted to first-order short periodic errors. However, larger errors may occur in special circumstances. On one occasion, it was possible to trace a multipath error, featuring 70 mm amplitude at 10 m and reducing to 13 mm at 80 m, to an optical part in a host theodolite, which did not carry an anti-reflex coating.

Statistically significant higher order short periodic errors are often present in test results. In some cases they can be explained by effects discussed in previous sections, in other cases not. In this context it should be noted that forced period Fourier series [as the one shown in Eq. (12.11)] can satisfactorily model most systematic errors within the interval of one unit length even if the latter are not sinusoidal. A physical explanation may therefore not always be possible. This follows also from the fact that the sum of a number of periodic errors is measured (and not the individual contributors) and that the sections above do not cover all possible errors.

12.2.5 Reduction of Short Periodic Errors

The manufacturers reduce the amplitude of these errors by rigorous electrical shielding of transmitter and receiver, separation of optical channels, special anti-reflex coating of common optical parts and/or electronic compensation. Compensation may be by adding to the received signal an analogue signal equal in phase and amplitude but with opposite sign of the latter. Alternatively, the short periodic error may be determined by the manufacturer numerically and the appropriate correction stored in the distance meter for on-line application. Both compensation mechanisms must assume that amplitude and phase of the short periodic error do not change with distance. Short periodic errors may also be reduced by an execution of the phase measurement at four different phases (90° apart) of the return signal (Hines 1976). As long as the crosstalk signal remains unaffected by the 90° shifts, the mean of the four phase measurements will be free of first- and second-order periodic errors. It has also been suggested that the crosstalk signal be measured (by disabling the internal and the external light path) and stored and then used to correct subsequent distance observations (Ohtomo 1983).

It will be shown later how a user of a distance meter can calibrate the short periodic error and then correct distance measurements by computation, if so required.

12.3 Scale Errors

Scale errors in EDM instruments are caused primarily by the oscillator and by the emitting and receiving diodes. It should be noted that a large number of other scale errors are caused by external effects, such as using an incorrect value of the velocity of light, using an incorrect first velocity correction, omission of humidity correction, limited resolution of ppm correction input, errors in temperature or pressure measurements/estimates, erroneous geometrical reductions (includes omission of geoid-spheroid separation), to name just a few.

12.3.1 Oscillator Errors

The most significant oscillator error is its dependency on temperature. As outlined in Section 4.1.1, oven-controlled crystal oscillators (OCXO's) perform best in this respect once they have reached their operational temperature. Most short range distance meters feature temperature-compensated crystal oscillators

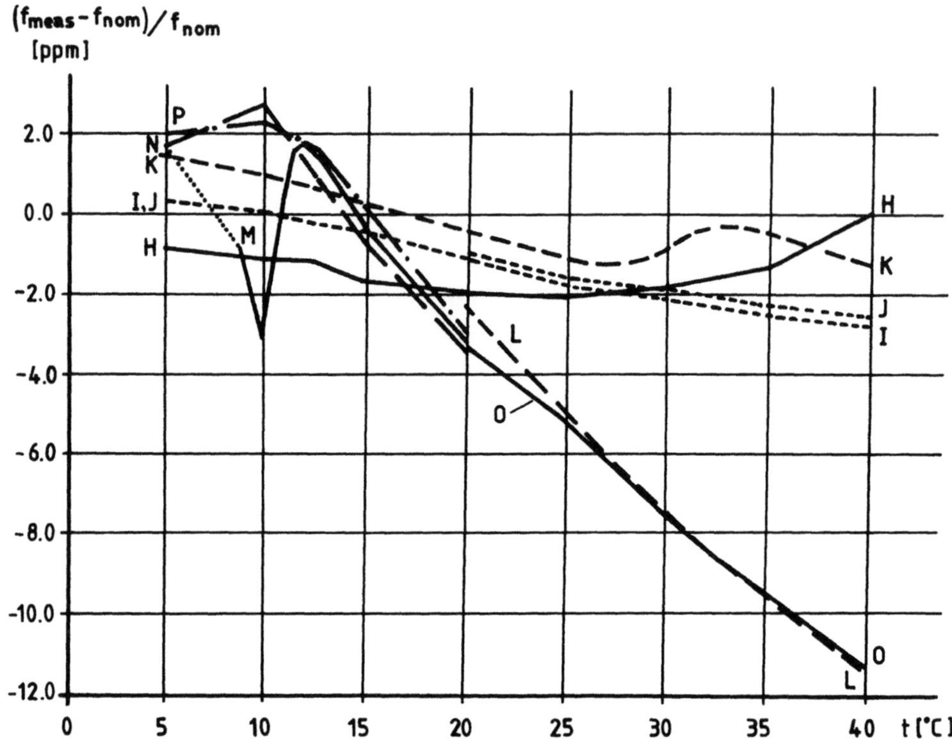

Fig. 12.2. Frequency versus temperature characteristics of some distance meters. The characteristics of the first and second instrument are labelled H and K. The curves I and J refer to the third instrument and show the good day-to-day repeatability of the frequency measurements. The curves L, M, N, O and P describe the characteristic of the fourth instrument. The discontinuity (± 3 ppm) at about 10 °C is real but not typical for distance meters. (After Rüeger 1982)

(TCXO's) as time bases. The frequency versus temperature characteristic of (AT-cut) quartz crystal oscillators can be modelled by a third degree polynomial. The inflection point is at about 25 °C. The frequency drops and rises steeply at low and high temperatures, respectively. The test results depicted in Fig. 12.2 show that not the full pattern may be evident within a relative narrow temperature range. Three out of four instruments shown are obviously equipped with TCXO's and perform very well. The fourth instrument is likely to employ a room temperature crystal oscillator (RTXO) and features a frequency versus temperature gradient of about 0.45 ppm/°C and an unusual discontinuity at about $+10$ °C (Rüeger 1982). Because of the temperature dependence of the oscillator, the scale of short range distance meters is also affected by warm-up effects. This means that the oscillator's frequency is dependent on how long the distance meter has been operating. During operation, the distance meter dissipates heat, which, in turn, warms up the oscillator. Different measuring modes lead to different warm-up effects. Warm-up effects may amount to as little as 1 ppm over the first hour or as much as 5 ppm, depending on instruments (Rüeger 1982).

The frequency versus time behavior of oscillators is described by the short-term and the long-term stability. The short-term stability is usually very good. The long-term stability (or ageing) is typically less than 1 ppm/year. Normally, the frequency drops with time as particles settle on the quartz slice and effectively increase its thickness (and decrease its resonance frequency). Because of the ageing pattern, the scale of distance meters must be checked periodically. Quartz crystal oscillators are also susceptible to a number of other (but usually negligible) effects, such as drive voltage changes, retrace, hysteresis and electro-magnetic interference (EMI). More details may be found in Rüeger (1982).

12.3.2 Diode Errors

Diode errors may affect the scale of distance meters in three ways. Firstly, the actual carrier wavelength of the GaAlAs diode's emission may be different from the nominal or published value. Errors arise because the latter rather than the former is used in the derivation of the first velocity correction. Secondly, the actual carrier wavelength may be temperature-dependent and affect scale in a similar way. Thirdly, the linear component of the effect of phase inhomogeneities across the emitting as well as the receiving diode also affects the scale of EDM instruments. Some earlier instruments exhibited scale errors of 10 to 30 ppm due to phase inhomogeneities, thus exceeding oscillator errors by at least a factor of 5 (Schwarz and Witte 1986)! Further details on the effects of phase inhomogeneities are given in the next section.

12.4 Non-Linear Distance-Dependent Errors

All distance-dependent systematic errors, which are repeatable and reproducible but do not fit the above three classes (Sects. 12.1, 12.2, 12.3) may be termed non-linear distance-dependent errors and include the so-called non-periodic errors as well as long-periodic errors. These errors are most likely caused by the phase inhomogeneities in both the emitting and receiving diodes. The effect has been discussed previously in Section 4.1.2.1 and Fig. 4.6. The aperture of a reflector moving further and further away from the EDM instrument will collect and return less and less of the non-uniform phase pattern across the measurement beam (see Fig. 4.6). The average phase information returned by the reflector may be imagined as the integral of the phase across the reflector's aperture weighted by the corresponding power (see Fig. 4.5). The weighted phase integral will be distance-dependent as soon as the power of phase pattern is not uniform. In consequence, the resulting errors will depend on the shape, size and number of the reflectors used.

Other contributing error sources may be: changes in diaphragm setting, stepwise changes in attenuator setting, changes of the operating point of the avalanche photodiode (e.g. stepwise variation of multiplication factor) and other signal strength effects.

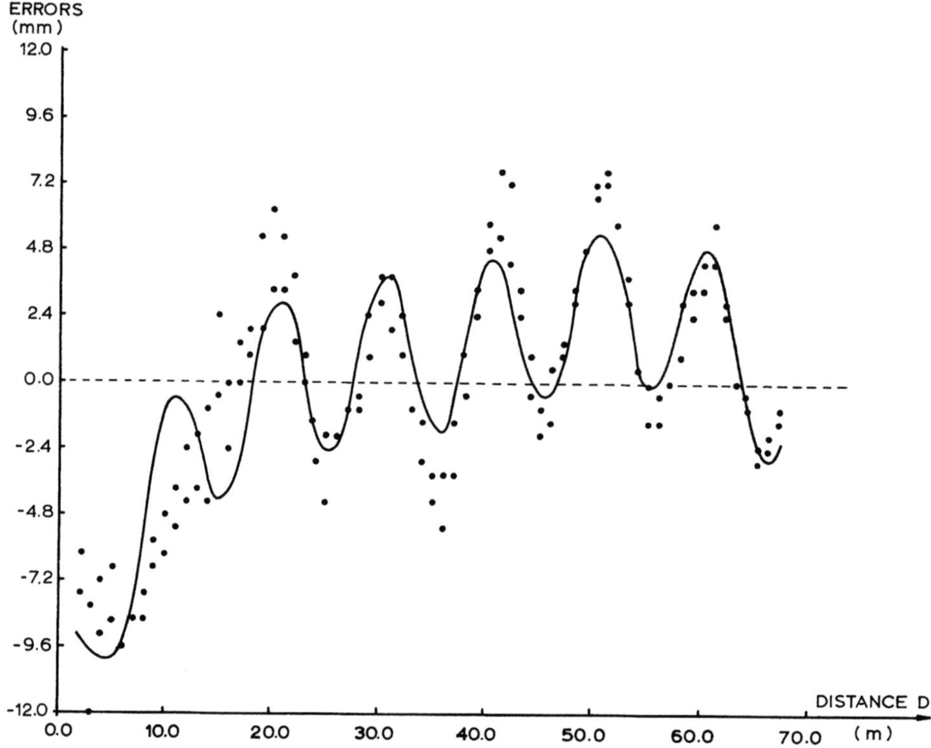

Fig. 12.3. Systematic errors of an infrared short range distance meter as determined in 1-m intervals between 3 and 68 m against a laser interferometer. Dominant are the 10 m (first-order) short periodic error and the non-linear distance-dependent error (change of 7 mm over first 20 m). The *ordinate values* represent measured distances minus true distances. The measurements are depicted by *dots* and the fitted 4th degree polynomial by a *continuous curve*. Note that the fitted curve has the sign of an error; the corresponding correction [in Eq. (12.12)] has the opposite sign

The non-linear distance-dependent errors are usually modelled as a polynomial expression of n-th degree and may then be called non-periodic errors (Covell 1979; Schwarz 1983). Test measurements indicate that non-periodic errors are most pronounced on close range, typically to about 20 m. Non-linear changes of 5 to 20 mm have been reported in some cases within this interval. Figure 12.3 shows an example of a non-periodic error on close range combined with a significant first-order short-periodic error. The equation for the fitted instrument correction (IC) in Fig. 12.3 is as follows:

$$IC = CONST - 167.8\,(D/100) + 679.0\,(D/100)^2 - 1239.9\,(D/100)^3$$

$$+ 820.1\,(D/100)^4 - 2.5\,SIN\,(2\pi D/10) - 1.6\,COS\,(2\pi D/10) \ , \qquad (12.12)$$

where   IC   = instrument correction (millimetre)
        D   = distance (metre)
        CONST = additive constant at zero distance (millimetre).

182

In some instances, the error patterns may be better represented by so-called long-periodic errors (Rüeger 1987; Covell and Rüeger 1982). The wavelengths of the latter sometimes agree with the coarse measurement wavelengths used by the distance meters (e.g. 1 km, 2 km). In other cases, the wavelengths of such long-periodic errors may be obtained with frequency analysis techniques. Long-periodic errors of 5 mm amplitude and 1000 m wavelength have been reported for some instruments.

The user may evaluate (but not calibrate) the magnitude of the problem of phase inhomogeneities by establishing so-called false pointing diagrams (Kahmen 1977; Kahmen and Zetsche 1974; Rüeger et al. 1975; Witte and Schwarz 1982). After an initial pointing and distance measurement to the centre of the reflector, the distance meter is deliberately mispointed in steps of several minutes of arc to the left, to the right, upwards and downwards, each time taking one or more distance measurement. The plots of distance measurement versus mispointing (separately for vertical and horizontal scan) are referred to as false pointing diagrams. The test may be improved by taking measurements in a grid pattern around the centre of the reflector. As mentioned before, these scans provide information on the magnitude of the problem only. Calibration has to occur elsewhere, usually on multistation EDM baselines, as discussed later in Section 13.

12.5 Summary and Mathematical Model of Errors

A brief summary of instrumental and non-instrumental errors is given in Table 12.1. The former have been discussed in the sections above. The latter are self-explanatory. The list of errors is by no means exhaustive nor have all independent variables been taken into account. For example, repair and servicing of instruments might affect quite a few error patterns. Also, non-tiltable reflectors lead to errors which depend on the zenith angle of the wave path. References for most listed errors may be found in Rüeger (1977).

Following from Table 12.1 and the previous sections, the mathematical model of most instrumental errors may be given as (Rüeger 1987):

$$IC = A_{00} + A_{01}T + A_{02}e^{(-A_{03}t)} + (A_{10} + A_{11}T + A_{12}T^2 + A_{13}T^3)D + (A_{14}Y)D$$

$$+ (A_{15}e^{(-A_{16}t)})D + A_{20}D^2 + A_{30}D^3 + A_{40}D^4 + A_{50}D^5 + \ldots$$

$$+ (B_{11} + B_{12}D)\cos\left(\frac{2\pi D}{U}\right) + (C_{11} + C_{12}D)\sin\left(\frac{2\pi D}{U}\right)$$

$$+ (B_{21} + B_{22}D)\cos\left(\frac{4\pi D}{U}\right) + (C_{21} + C_{22}D)\sin\left(\frac{4\pi D}{U}\right)$$

$$+ (B_{31} + B_{32}D)\cos\left(\frac{6\pi D}{U}\right) + (C_{31} + C_{32}D)\sin\left(\frac{6\pi D}{U}\right)$$

$$+ (B_{41} + B_{42}D)\cos\left(\frac{8\pi D}{U}\right) + (C_{41} + C_{42}D)\sin\left(\frac{8\pi D}{U}\right) + \ldots \quad (12.13)$$

Table 12.1. List of errors occurring in distance measurements with EDM instruments using IR emitting diodes and photodiodes

Error	Type of error							
	C	D	E	F	G	H	I	K
Instrumental errors								
1. Additive constant		×				×	(×)	
2. Electrical and optical cross-talk and stray coupling			×		(×)			
3. Multipath errors			×		×			
4. Phase inhomogeneities in IR emitting diode and photodiode		×		×				
5. Errors in frequency				×		×	×	×
6. Phase measurement								
– Digital	×							
– Analog		×	×					
7. Phase drift	×	×					(×)	
8. Pointing:								
– Electronic pointing	×							
– Optical pointing only with a maladjusted telescope-mounted EDM instrument		×		×	×			
9. Effect of signal strength			×	(×)	×			
10. Adjustment of diaphragm, attenuator		×			×			
11. Variability of reflector constant (for one make and shape)	×	(×)						
12. Variability of reflector constant (for other makes and shapes)		×		×	×			
13. Error in given carrier wavelength				×			×	
14. Resolution of ppm input	×			×				
Non-instrumental errors								
15. Centring errors of reflector and EDM instrument	×	(×)						
16. Levelling errors of reflector and EDM instrument	×	(×)						
17. Reflector pointing	×	(×)						
18. Measurement of temperature	×			×				
19. Measurement of atmospheric pressure	×			×				
20. Neglect of humidity	×			×				
21. Spurious reflections from illuminated objects other than reflector				×				
22. Atmospheric turbulence	×							

The following types of errors are distinguished: Type C: random error; Type D: systematic, but constant error; Type E: systematic error, periodic with distance; Type F: systematic error, linear with distance; Type G: systematic error, non-linear, dependent on distance; Type H: temperature-dependent; Type I: time-dependent; Type K: voltage-dependent

where IC = instrument correction
A, B, C = unknown coefficients (to be calibrated)
D = distance
t = elapsed time since the instrument was switched on
T = temperature
U = unit length of distance meter
Y = years (elapsed since purchase of instrument).

The first three terms in Eq. (12.13) represent the three components of the additive constant, namely the constant, temperature-dependent and warm-up terms. The coefficients A_{10} to A_{15} model the scale corrections. The coefficients A_{14} and A_{15} refer to the ageing and warm-up effects. The fifth degree polynomial (coefficients A_{20} to A_{50}) expresses the non-linear distance-dependent errors or non-linearity of the instrument correction. The last four lines of Eq. (12.13) constitute a fourth-order Fourier series with forced period U and model the short periodic errors. The amplitudes of the periodic errors cater for a change with distance (e.g. coefficient B_{12}). The wavelengths of the first-, second-, third- and fourth-order terms are U, U/2, U/3 and U/4, respectively.

It is not suggested that Eq. (12.13) be used routinely for calibration purposes. All instrument users are advised to periodically test the coefficients A_{00} (additive constant), A_{10} (scale correction) as well as B_{11} and C_{11} (first-order short periodic error correction). In some cases it might also be necessary to determine the coefficients A_{20} to A_{50} (non-linear distance-dependent errors) as well as B_{21} and C_{21} (second-order short periodic errors). When operating distance meters within their claimed accuracy bands, it should not be necessary to calibrate any further terms. For improved accuracy performance, further terms might have to be considered. Some appropriate calibration procedures are discussed in the following chapter.

13 Calibration of Electro-Optical Distance Meters

13.1 Introduction

In this context, the calibration of a distance meter is defined as the determination of its instrument correction and associated precision. The instrument correction IC [see Eq. (12.13), for example] is added to distance measurements to obtain the correct distance. It has been shown previously that the instrument correction is a function of a number of independent variables, the most important being distance, temperature and time. The instrument correction is determined for a particular instrument-reflector combination. All distance-dependent and the constant terms require re-evaluation when using the distance meter with another type of reflector.

13.1.1 Reasons for Calibration

Electro-optical distance meters may be calibrated for a number of reasons. The most important ones are:

- quality control
    - at time of purchase
    - periodically thereafter
- improvement of accuracy
- legal metrology.

The quality control measurements at the time of purchase have to establish if an instrument fulfils the manufacturer's specifications. The most important specifications are those of accuracy, temperature range and distance range. The accuracy specification can be tested in the context of EDM instrument calibration. Possible interpretations of accuracy specifications are discussed later in Section 13.6. The specified temperature range for the operation of distance meters cannot be easily verified by users. If operation of an instrument is expected to be predominantly at the upper (or lower) end of the specified temperature range, it is advisable to consult the manufacturer prior to ordering. [Rüeger (1987) reported the failure of two out of eight distance meters at $-20\,^\circ$C.] The verification of the range specifications is possible, but very time-consuming. It should be noted that range specifications usually imply the use of specific reflectors. Sect. 5.2 provides some information on range formulae.

After the initial quality control measurements, it is advisable to schedule periodic quality control measurements, typically at half yearly or yearly intervals. General purpose distance meters require the periodic determination of precision and instrument correction. The latter can usually be restricted to the constant, linear and short periodic terms or, in other words, to additive constant, scale correction and short periodic error corrections. Calibrations should also be carried out after any service or repair work on instruments. It is strongly recommended to maintain a file on all calibration measurements carried out with a particular distance meter. If the same quality control tests are carried out each time, the file will indicate any variations and deterioration of the instrument constant and the instrument's precision. The key parameters of each test should be summarized in a table for easier detection of such changes with time. It is advisable to include ambient temperature and prevailing weather conditions in such tables as some parameters are, for example, temperature-dependent.

The quality control measurements discussed above are a necessity rather than an option and are essential if the specified accuracy of a distance meter is to be realized. In consequence, most manufacturers prescribe the determination by the user of the additive constant and scale (modulation frequency). If a distance meter features additional and repeatable systematic errors, it may be possible to calibrate such errors and to compensate them by the application of corrections. If relatively large errors are detected and compensated, a sizable improvement in accuracy may be possible. Often, the improvement in accuracy by calibration requires some changes in the field procedures as the measuring conditions during calibration and field measurements should be as equal as possible. Some systematic errors can be eliminated by suitable field procedures alone.

Corrections should be applied for statistically significant systematic errors only. Economic reasons usually dictate a further restriction to the largest significant errors which provide maximum improvement in accuracy. To establish these, a comprehensive calibration similar to Eq. (12.13) needs to be executed initially. Subsequent periodic calibrations are then restricted to the verifications of the significant terms. The accuracy of an EDM instrument can typically be improved by a factor 2 to 4 through a more extensive calibration and improved field procedures (Rüeger 1987). The cost of extensive calibrations should always be compared with the cost of hiring or purchasing a distance meter with better accuracy specifications.

In some countries (such as Australia), the length measurements carried out by surveyors in the context of "legal" (usually cadastral) surveys are governed by national "weights and measures" legislation. This requirement for legal metrology may provide a third reason for the calibration of EDM instruments. The requirements of legal metrology necessarily differ from country to country. In Australia, the following coefficients of Eq. (12.13) are evaluated and incorporated in the "legal instrument correction" if found to be significant: A_{00}, A_{10}, A_{20}, A_{30}, A_{40}, A_{50}, B_{11}, C_{11}, B_{21}, C_{21}. The coefficients A_{00} (additive constant) and A_{10} (scale correction) are always stated by the legal instrument correction. More information on legal EDM instrument calibration in Australia may be found in Rüeger (1985) and Norton (1986). Some results of routine legal EDM calibration may be found in Benwell et al. (1985).

13.1.2 Concept of Calibration

As indicated before, the calibration of a distance meter consists in the determination of its instrument correction and associated precision. It follows from the discussions in Chapter 12 that this instrument correction is dependent on many variables, such as distance, temperature, time, supply voltage and ambient atmospheric conditions, to name the most important relationships. The change of the IC with distance could be determined by mounting the reflector on a carriage and moving this on a rail of one or more kilometre(s) length. With the distance meter mounted at the end of the rail, distances of different lengths could be measured over the total working range of the distance meter and instantly compared with a working standard of length. The temperature effects could be determined by placing the facility (with a length of a few kilometres) in a temperature chamber and by repeating the test versus distance at equally spaced temperatures over the total specified temperature range of the instrument. The long-term component of the time effect could be established by repeating the distance/temperature tests periodically, say every year. The voltage effect could be calibrated by carrying out all distance measurements in the above tests at selected input voltages.

It is needless to say that the costs of such an approach would be prohibitive. However, the concept could be realized by using an underground tunnel as laboratory (and by air-conditioning it) and the Kern Mekometer ME 5000 as length standard. On a smaller scale and at one temperature only, the rail concept is used by most manufacturers as well as some universities and national standard laboratories for general-purpose EDM instrument testing on close range. The range of these facilities is typically restricted by the working range of the laser interferometers used as length standards (presently 50 to 70 m).

As it is generally not practical to test the instrument correction simultaneously against all variables, separate tests are used to determine one coefficient or groups of coefficients of Eq. (12.13) at a time. As the largest number of (known) errors are distance-dependent, procedures have been developed to efficiently determine the distance-dependent coefficients of the instrument correction (IC). For practical reasons, the ideal rail approach is usually replaced by a finite number of survey marks (or pillars) along a line. Such testlines are typically called EDM (calibration) baselines. Three basic baseline designs will be discussed later. They all permit determination of most if not all distance-dependent terms of the instrument correction as long as the true baseline lengths are known. Short periodic errors are sometimes determined on specialized testlines, either outdoors or in a laboratory.

Additional laboratory tests may be used to efficiently determine the temperature dependence of an instrument's scale as well as its additive constant. The warm-up effects on additive constant and scale can also be established in the laboratory. It should be noted that repeated baseline observations also provide data on temperature variations of additive constant, scale and, possibly, other coefficients of Eq. (12.13), if the ambient temperatures experienced during baseline observations vary greatly between baseline measurements.

The change with time of some coefficients of the instrument correction can only be established through repeated measurements on baselines (and, possibly, in the laboratory) over long periods of time.

The step-by-step approach to instrument calibration is less rigorous than the ideal approach discussed before as some potential dependencies remain untested. Table 12.1 and Eq. (12.13) indicate the most widely (but not all) reported error patterns. This should always be kept in mind when analyzing calibration measurements.

13.2 Calibration on EDM Baselines

EDM calibration baselines have been the preferred approach to general-purpose EDM instrument calibration since the introduction of infrared distance meters. A first publication on distance measurements in all combinations on EDM baselines dates back to 1968 (Pauli 1968). Alternative calibration schemes, such as radial patterns and calibration networks, are rarely employed by users. Measurements in a radial pattern to fixed reflectors (e.g. from an elevated window) are used by most manufacturers and their agents for routine tests on longer lines. The properties of the baseline design compare favourably with the two other designs mentioned:

- (relatively) low cost of installation
- few stations, many observations
- little space needed (but linear)
- high precision for additive constant, even if known distances are not available or are out of date
- distances (evenly) spread over the whole range of the instrument
- easy computation, with or without known distances.

The discussions below refer to EDM baselines only.

13.2.1 Geometric Design of EDM Baselines

13.2.1.1 Golomb Rulers in EDM Baseline Design

All baseline designs discussed below aim at an equal distribution of all measured distances between the shortest and longest line on the baseline, with no repetitions. Some designs require each length to be a multiple of a basic unit length. In two of the three cases, this requirement must be fulfilled by distances measured in all combinations. Staiger (1987) pointed out that the equivalent mathematical problem may be found in Golomb's Graceful Graphs and in Golomb's Rulers (Gardner 1983; Bloom and Golomb 1976). A ruler marked at 0, 1, 4, 6 (cm) is called gracefully labelled because such a ruler can measure all integral distances from zero to the length of the ruler, namely 1, 2, 3, 4, 5, 6 (cm). In this case, all distances between pairs of marks are different and run consecutively from zero to the ruler's total length. Golomb proved that gracefully numbered rulers can be found for rulers of four marks or less. This means that it is impossible to design baselines of five or more stations with truly equal spacing of distances measured in all combinations. For rulers with more than four marks, non-redundant, minimum-length rulers "with best numbering" have been found. These semi-graceful rulers are the shortest possible rulers of a given number of marks (end points included) with all distances between pairs of marks being different but not featuring all consecutive distances from zero to the total length of the ruler. Tables of possible rulers with 2 to 11 marks may be found in Bloom and Golomb (1976) and Staiger (1987). These semi-graceful Golomb rulers may be used in lieu of the design equations given below for the Heerbrugg-type and Aarau-type baselines. Golomb rules provide four design options for six station baselines and five options for seven station baselines. Only one design option each is available for baselines with 8 to 11 stations.

13.2.1.2 Heerbrugg Design

Schwendener (1971, 1972) published a baseline design which features an (almost) equal distribution of the distances measured in all combinations over the baseline length as well as over the unit length of the distance meter. The original design has seven stations, provides 21 observations in all combinations and a total length of 1021.5 m. It is based on a unit length of 10 m and has the following section lengths (in metres): 19.5, 39.0, 68.0, 127.5, 256.0, 511.5. Some fractions of the unit length are repeated: 1.5 m three times and 7.5 m two times. The design of this baseline permits the detection of all distance-dependent errors, including the short periodic errors.

The original publication gives the least-squares solution for the additive constant and six unknown baseline distances. The short periodic errors have to be checked by a plot of the residuals versus the corresponding fraction of the unit length. If present, the adjustment has to be repeated with data corrected for the short periodic error. If the latter is not done, the additive constant may be in error by up to 75% of the short periodic error's amplitude (Rüeger 1976).

A design equations for Heerbrugg-type (Schwendener-type) baselines is given in Tables 13.1 and 13.2. The baseline design is based on four input parameters:

U = unit length of distance meter(s) to be tested
A = shortest distance on baseline (multiple of U)
C_0 = desired total length of baseline
n = number of baseline stations

The value C_0 should be selected according to the maximum distance which can be measured to a single reflector in worse than average atmospheric conditions. The range between parameters A and C_0 should include the lengths of lines experienced in everyday survey work with the distance meter in question. The number of stations has to be decided on the basis of the required precision of the additive constant, the availability of true distances and the expected field effort per calibration. A reasonable cost versus precision balance is obtained with six to seven stations. The following parameters are determined in the design process:

C = final total length of baseline
B_0 = estimate of first design parameter
B = final value of first design parameter (B_0 rounded to the nearest multiple of A)
D = second design parameter.

190

Table 13.1. Values of the design parameters B_0 and D for Heerbrugg-type EDM baselines. (Rüeger 1978)

Number of stations	B_0	D
5	$\frac{1}{6}(C_0 - 4A - U)$	$\frac{1}{16}U$
6	$\frac{1}{10}(C_0 - 5A - U)$	$\frac{1}{25}U$
7	$\frac{1}{15}(C_0 - 6A - U)$	$\frac{1}{36}U$
8	$\frac{1}{21}(C_0 - 7A - U)$	$\frac{1}{49}U$

Table 13.2. Baseline sections and total baseline length versus number of stations for Heerbrugg-type EDM baselines. (Rüeger 1978)

Section	Baseline with			
	5 Stations	6 Stations	7 Stations	8 Stations
1st	A+ B+ 3D	A+ B+ 3D	A+ B+ 3D	A+ B+ 3D
2nd	A+3B+ 7D	A+ 3B+ 7D	A+ 3B+ 7D	A+ 3B+ 7D
3rd	A+2B+ 5D	A+ 4B+ 9D	A+ 5B+11D	A+ 5B+11D
4th	A+ D	A+ 2B+ 5D	A+ 4B+ 9D	A+ 6B+13D
5th	–	A + D	A+ 2B+ 5D	A+ 4B+ 9D
6th	–	–	A+ D	A+ 2B+ 5D
7th	–	–	–	A+ D
C =	4A+6B+16D	5A+10B+25D	6A+15B+36D	7A+21B+49D

Firstly, B_0 is determined according to Table 13.1 and then rounded downwards to the next multiple of the unit length U. The second design parameter D follows directly from the number of stations and the unit length U. Its value is listed in Table 13.1. Based on the values of A, B and D, the baseline sections as well as the final baseline length C may be computed from the equations given in Table 13.2. With the section lengths known, all combinations of distances can be computed. It is advisable to plot all combinations of distances versus distance to verify the quality of the design. As the distribution of the distances within the unit length is exactly the same as the distribution over the total baseline length, a plot of the former is not required. If the design is not totally satisfactory, it may be varied by small changes to A or B or both. B should not be equal to A if repetitions of distances are to be avoided. Good designs result in cases where B is equivalent to 2A.

Table 13.3. Example of the design of a Heerbrugg-type baseline

Stations	Sections	All combinations of distances						
1		0						
	20+ 40+0.61 = 60.61							
2		60.61	0					
	20+120+1.43 = 141.43							
3		202.04	141.43	0				
	20+200+2.24 = 222.24							
4		424.28	363.67	222.24	0			
	20+240+2.65 = 262.65							
5		686.93	626.32	484.89	262.65	0		
	20+160+1.84 = 181.84							
6		868.77	808.16	666.73	444.49	181.84	0	
	20+ 80+1.02 = 101.02							
7		969.79	909.18	767.75	545.51	282.86	101.02	0
	20+ 0.20 = 20.20							
8		989.99	929.38	787.95	565.71	303.06	121.22	20.20

Input values: $U = 10$ m, $A = 2U = 20$ m, desired baseline length (C_0) = 1000 m, eight stations. $B_0 = 40.47$ m, rounded to $B = 40$ m. Final total length (C) = 990 m. D = 0.204 m.

A numerical design example is given in Table 13.3. The general design formula leads to an equal distribution of distances over the baseline length as well as over the unit length. Distances or their fractions of the unit length do not repeat.

Because all combinations of distances are distributed in exactly the same way over the unit length as over the baseline length, short periodic errors will be mapped as long periodic errors in a plot of the residuals of a baseline adjustment versus distance unless short periodic errors are solved for simultaneously. It should also be noted that additive constants obtained from baseline adjustments with solution for additive constants and scale (or baseline distances) only may be in error by as much as 75% of the amplitude of the first-order short periodic error. To remove the error from the additive constant, the short periodic error may be determined and the baseline measurement corrected prior to the adjustment. Generally it is preferable to always include the short periodic errors as unknown parameters in the least-squares adjustment.

13.2.1.3 Aarau Design

Although a first multiple-of-unit-length design was reported by Pauli in 1968, the concept of all combinations of distances being equally distributed over the total baseline distance in combination with a separate determination of the short periodic errors was first reported by Aeschlimann and Stocker (1974) and Kern (1974). The particular baseline in Aarau has a total length of 520 m with sections of 30, 70, 80, 120, 170 and 50 m length. This design features baseline sections which are exact multiples of the unit length U. Assuming that the amplitudes and phases of the short periodic errors do not change over the length of the baseline, the short periodic errors will affect each combination measured on the baseline by an equal amount. Any short periodic errors will therefore map into the additive

Table 13.4. Equations for the design parameter B_0 for Aarau-type EDM baselines. (Rüeger 1977)

Number of stations	B_0
5	$\dfrac{1}{6}(C_0 - 4A)$
6	$\dfrac{1}{10}(C_0 - 5A)$
7	$\dfrac{1}{15}(C_0 - 6A)$
8	$\dfrac{1}{21}(C_0 - 7A)$

Table 13.5. Baseline sections and total baseline length versus number of stations for Aarau-type EDM baselines. (Rüeger 1977)

Section	Baseline with			
	5 Stations	6 Stations	7 Stations	8 Stations
1st	A + B	A + B	A + B	A + B
2nd	A + 3B	A + 3B	A + 3B	A + 3B
3rd	A + 2B	A + 4B	A + 5B	A + 5B
4th	A	A + 2B	A + 4B	A + 6B
5th	–	A	A + 2B	A + 4B
6th	–	–	A	A + 2B
7th	–	–	–	A
C =	4A + 6B	5A + 10B	6A + 15B	7A + 21B

constant which is derived from the baseline measurements. The determined scale or unknown baseline distances will not be affected. Because of the effect of short periodic errors on the additive constant, the short periodic errors must be determined separately on special "cyclic error testlines" and the bias of the additive constant removed. This aspect will be dealt with in Section 13.3.

A generalized design equation for Aarau-type baselines is given in Tables 13.4 and 13.5. The definitions of the parameters are the same as in the previous section. With all combinations being multiples of the unit length, the parameter D is no longer required. A numerical example is given in Table 13.6. Because of their separate determination, the short periodic errors as well as non-linear distance-dependent or long periodic errors are obtained with a higher precision than in the case of Heerbrugg-type baselines. The measuring effort increases accordingly. Should the assumption of the short periodic errors remaining constant over the baseline length prove to be wrong, biased additive constants and distance-depen-

Table 13.6. Numerical example of the design of a Aarau-type baseline

Stations	Sections	All combinations of distances							
1		0							
	$20 + 40 = 60$								
2		60	0						
	$20 + 120 = 140$								
3		200	140	0					
	$20 + 200 = 220$								
4		420	360	220	0				
	$20 + 240 = 260$								
5		680	620	480	260	0			
	$20 + 160 = 180$								
6		860	800	660	440	180	0		
	$20 + 80 = 100$								
7		960	900	760	540	280	100	0	
	$20 = 20$								
8		980	920	780	560	300	120	20	0

Input values: $U = 10\,m$, $A = 2U = 20\,m$, desired baseline length $(C_0) = 1000\,m$, eight stations. B_0 is obtained as 41 m and rounded to 40 m, leading to a total length of 980 m.

dent errors might result. This problem may be overcome by determining short periodic errors at different ranges (Rüeger 1986).

13.2.1.4 Hobart Design

This third baseline design, as proposed by Sprent and Zwart (1978), is not based on the measurement of all combinations of distances like the two designs above. It also differs from the two other designs by relying on true baseline distances being known at all times. The stations are spaced at equal intervals over the total baseline range. The distance meter to be calibrated occupies only one endpoint of the baseline as well as an auxiliary station half a unit length apart. The pillar coordinates of the original Hobart EDM baseline are given as 0, 5, 10, 123, 231, 359, 477 and 595 m (Sprent 1980), with the 5 m pillar being used for 10 m unit length distance meters and the 10 m pillar for 20 m unit length instruments (e.g. Wild Distomat DI3). This specific baseline features six fundamental stations as well as two auxiliary stations.

The test measurements are completed in minimum time as the distance meter does not move more than a few meters. The design permits the computation of the additive constant and the scale correction by simple linear regression techniques. Both parameters are not affected by any first-order short periodic errors present and minimally by any third-order short periodic errors. Their estimated standard deviations, however, are. Both parameters are affected by any second-order short periodic errors. The poor redundancy of the original Hobart design can be improved by an increased number of fundamental stations and/or additional auxiliary stations at 1/4 and 3/4 unit length from the first station.

Table 13.7. Equations for all design parameters for Hobart-type EDM baselines

Number of stations	Number of sections	Number of measurements	C	B_0	D
6 (+1)	5	10	$5B+U$	$(C_0-U)/5$	U/5
7*(+1)	6	12	$6B+U$	$(C_0-U)/6$	U/6
8 (+1)	7	14	$7B+U$	$(C_0-U)/7$	U/7
9*(+1)	8	16	$8B+U$	$(C_0-U)/8$	U/8
10 (+1)	9	18	$9B+U$	$(C_0-U)/9$	U/9
11*(+1)	10	20	$10B+U$	$(C_0-U)/10$	U/10
12 (+1)	11	22	$11B+U$	$(C_0-U)/11$	U/11

Baseline options marked by asterisks are not recommended as the distribution of measurements from the auxiliary station over the unit length is exactly the same as from the first station. The ability to resolve higher-order short periodic is reduced in these cases. The section lengths between all fundamental stations are constant and defined by the sum of B and D.

Table 13.7 gives generalized design equations for Hobart-type EDM baselines. The description of the parameters is as follows:

U $\quad$ = unit length of EDM instrument
C_0 $\quad$ = desired total length of baseline
C $\quad$ = final total length of baseline
B_0 $\quad$ = approximate value of first design parameter
B $\quad$ = final value of first design parameter ($= B_0$ rounded downwards to nearest multiple of U)
D $\quad$ = second design parameter
$B+D$ = section length between fundamental stations.

A numerical example may now be given. Assuming a desired baseline length C_0 of 800 m, a unit length of 10 m and eight fundamental stations, values of 110 m and 1.43 m are obtained for B and D, respectively. The coordinates of the stations compute then as: 0.00, 5.00, 111.43, 222.86, 334.29, 445.71, 557.14, 668.57, 780.00 m.

13.2.2 Physical Design

An ideal baseline site should have the following properties:

1. be of required length
2. even (or slightly concave) ground surface (horizontal or even slope)
3. equal irradiation by sunlight on entire line (no change between shaded and unshaded sections)
4. even vegetation, small growth (e.g. open grassland)
5. limited or no public access (vandalism, beam interruptions)
6. bedrock not too far below ground surface
7. easy access (road)

8. access road along base (to speed up transport of reflectors and instruments between stations)
9. North-South orientation (allowing measurement with and against the sun)
10. sympathetic owner.

If the baseline is to be used for one specific instrument only, it should have a length equal to the maximum distance normally measured in practice. This distance should be measurable with the normal number of prisms (1 or 3) under average atmospheric conditions.

If the baseline is to be used for instruments of varying range, a length may be chosen corresponding to the instrument having the longest range. In such a case more stations than normal should be considered in order to be able to accurately determine the instrument correction of the instruments with shorter ranges, even if this means that only a selection of all distances can be measured with the latter. Alternatively, the base length may be chosen to correspond to the range of the instrument with the shortest range, the whole length of the base then being used for all instruments (easier solution for computations).

Concrete pillars have some invaluable advantages over ground marks, and consequently are commonly used where high precision is required. The advantages of pillars are:

1. Stability during measurement (effects of operator and sun)
2. no danger of movement when interchanging reflector and heavy instruments
3. fast instrument and reflector set-up resulting from
   − constrained centring
   − pre-levelled centring plates
4. precise centring to 0.1 mm or better
5. constant height of instrument and reflector resulting in standardized computations.

Ground marks on the other hand require many tripods to provide constrained centring. The centring of these tripods and the measurement of zenith angle, every time a calibration is carried out, is very time-consuming. In addition, it is not possible to achieve the same accuracy in centring as on pillars and there is a degree of uncertainty about the stability of tripods due to the effects of operators, sun and interchange of reflectors, EDM instruments and theodolites. Therefore a pillared baseline is always preferable to one with ground marks. The costs of establishing pillars will soon be compensated by savings of time in the field and during computation.

For the pillar design, different solutions are available. The pillar centring system is a very crucial point and full attention should be paid to this problem. Suitable pillar designs are described in Kern (1974); Raphael (1983) and Kääriäinen et al. (1986), for example.

Although pillared EDM baselines are the preferred option, it should be noted that pillars may exhibit diurnal movements due to the daily change in sun exposure. Kobold (1958, 1961) and Borutta and Maass (1986) report on diurnal peak-to-peak pillar movements of 1.0 mm. Slightly smaller values of 0.6 mm have been measured for roof pillars at the University of New South Wales (Rüeger

1983). The effect of such short-term movements may have to be considered when deciding on a priori centring errors unless the pillars are protected from direct sunlight. Pillars may also exhibit long-term movements. Such movements may be caused by settlement due to the dead weight of the mark, slope creep, shrinkage and swelling of cohesive soils following changes in the water table as well as by underground mining, to name just a few.

Because of the possible long-term movements of pillars, national standard baselines measured with the Väisälä compensator are traditionally marked by both primary underground marks and secondary observation pillars (about 2 m apart). On occasions, the short-term pillar movements did adversely affect the Väisälä comparator measurements on such standard baselines. The marking of two recent Väisälä baselines is described by Kääriäinen et al. (1986, 1988).

13.2.3 Measurements on EDM Baselines

The procedures listed below are suggested for routine baseline measurements for the purpose of legal calibration of EDM instruments in Australia. They may be taken as a guide. More elaborate procedures are required when calibrating precision or high precision distance meters.

1. The observation sequence should be chosen in such a way that short lines are measured first and long lines later. (For example, if all 21 combinations on a 7-station baseline are to be measured: 6→7, 5→7, 5→6, 4→6, 4→5, 4→7, 3→7, 3→6, ..., 1→5, 1→4, 1→3, 1→2. (On Hobart-type baselines, the reflector will move from the closest to the most distant station and back.)
2. The EDM instrument station must be shaded by an umbrella.
3. Temperature and pressure is measured (in the shade) at the instrument station at the beginning of baseline measurements, upon completion of these and at half-hourly intervals between the two times and at other times when a change in weather is experienced.
4. Partial water vapour pressure or relative humidity is not measured. A baseline specific yearly average should be taken into account by the analysis.
5. On the first EDM instrument station to be occupied, the EDM instrument is set up and shaded (without switching on) at least 15 min prior to measurement of the first line on the baseline.
6. The "ppm-knob" is set to the neutral position, usually 0 ppm.
7. On each line, four distance measurements are taken, with repointing after each measurement. Pointing is "optically" or "electronically" (maximum signal strength) as prescribed by the manufacturer of the instrument concerned.
8. If possible, all distances are measured with one single prism. If necessary, longer distances may be measured to one triple prism. This (these) prism(s) should carry a permanent and unique identification label(s).
9. The measurements should be executed with the attenuator or aperture setting as prescribed by the instrument's manufacturer for a particular distance range.
10. Between the first and the last measurement on the baseline, the EDM instrument should be kept in the open air and in the shade.

11. EDM instruments should be operated according to the manufacturer's instruction and/or according to the measuring procedures followed during field surveys. Should the manufacturer suggest a number of alternative procedures, the field notes and the verification certificate should clearly indicate which procedure was followed.

12. The EDM instrument should be switched on immediately prior to the four measurements of a line and it should be turned off after these four measurements.

13. On pillared baselines, the height of the reflector, of any tilting axis on target/reflector assembly, of the EDM instrument and of the height of theodolite (if EDM equipment mounted on such) above the bottom plate of the tribrach used should be measured and booked to millimetres before *and* after the baseline measurements. This is best done in the office on a table, by measuring the heights from the table top with a pocket tape in four cardinal directions. The footscrews should be in mid-position during these measurements. Before levelling theodolite, EDM instrument and reflector on any station, the footscrews should be returned to the mid-position.

14. Whenever spot bubbles (circular levels) are used for levelling purposes, which cannot be rotated through 180° about the vertical axis of the respective equipment, they need to be checked and adjusted, if necessary, against a plate level of a theodolite before commencing the first measurement on a baseline and again after completion of all measurements on baseline.

15. The temperatures should be measured at instrument height and in the shade with good quality mercury thermometers or with platinum resistance or thermistor thermometers designed for temperature measurements of gases. Temperatures should be read to one degree celsius. The error of the thermometer used, at the points of the scale used during the baseline measurements, should not exceed $\pm 1.0\,°C$. The thermometer(s) used should carry a permanent and unique identification label.

16. The pressures should be measured in the shade and with a horizontal barometer carrying a permanent and unique identification label. Aneroid barometers should be graduated to at least 2 mb and should be gently tapped prior to reading. Pressures should be read to at least 2 mb.

    The barometers should be calibrated against a mercury column barometer at least prior to baseline measurements. If a malfunction is detected on the baseline, a second calibration after the baseline measurements should be executed.

    Mercury column barometers may usually be found at weather stations operated by the local Bureau of Meterology. Comparisons should be made against *station level pressure* and <u>NOT</u> against *sea level pressure*.

17. In the case of telescope- or theodolite-mounted EDM instruments, the axis of the EDM beam should be adjusted according to the manufacturers instructions. Under no circumstances should this adjustment be changed during the baseline test.

18. For theodolite (standard)-mounted instruments the field form should clearly state, if the EDM instrument was attached in the face left or face right position of the theodolite (when EDM instrument and theodolite telescope are pointing to reflector).

19. The levelling of theodolite, EDM instrument and reflector is critical and should be done with utmost care. The field form should feature a field, which can be ticked upon completion of levelling.
20. Baseline measurements should be carried out either fully during day-time or fully during night-time.

13.2.4 Analysis of Baseline Measurements

13.2.4.1 Preprocessing of Data

The following information is required for the preprocessing of the measured data:

- elevation of all stations above a reference level
- offset of all stations from a vertical plane through the first and last station
- calibration value (additive constant) of barometer(s) and thermometer(s) used.

The measured distances may then be processed as shown below:

1. For each line measured the mean (of four) distance measurements is computed as well as the standard deviation of a single distance measurement (S_D).
2. The standard deviations (of single distance observations) from all lines are plotted versus distance. A simple linear regression leads a reasonable a priori standard deviation of distances for subsequent least-squares adjustments:

$$S_D = \pm(A\ \text{mm} + B\ \text{ppm}) \ . \tag{13.1}$$

As the manufacturer's specifications are totally unsuitable for the estimation of the precision of distance measurements on baselines, this is done on the basis of the actual measurements. Using the propagation law of variances, A may be increased to account for centring errors and B for temperature/pressure reading errors, for example.

3. For each line, the corresponding temperature is interpolated (according to time) from the temperatures measured, at, typically, 30 minutes intervals and corrected for the calibration value.
4. After applying the calibration constant to all barometer readings, the pressure readings for all lines may be obtained by interpolation according to time. On sloping baselines, the change of pressure with height may have to be taken into account (see Appendix C).
5. A first velocity correction K′ of the form given by Eq. (6.11), is then added to all mean distances. An early average value of the partial water vapour pressure e is used as determined for a particular baseline site.
6. Where necessary, lines are corrected for any eccentricities their respective terminals may have from the straight line between the first and last station.
7. Where necessary, an additional correction has to be applied for EDM instruments which are attached to theodolites. However, no correction is required if:
   a) the EDM instrument is mounted on the standards of the theodolite and features its own trunnion axis provided that the tilting axis of the reflector is vertically above the mark. h_{EDM} and h_{REF} are measured according to the definition in Eq. (13.2).

b) the EDM instrument is mounted on the telescope of the theodolite at an offset and the reflector is mounted at an offset above a target, with the reflector tilting about an axis which goes through the centre of the target, provided that the reflector/target is always tilted to point at the theodolite, that the theodolite height is entered as h_{EDM} and the target's tilting axis as h_{REF} in Eq. (13.2) below. Refer to Sections 8.1.2 and 8.1.3 for details.

8. Finally, all lines are reduced on the same horizontal reference elevation E_R:

$$HD_{i,j} = \left(D - \frac{\Delta H^2}{2D} - \frac{\Delta H^4}{8D^3} - \frac{\Delta H^6}{16D^5} + \frac{\bar{H}\,\Delta H^2}{2DR} + \frac{\bar{H}\,\Delta H^4}{8D^3R} + \frac{\bar{H}\,\Delta H^6}{16D^5R} - \frac{\bar{H}\,D}{R} \right)$$
$$\times \left(1.0 + \frac{E_R}{R} \right) , \tag{13.2}$$

where
$$
\begin{aligned}
D &= \text{corrected mean distance of line i to j} \\
\bar{H} &= 0.5\,(H_i + h_{EDM} + H_j + h_{REF}) \\
\Delta H &= (H_i + h_{EDM}) - (H_j + h_{REF}) \\
H_i &= \text{elevation of instrument station "i" (in m)} \\
H_j &= \text{elevation of reflector station "j" (in m)} \\
h_{EDM} &= \text{height of EDM instrument (in m)} \\
h_{REF} &= \text{height of reflector (in m)} \\
R &= \text{mean radius of curvature of the earth (in m)} \\
HD_{ij} &= \text{horizontal distance from i to j at elevation } E_R.
\end{aligned}
$$

All horizontal distances HD are colinear as far as the horizontal component is concerned and at exactly the same elevation (E_R). The above equation can be derived easily from Eqs. (7.1) and (7.28).

13.2.4.2 Review of Baseline Adjustment Options

The available adjustment options depend on whether "true" baseline distances are known or not, whether a complete or reduced instrument correction is required, and on the computers to be used. It has been discussed before that the Hobart baseline design requires true distances at all times. The other two baseline designs permit the determination of most coefficients of the instrument correction with or without known distances. Table 13.8 outlines the four major least-squares adjustment options. Solutions for unknown baseline distances are only possible if distance measurements in all (or most) combinations are available. This case is restricted to the Heerbrugg and Aarau baseline designs.

The two pocket calculator solutions are described in the next two sections in more detail. For the general least-squares solutions, the reader is referred to text books on least-squares estimation techniques as well as to Eqs. (13.1) and (12.13) for the stochastic and mathematical models, respectively.

The residuals and the standard deviations of all determined coefficients of the instrument correction IC should be computed in all four cases shown in Table 13.8. In the case of the closed pocket calculator solutions, the former should be plotted against distance as well as unit length to check for remaining systematic (non-linear distance-dependent or long periodic) errors. Both plots of residuals

Table 13.8. Major least-squares adjustment options for the determination of the instrument correction from distance measurements on EDM calibration baselines

	Baseline of known length	Baseline of unknown length
Computation with pocket calculator	Method: linear regression (= least-squares adjustment) (see Sect. 13.2.4.3) Unknowns: − additive constant − scale correction	Method: least-squares adjustment (equal weight for all observations) (see Sect. 13.2.4.4) Unknowns: − additive constant − $(N-1)$ baseline distances (N = number of stations on baseline)
Computation with personal or main-frame computer	Method: least-squares adjustment with correct weighting of observations Unknowns: − additive constant − scale correction − short periodic error corrections − other parameters	Method: least-squares adjustment with correct weighting of observations Unknowns: − additive constant − $(N-1)$ baseline distances − short periodic error corrections − other parameters

are also recommended when using general least-squares solutions unless short periodic errors are solved for. In this case the plot versus unit length may be omitted. In general least-squares solutions, a large number of coefficients of the IC are solved for initially. After testing the statistical significance of the coefficients, the least-squares adjustment is repeated with the significant terms only. The pocket calculator solutions assume equally weighted data. Correctly weighted data are used in general least-squares solutions. It has been shown that correctly weighted data provide more consistent coefficients of the instrument correction in repeated determination (Elmiger and Siegerist 1977).

13.2.4.3 Determination of Additive Constant and Scale Correction on Known Baselines

Linear regression solutions may be used to resolve additive constant and scale correction from measurements obtained on any of the three baselines designs whenever the "true" lengths of the baselines are known.

For y = measured distance [see Eq. (13.2)]
and x = given or "true" distances
the equation of the linear regression line is given as

$$\hat{y} = a + bx \, , \tag{13.3}$$

where $\hat{y}$ = estimated y for given x.

The closed least-squares solutions for the unknown parameters a and b are as follows:

$$a = \frac{(\Sigma x^2)(\Sigma y) - (\Sigma x)(\Sigma xy)}{n(\Sigma x^2) - (\Sigma x)^2} \tag{13.4}$$

$$b = \frac{n(\Sigma xy)-(\Sigma x)(\Sigma y)}{n(\Sigma x^2)-(\Sigma x)^2} \tag{13.5}$$

$$s_0^2 = \frac{\Sigma(a+bx-y)^2}{n-2} \tag{13.6}$$

$$s_a = s_0 \sqrt{\frac{(\Sigma x^2)}{n(\Sigma x^2)-(\Sigma x)^2}} \tag{13.7}$$

$$s_b = \frac{s_0}{\sqrt{(\Sigma x^2-(\Sigma x)^2/n)}} \; . \tag{13.8}$$

To derive the instrument correction IC, Eq. (13.3) is solved for x:

$$x = -a/b+(1/b)\hat{y} \; . \tag{13.9}$$

The first and second terms on the right hand side represent the additive constant and the scale correction, respectively.

The following example may demonstrate the necessary steps. A distance meter was calibrated on a seven-station baseline. Twenty-one measured and true distances are available. Table 13.9 lists the corresponding data pairs. Defining the given distances as independent variables x and the measured distances as dependent variables y, the following solutions are obtained:

$$\hat{y} \;\; = \; +0.0057\,\mathrm{m}+1.000\,002\,18\,x \tag{13.10}$$

$$x \;\;\; = \; -5.7\,\mathrm{mm}+0.999\,997\,82\,\hat{y} \tag{13.11}$$

$$IC = \; -5.7\,\mathrm{mm}-2.18\,(D/1000) \; , \tag{13.12}$$

where D = distance (in m)
        IC = instrument correction (in mm)

The following standard deviations are calculated

Table 13.9. Pairs of measured and true distances obtained on an Aarau-type EDM calibration baseline. The baseline stations are numbered consecutively from 1 to 7

Line	y(m)	x(m)	Line	y(m)	x(m)
12	30.0235	30.0191	67	50.0094	50.0022
23	69.9931	69.9882	34	79.9959	79.9913
13	100.0121	100.0073	45	120.0060	119.9985
24	149.9844	149.9795	56	170.0124	170.0029
14	180.0029	179.9986	35	199.9959	199.9898
57	220.0147	220.0051	25	269.9823	269.9780
46	290.0109	290.0014	15	300.0033	299.9971
47	340.0088	340.0036	36	369.9978	369.9927
37	420.0031	419.9949	26	439.9865	439.9809
16	470.0070	470.0000	27	489.9883	489.9831
17	520.0098	520.0022			

$s_0 = \pm 1.8$ mm
 = standard deviation of an measured (input) distance

$s_b = \pm 2.6$ ppm
 = standard deviation of the scale correction

$s_a = \pm 0.8$ mm
 = standard deviation of the additive constant.

The corresponding 95% confidence intervals are obtained by multiplication with 2.093 (two-tailed t-distribution, degree of freedom $= 21 - 2 = 19$). It follows that the additive constant is statistically significant and the scale correction not. As the data were obtained on an Aarau-type baseline (all distances being multiples of 10 m), the short periodic errors would need to be determined separately and the additive constant corrected accordingly. The plot of residuals versus distance is omitted here.

13.2.4.4 Determination of Additive Constant and Unknown Baseline Lengths

Closed least-squares solutions for additive constant and unknown baseline distances from distance measurements in all combinations are available. In this context, the solution given by Halmos and Kadar (1972) is used, which provides the additive constant as well as the adjusted lengths between the first and any other station. This solution gives the precision of the total length of the baseline directly. [A number of other authors (Schlichting 1980; Heck and Heil 1981; Reissmann 1982; Emenike 1982a) report solutions for the baseline sections.] The method suits pocket calculators well. No approximate values are required as the measured values are employed directly.

The Halmos and Kadar (1972) solution is first presented for a five-station baseline. The additive constant (here denoted by c) and the four unknown baseline distances are calculated as:

$$c = -\frac{1}{10}\,[3\,(S_{12}+S_{23}+S_{34}+S_{45}-S_{15})+(S_{13}+S_{35}-S_{25}+S_{24}-S_{14})]\,, \qquad (13.13)$$

where S_{ij} = observed (EDM) distance from point i to point j.

$$\bar{S}_{12} = \tfrac{1}{5}[2c+2S_{12}+(S_{13}-S_{23}+S_{14}-S_{24}+S_{15}-S_{25})] \qquad (13.14a)$$

$$\bar{S}_{13} = \tfrac{1}{5}[4c+2S_{13}+(S_{12}+S_{23}+S_{14}-S_{34}+S_{15}-S_{35})] \qquad (13.14b)$$

$$\bar{S}_{14} = \tfrac{1}{5}[6c+2S_{14}+(S_{12}+S_{24}+S_{13}+S_{34}+S_{15}-S_{45})] \qquad (13.14c)$$

$$\bar{S}_{15} = \tfrac{1}{5}[8c+2S_{15}+(S_{12}+S_{25}+S_{13}+S_{35}+S_{14}+S_{45})]\,, \qquad (13.14d)$$

where $\bar{S}_{1j}$ = adjusted value of distance between first point of baseline and point j of baseline.

The residuals may then be computed from the following relationship:

$$\bar{S}_{ij} = S_{ij}+c+v_{S_{ij}}\,. \qquad (13.15)$$

The standard deviation of a measured (input) distance is obtained in the usual way from

$$s_0 = \pm \sqrt{\frac{\Sigma v^2}{n-u}} \ , \tag{13.16}$$

where   n = number of observations
   u = number of unknowns
   u = N

$$n = \frac{N!}{2(N-2)!} \ . \tag{13.17}$$

For baselines with N stations, the formulae for the additive constant c and the $(N-1)$ unknown distances may be given as (Halmos and Kadar 1972, 1976; Halmos 1977, 1980):

$$\bar{S}_{1j} = \frac{1}{N} [2(j-1)c + 2S_{1j} + \Sigma S_{ij}] \ , \tag{13.18}$$

where $2 \leqslant j \leqslant N$
   $\bar{S}_{1j}$ = adjusted value of distance between first and j-th point
   S_{1j} = measured value of distance between first and j-th point
   N = number of baseline stations

and where ΣS_{ij} is determined as follows:

$$\Sigma S_{ij} = \sum_{k=2}^{N} (S_{1k} + S_{kj}) \quad k \neq j \tag{13.19}$$

where S_{1k} = measured distance between first and k-th point
   S_{kj} = measured distance between k-th and j-th point.
   (If $k > j$, replace S_{kj} by $(-S_{jk}.)$

$$c = -\frac{\sum_{i=1}^{m} \{N - 2i\}\Delta_i}{N(N-1)(N-2)/6} \ , \tag{13.20}$$

where $m = (N-2)/2$ for even number of stations (N)
   $m = (N-1)/2$ for odd number of stations (N)

and where the polygonal closing errors are denoted by Δ_i. The distance misclosures are computed first $(m = 1)$ for neighbouring points, then for points spaced by two sections $(m = 2)$, then for points separated by three sections $(m = 3)$ and so forth. Rather than trying to define the misclosures Δ_i any further, the equations for baselines with 6, 7 and 8 stations are listed as examples.

6-station-baseline:

$$c = -\frac{1}{10} [2(S_{12} + S_{23} + S_{34} + S_{45} + S_{56} - S_{16})$$
$$+ (S_{13} + S_{35} - S_{15} + S_{24} + S_{46} - S_{26})] \ . \tag{13.21}$$

7-station baseline:

$$c = -\frac{1}{35} \, [5\,(S_{12}+S_{23}+S_{34}+S_{45}+S_{56}+S_{67}-S_{17})$$
$$+3\,(S_{13}+S_{35}+S_{57}-S_{27}+S_{24}+S_{46}-S_{16})$$
$$+(S_{14}+S_{47}-S_{37}+S_{36}-S_{26}+S_{25}-S_{15})] \; . \tag{13.22}$$

8-station baseline:

$$c = -\frac{1}{28} \, [3\,(S_{12}+S_{23}+S_{34}+S_{45}+S_{56}+S_{67}+S_{78}-S_{18})$$
$$+2\,(S_{13}+S_{35}+S_{57}-S_{17}+S_{24}+S_{46}+S_{68}-S_{28})$$
$$+(S_{14}+S_{47}-S_{27}+S_{25}+S_{58}-S_{38}+S_{36}-S_{16})] \; . \tag{13.23}$$

Closed solutions for the standard deviations of the additive constant (s_c) and the adjusted baseline line lengths (s_{1j}) are given by Pauli (1977a) as:

$$s_c = s_0 (Q_{cc})^{0.5} \tag{13.24}$$

$$s_{1j} = s_0 (Q_{1j})^{0.5} \; , \tag{13.25}$$

where s_0 is defined by Eq. (13.16). The cofactors Q are given as:

$$Q_{cc} = \frac{6}{(N-1)(N-2)} \tag{13.26}$$

$$Q_{1j} = \frac{2}{N} \left\{ 1 + \frac{12(j-1)^2}{N(N-1)(N-2)} \right\} \tag{13.27}$$

with $j = 2, \ldots, N$.

Two additional cofactors (for distances between any two stations and between additive constant and any adjusted baseline distance s_{1j}) are also listed although they are generally not required:

$$Q_{ij} = \frac{1}{N} \left\{ 1 + \frac{24(i-1)(j-1)}{N(N-1)(N-2)} \right\} \tag{13.28}$$

for $i \neq j$; $j = 2, \ldots, N-1$; $i = j+1, \ldots, N$

$$Q_{cj} = \frac{12(j-1)}{N(N-1)(N-2)} \tag{13.29}$$

with $j = 2, \ldots, N$.

An example of all combination measurements on an eight-station Aarau-type EDM calibration baseline is now given. The input distances as well as their residuals are listed in Table 13.10. The necessary equation for the additive constant c is given by Eq. (13.23). The equations for the unknown baseline distances may be derived from Eqs. (13.18) and (13.19) as follows:

Table 13.10. Measurements in all combinations on an eight-station, Aarau-type EDM calibration baseline. The residuals are the result of a closed least-squares solution with equally weighted observations

Line	Measured distance (m)	Residual (mm)
12	139.99750	-1.930
13	199.99670	0.609
14	310.00130	-0.764
15	459.99660	-0.463
16	650.00040	1.902
17	879.99730	0.154
18	980.00220	0.493
23	60.00090	-0.068
24	170.00320	0.859
25	319.99990	-0.239
26	510.00870	-2.875
27	740.00050	0.477
28	840.00630	-0.084
34	110.00120	1.120
35	259.99790	0.021
36	450.00370	0.386
37	679.99920	0.038
38	780.00550	-1.023
45	149.99370	0.995
46	340.00110	-0.241
47	569.99680	-0.789
48	670.00000	1.250
56	190.00420	1.057
57	420.00040	0.009
58	520.00640	-0.752
67	229.99450	-0.255
68	329.99900	0.484
78	100.00470	-0.368

$j = 2$

$$\bar{S}_{12} = \tfrac{1}{8}[2c + 2S_{12} + (S_{13} - S_{23} + S_{14} - S_{24} + S_{15} - S_{25} + S_{16} - S_{26} + S_{17} - S_{27} + S_{18} - S_{28})] \tag{13.30a}$$

$j = 3$

$$\bar{S}_{13} = \tfrac{1}{8}[4c + 2S_{13} + (S_{12} + S_{23} + S_{14} - S_{34} + S_{15} - S_{35} + S_{16} - S_{36} + S_{17} - S_{37} + S_{18} - S_{38})] \tag{13.30b}$$

$j = 4$

$$\bar{S}_{14} = \tfrac{1}{8}[6c + 2S_{14} + (S_{12} + S_{24} + S_{13} + S_{34} + S_{15} - S_{45} + S_{16} - S_{46} + S_{17} - S_{47} + S_{18} - S_{48})] \tag{13.30c}$$

$j = 5$

$$\bar{S}_{15} = \tfrac{1}{8}[8c + 2S_{15} + (S_{12} + S_{25} + S_{13} + S_{35} + S_{14} + S_{45} + S_{16} - S_{56} + S_{17} - S_{57} + S_{18} - S_{58})] \tag{13.30d}$$

$j = 6$

$$\bar{S}_{16} = \tfrac{1}{8}[10c + 2S_{16} + (S_{12} + S_{26} + S_{13} + S_{36} + S_{14} + S_{46}$$
$$+ S_{15} + S_{56} + S_{17} - S_{67} + S_{18} - S_{68})] \tag{13.30e}$$

$j = 7$

$$\bar{S}_{17} = \tfrac{1}{8}[12c + 2S_{17} + (S_{12} + S_{27} + S_{13} + S_{37} + S_{14} + S_{47}$$
$$+ S_{15} + S_{57} + S_{16} + S_{67} + S_{18} - S_{78})] \tag{13.30f}$$

$j = 8$

$$\bar{S}_{18} = \tfrac{1}{8}[14c + 2S_{18} + (S_{12} + S_{28} + S_{13} + S_{38} + S_{14} + S_{48}$$
$$+ S_{15} + S_{58} + S_{16} + S_{68} + S_{17} + S_{78})] \tag{13.30g}$$

Substitution of the measured data of Table 13.10 into the above equations leads to the following values of the adjusted parameters and their standard deviations:

Additive Constant	= +0.91 mm	±0.43 mm
Line 12	= 139.9965 m	±0.58 mm
Line 13	= 199.9982 m	±0.61 mm
Line 14	= 310.0014 m	±0.65 mm
Line 15	= 459.9970 m	±0.71 mm
Line 16	= 650.0032 m	±0.78 mm
Line 17	= 879.9984 m	±0.86 mm
Line 18	= 980.0036 m	±0.94 mm

It should be noted that the given additive constant is provisional pending separate tests for short periodic errors. The standard deviation of a measured (input) distance follows from Eq. (13.16) as ±1.14 mm. The standard deviations of the adjusted parameters are computed with the aid of Eqs. (13.32) and (13.33). The above least-squares adjustment problem includes 8 unknown parameters and 28 observations, leading to a degree of freedom of 20. The 95% confidence intervals of the parameter may be obtained by multiplication of the standard deviations with a factor of 2.086 (two-tailed t-distribution). It follows that the additive constant is (just) statistically significant.

13.2.5 Determination of Baseline Lengths

EDM calibration baselines of known length have at least two distinct advantages over baselines of unknown length. Firstly, they permit the establishment of the absolute scale [coefficient A_{10} in Eq. (12.13)] of a distance meter. Secondly, they provide a much larger degree of freedom in the least-squares adjustment as the baseline lengths need not to be solved for. This leads to a better accuracy of the derived instrument correction and/or permits the solution for a larger number of terms of Eq. (12.13). (For example, it is unlikely that the solution for the fifth degree polynomial (parameters A_{20}, A_{30}, A_{40}, A_{50}) is feasible in the case of baselines of unknown length.) In the case of legal metrology, the baseline distances must be known as the determination of the scale of an instrument is of prime importance.

The "true" lengths of EDM calibration baselines may be established by:

- invar taping
- Väisälä interference comparator
- high precision EDM
- length transfer between baselines
- multiwavelength interferometry.

The first two options are no longer economically viable but were used successfully in the past (see Sect. 3.4.3, for example). High precision EDM instruments might be used in an absolute mode or in a relative mode. In the former, they provide a direct determination of baseline through self-calibration of the additive constant and through laboratory calibration of the on-board master oscillator. Single-colour instruments are likely to be used on relative short baselines and two-colour instruments on baselines in excess of a few kilometres lengths. In the relative mode, they are employed as transfer standards between known (e.g. Väisälä standard baselines) and unknown baselines. In this case, calibration of the transfer instrument occurs on the reference baseline. (Because of the more comprehensive field calibration involved, less sophisticated precision distance meters may also be used as transfer instruments between baselines of known and unknown lengths.)

Multiwavelength interferometers are in development and may eventually permit measuring 100 m directly to interferometric accuracies [refer to Walsh (1987), for example]. With such instruments, some sections of baselines may be measured directly, with the remainder extrapolated by high precision (or precision) distance meters (Hölscher 1980; Konttinen 1988) and use of the local scale parameter method (see Sect. 5.9.3.5). The use of commercial laser interferometers is a less practical alternative, as a rail is required to shift the reflector between stations. A commercial laser interferometer was originally used on at least one baseline to determine a (41 m) section (Dodson et al. 1981).

The maintenance of the true lengths of known baselines is a major task, in particular, if one or more stations prove to be unstable. Pillar movements of 1 mm per year are not uncommon. (The author has even recorded cyclic pillar movements of 15 mm peak-to-peak over one year and for a particular baseline!) It is strongly recommended to select the site for known baselines very carefully and to analyze all data from such known baselines in two ways, namely with and without the known lengths. Comparisons of computed baseline lengths with given lengths will provide a first indication of the stability of a baseline or otherwise. Unstable baselines need to be remeasured more frequently than stable ones with the techniques listed above. Proper deformation analysis techniques may be used to track station movements between periodic remeasurements on the basis of data gathered with normal EDM instruments (Caspary 1987; Rüeger 1983).

13.3 Calibration on Cyclic Error Testlines

The calibration of short periodic errors on "cyclic error" testlines is complementary (a must) or supplementary (an option) depending on the type of EDM calibration baseline used for the calibration of the distance-dependent terms of the instrument correction.

Complementary measurements on cyclic error testlines are required whenever:

- the instrument correction has been determined on an EDM calibration baseline of the Aarau design
- the instrument correction is determined on an EDM calibration baseline of the Heerbrugg or Hobart design, which was not designed for the unit length of the distance meter in question.

In the former case, the independent determination of the short periodic errors is required to remove any bias from the additive constant. As most Heerbrugg- or Hobart-type baselines are designed for unit lengths of 10 m, distance meters with "exotic" unit lengths such as 33.33 m, 30.77 m and 10.10 m will often require separate testing for short periodic errors. However, EDM calibration baselines designed for a unit length of 10 m may also accommodate instruments with unit lengths of 5 m and 20 m.

Supplementary measurements on cyclic error testlines may be appropriate whenever:

- the change of the short periodic errors with distance needs to be evaluated,
- the accuracy of the short periodic errors determined on Heerbrugg- or Hobart-type baselines is not sufficient.

Separate calibrations on cyclic error testlines are therefore required if the variation with distance of the short periodic errors needs to be evaluated. This might be required to improve the precision of a distance meter or to resolve suspected multipath errors (see Sect. 12.2.3). Heerbrugg- and Hobart-type baselines permit the detection of short periodic errors but often do not provide sufficient redundant data to determine the short periodic errors accurately enough or with a sufficiently large number of higher order terms. Separate tests may be required in such cases.

13.3.1 Design of and Measurements on Cyclic Error Testlines

In this context the discussion on testlines for the calibration of short periodic errors is restricted to simple outdoor facilities which can be easily established. More sophisticated indoor facilities are usually based on rails for the movement of reflectors and on laser interferometers as length standards (see Sect. 13.1.2). Details of an indoor facility may be found in Ciddor (1987) and Rüeger and Ciddor (1987), for example.

The general layout for such a test is depicted in Fig. 13.1. The wall should have a flat but not necessarily horizontal top. A good quality 30-m steel tape (graduated at centimetre intervals) is fixed at standard tension (usually 50 newtons) and fully supported on top of the wall. A reliable anchorage at the zero end of the tape must be assured. For general purpose testing, the EDM instrument should be set up on line about 50 to 100 m from the wall and at a height which

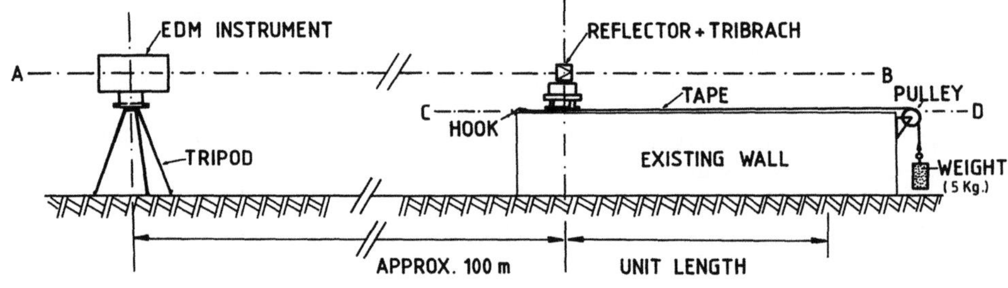

Fig. 13.1. Longitudinal section through a cyclic error testline. The pulley and 5 kg weight ensure proper tension at changing temperatures. The unit length is equivalent to half of the modulation wavelength. The axis of the EDM beam ($A-B$) must be parallel in three dimensions to the axis of the tape ($C-D$)

makes the line of sight to the prism parallel to the steel tape and thus the top of the wall.

To test an EDM instrument with a unit length of 10 m (modulation wavelength of 20 m), the reflector is placed on the 1.0-m mark in such a way that one footscrew of the tribrach points to the distance meter. The edge of the tribrach's footplate pointing to the instrument is aligned precisely with the tape mark. The tribrach is levelled and the distance measured (e.g. four times). The reflector assembly is then moved to the tape marks 3.0, 5.0, 7.0, 9.0, 10.0, 8.0, 6.0, 4.0, 2.0 in turn, measuring distances to the reflector at each of these set-ups. This procedure yields ten distances equally spaced over the 10-m unit length. For higher resolution, the reflector may be shifted in 0.5-m or even 0.25-m intervals instead of the 1.0-m interval described here.

Critical are, first, the accuracy with which the reflector's tribrach can be positioned on the tape and the accuracy of the tape graduations as such. The latter is specified by the German standard for steel tapes as $\pm(0.2\ \text{mm}+100\ \text{ppm})$, being the maximum deviation from nominal (in mm) between any two marks at 20 °C and 50 newtons tension. A distance between two marks 10 m apart would therefore be accurate to at least 1.2 mm. A corresponding setting accuracy of the reflector's tribrach on the tape of ±0.2 mm can be achieved.

To match the tape's scale uncertainty of 100 ppm may prove more difficult. The appropriate ppm correction for the distance meter is usually less than 50 ppm, and may be ignored. Otherwise, the ppm correction can be easily set on most distance meters. Changes in air temperature are more critical. Assuming the distance meter at a distance of 100 m from the reflector, a change of 10 °C during the test will lead to a change of the measured distance of 1 mm and thus to a *scale error* of 100 ppm over the 10 m. Should the temperature change more than 5 °C during a short periodic error test over 100 m, the distances should be corrected or the ppm-dial reset whenever necessary.

Furthermore, the average scale of the tape during the test should not differ significantly from true scale. Quoting the German standard, the difference will be less than 100 ppm at 50 newtons and 20 °C. It can be assumed that this value could be reduced considerably by calibration of the steel tape. If a tape is not at

210

20 °C (or standard temperature), a scale error of about 10 ppm per degree Celsius will result. It is therefore suggested to apply an average correction to the tape measurements whenever the average field tape temperature differs by more than 5 °C from the standard temperature.

It is suggested that the tape (and its support) be shaded on clear days and the tape temperature measured directly, for example, with contact thermistor thermometers. On overcast days, the air temperature may be substituted for the tape temperature; therefore overcast days are best suited for short periodic error tests.

A last effect to be considered is the change of the tape temperature during the test. Again, should this temperature change by more than 5 °C, it should be taken into account. Such changes are likely to occur only on clear days and for unshaded tapes if conditions change from wind-still to windy and/or from clear to overcast (Angus-Leppan 1979).

The measures to be observed to ensure correct EDM and tape scale should be selected on the basis of the precision required for the short periodic error determination. For example, a residual scale error of 100 ppm or 1 mm/10 m produces a fictitious error of 0.3 mm amplitude and 10-m wavelength, as well as errors of shorter wavelengths and smaller amplitudes.

This induced error is well below the intrinsic resolution of standard EDM instruments. Scale is therefore not a serious problem whenever induced periodic errors of up to 1 mm amplitude are acceptable. Also, the measurement sequence suggested above, namely, to measure every second data point in a forward and the remaining points in a reverse sense, tends to compensate linear changes in air and steel tape temperatures which occur during the test.

Cyclic error testlines can be integrated with EDM calibration baselines by building of a, say, 35-m-long wall (or other structure) beyond one of the two terminal stations. This then permits the determination of short periodic errors from each of the (typically seven) baseline stations and at increasing distances to the full length of the baseline.

13.3.2 Semi-Graphic Determination of Short Periodic Errors

The reductions are made according to Table 13.11. The resulting s* values are subsequently plotted as a function of distance as shown in Fig. 13.2 (unit length in this example is 10 m).

In Fig. 13.2, s* is plotted as a function of s and the mean of s* is drawn as a line. A sine curve of 10 m wavelength (unit length in this example) with its axis as s* mean is now fitted to the data. The correction to the original additive constant dc for an exact multiple of the unit length (here 10 m) is obtained as the difference, in ordinate s*, between the estimated sine curve of the short periodic error and s* mean. The correction dc, which is required to remove the bias from an additive constant c (or the instrument correction IC) established on an Aarau-type EDM calibration baseline, is taken as positive if the curve for this abscissa is above the line of s* mean and negative if the curve at this point is below s* mean. In this example (see Fig. 13.2), dc is equal to $+8$ mm for $s = 100$ m or for $s = 110$ m (multiple of unit length $= 10$ m). Together with the additive constant c

Table 13.11. Original measurements and reduced data for the determination of short periodic errors

Tape mark	s	Subtract	s*	
1 m	100.032 m	0 m	100.032	m
2 m	101.032 m	1 m	100.032	m
3 m	102.026 m	2 m	100.026	m
4 m	103.022 m	3 m	100.022	m
5 m	104.018 m	4 m	100.018	m
6 m	105.016 m	5 m	100.016	m
7 m	106.016 m	6 m	100.016	m
8 m	107.020 m	7 m	100.020	m
9 m	108.026 m	8 m	100.026	m
10 m	109.030 m	9 m	100.030	m
		Mean s* =	100.0238 m	

s = observed distance; s^* = observed distance, reduced to first observed tape mark.

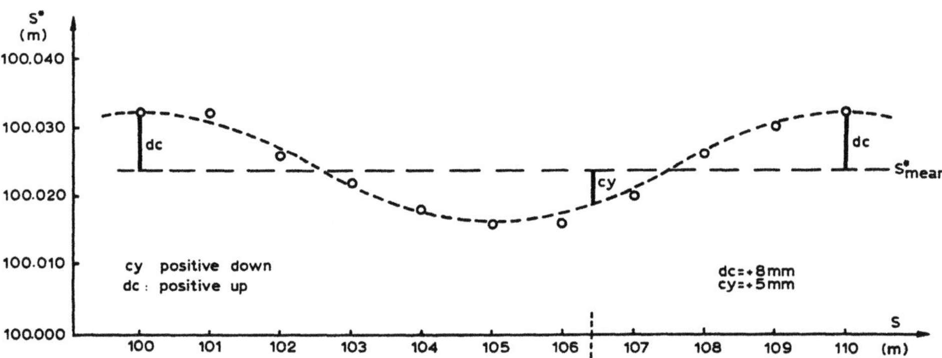

Fig. 13.2. Plot of reduced cyclic error testline data (s^*) versus observed distances (s). The supplementary additive constant and the short periodic error correction (at 106.4 m) are denoted by dc and cy. The former is taken positive when above the level of s^*_{mean}, the latter when below s^*_{mean}

determined on a baseline of the Aarau design the total additive constant c_{tot} then yields

$$c_{tot} = c + dc \ . \tag{13.31}$$

The total additive constant is equivalent to the first parameter A_{00} of the instrument correction as defined by Eq. (12.13). It is important that the supplementary correction dc is required only if the corresponding additive constant is derived from measurements on Aarau-type EDM calibration baselines. There is no such simple way of correcting for the bias caused by short periodic errors in additive constants derived from Heerbrugg- (and Hobart-) type baselines. Refer to Sections 13.2.1.2 and 13.2.1.4 for a discussion of the effect (and its removal) of short periodic errors on additive constant (and scale) in the case of Heerbrugg and Hobart baseline designs.

212

The correction (cy) for short periodic errors may be derived graphically from Fig. 13.2, irrespective of the design of EDM calibration baselines used. cy is defined as the interval between the estimated cyclic error curve and s* mean for a certain fraction of the unit length. cy is taken positive if the curve is below s* mean and negative if the curve is above the s* mean line.

The short periodic error correction (cy) is equivalent to the B_{i1} and C_{i1} terms in the general formula for the instrument correction in Eq. (12.13). In practice, short periodic error corrections (cy) may be omitted whenever their magnitude is below the stated accuracy of the distance meter or below the accuracy requirements of a particular survey task. The accuracy of the determination of the short periodic error correction can be improved by moving the reflector in shorter steps (rather than the 1.0-m intervals shown in Table 13.11 and Fig. 13.2, steps of 0.5 or 0.25 m may be selected) and by replacing the semi-graphic solution by a rigorous least-squares adjustment. The latter option is discussed below.

13.3.3 Analytical Determination of Short Periodic Errors

First-order short periodic errors (cyclic errors) may be described by the following expression (Pauli 1977b; Halmos 1980; Rüeger 1978)

$$CE = A_1 \sin [(2\pi)/U(S+B_1)] \, , \tag{13.32}$$

where A_1 is the amplitude of the error, U is the unit length, S is the measured distance, B_1 is the phase of the error, and CE is the error caused at the distance S.

The above equation may be developed into the following form, using well-known identities:

$$CE = A_1 [\sin (2\pi S/U) \cos (2\pi B_1/U) + \cos (2\pi S/U) \sin (2\pi B_1/U)] \, . \tag{13.33}$$

Considering the substitutions

$$b_1 = A_1 \cos (2\pi B_1/U) \; ; \quad a_1 = A_1 \sin (2\pi B_1/U) \, , \tag{13.34}$$

the first-order short periodic error can be given in a form better suited for computations (Couchman 1974; Halmos 1980; Sprent and Zwart 1978; Witte and Schwarz 1982; Emenike 1982a, b)

$$CE = a_1 \cos (2\pi S/U) + b_1 \sin (2\pi S/U) \, . \tag{13.35}$$

Considering all short periodic errors to the n-th order, the combined error (CE') yields:

$$CE' = \sum_{j=1}^{n} a_j \cos [(2\pi S/U)j] + \sum_{j=1}^{n} b_j \sin [(2\pi S/U)j] \, . \tag{13.36}$$

The original form of periodic errors [see Eq. (13.32)] may be reproduced using the equations below, which follow directly from Eq. (13.34):

$$A_j = (a_j^2 + b_j^2)^{1/2} \; ; \quad B_j = (U/2\pi) \arctan (a_j/b_j) \, . \tag{13.37}$$

It is now assumed that the total number of m observations (equally distributed over a unit length U) has been made during a short periodic error test with:

$$mD = U ,\tag{13.38}$$

where the spacing of the measuring points is denoted by D. Therefore, in terms of fractions of the unit length, the measurements have been made at the points:

$$0, D, 2D, \ldots (m-2)D, (m-1)D .\tag{13.39}$$

For each of the m measurements, an observation equation of the following form can be formulated:

$$s_i + v_i = S_i + \sum_{j=1}^{n} a_j \cos [(2\pi j/U)S_i] + \sum_{j=1}^{n} b_j \sin [(2\pi j/U)S_i] ,\tag{13.40}$$

where
s_i = i-th distance measurement
v_i = residual of i-th measurement
S_i = adjusted distance to i-th tape mark
i = 0, 1, 2, 3, ..., (m−1)
j = degree of periodic error = 1, 2, 3, 4, ..., (n)
a_j = adjusted amplitude of j-th cosine-term
b_j = adjusted amplitude of j-th sine-term
m = total number of observations
n = highest order of periodic error considered where $m > (2n+1)$.

Assuming that the tape graduations and thus the increments D are free of error, the parameter S_i may be replaced by:

$$S_i = S_0 + iD ,\tag{13.41}$$

where D is equal to U/m and S_0 is equal to the adjusted distance to the nullth reflector position (i = 0). The latter should ideally be an exact multiple of the unit length U.

Defining as in Table 13.11:

$$s_i^* = s_i - iD \quad \text{and} \quad \bar{s}^* = (1/m) \sum_{i=0}^{m-1} s_i^*\tag{13.42}$$

the adjusted distance S_0 may be expressed as:

$$S_0 = \bar{s}^* + ds .\tag{13.43}$$

Substitution of Eqs. (13.42) and (13.43) into Eq. (13.40) and considering the substitution

$$E_i = (2\pi/U)(\bar{s}^* + ds + iD)\tag{13.44}$$

yields the following observation equations:

$$l_i + v_i = ds + \sum_{j=1}^{n} a_j \cos (jE_i) + \sum_{j=1}^{n} b_j \sin (jE_i) ,\tag{13.45}$$

where

214

$$l_i = s_i^* - \bar{s}^* \ . \tag{13.46}$$

Equation (13.45) leads to the classical case of a harmonic analysis where m values y_i (here l_i) are known for the corresponding m abscissa values x (here E_i), with the x_i equally distributed from 0 to 2π. The solution for all the unknown parameters ds, a_j, and b_j are given by Carslaw (1950) as follows:

$$ds = (1/m) \sum_{i=0}^{m-1} l_i \tag{13.47}$$

$$a_j = (2/m) \sum_{i=0}^{m-1} [l_i \cos(j E_i)] \ ; \quad j \neq (m/2) \tag{13.48}$$

$$b_j = (2/m) \sum_{i=0}^{m-1} [l_i \sin(j E_i)] \ , \tag{13.49}$$

where $m > 2j+1$.

Substitution of Eqs. (13.46) and (13.42) in Eq. (13.47) leads to

$$ds = 0 \ , \tag{13.50}$$

which means that E_i reduces to:

$$E_i = (2\pi/U)(\bar{s}^* + iD) \ . \tag{13.51}$$

The standard deviation s_{EDM} of the observed distances (in fact, l_i) about the adjusted cyclic error curve CE' is computed as:

$$s_{EDM} = \pm (\Sigma v_i^2/f)^{1/2} \ , \tag{13.52}$$

where f = degrees of freedom = $m - 2j - 1$.

The standard deviations of all Fourier coefficients a_j, b_j are equal to

$$s_{a_j} = s_{b_j} = s_{EDM}(2/m)^{1/2} \ . \tag{13.53}$$

The significance of each individual coefficient is investigated by testing its difference from zero statistically. A coefficient is not significantly different from zero and may be ignored for practical purposes if the following unequality is fulfilled

$$(a_j/s_{a_j}) \leq t_{f,\alpha} \ , \quad (b_j/s_{b_j}) \leq t_{f,\alpha} \ , \tag{13.54}$$

where t denotes the test statistic (t-distribution)
       α = level of significance (two-tailed test) and
       f = degree of freedom.

Application of the propagation law of variances to Eq. (13.36) yields the standard deviation of short periodic error correction (cy):

$$s_{cy} = s_{EDM}(2j/m)^{1/2} \tag{13.55}$$

where j = number of wavelengths represented in the short periodic error correction (cy).

As an example, the data of Table 13.11 are evaluated using the analytical method described above. Only the results are given [a fully worked example may be found in Rüeger (1986)]. A first solution for short periodic errors of orders one to four establishes the two components of the first-order short periodic error as the largest terms. (The sine term of the third-order short periodic error featured the next largest amplitude of 0.5 mm.) A second solution for the first-order terms only of the cyclic error leads to

$$CE' \text{ (mm)} = 8.34 \cos (E) + 0.98 \sin (E) \, , \tag{13.56}$$

where $\quad E \text{ (rad)} = \dfrac{D 2\pi}{U}$

and $\quad D \quad = $ measured distance (m)

$\quad\quad\quad U \quad = $ unit length (m)

number of unknown parameters $= 3$

number of measurements $= 10$

standard deviation of input distance $(s_{EDM}) = \pm 0.71$ mm

standard deviation of Fourier coefficients $(s_{a_j}, s_{b_j}) = \pm 0.32$ mm.

The statistical significance of the coefficients is then tested on the basis of Eq. (13.54). The critical t-value amounts to 2.36 in this case. Both coefficients are found to be significantly different from zero.

The short periodic error correction $(cy = -CE')$ may then be written as

$$cy = -8.34 \cos (E) - 0.98 \sin (E) \, , \tag{13.57}$$

where the first and second coefficients are equivalent to B_{11} and C_{11} of Eq. (12.13).

Assuming firstly that the same distance meter was calibrated on an Aarau-type EDM baseline and secondly, that the data of Table 13.9 as well as the instrument correction of Eq. (13.12) refer to the same distance meter, the supplementary additive correction dc may be computed and both dc and cy added to the instrument correction. The instrument correction as computed in Section 13.2.4.3 is repeated here:

$$IC = -5.7 \text{ mm} - 2.18 (D/1000) \, . \tag{13.12}$$

The supplementary correction dc computes easily as the short periodic error at an exact multiple of the unit length. At multiples of the unit length (here 10.0 m) the cosine and sine in Eq. (13.56) become 1.0 and 0.0 exactly. Therefore:

$$dc = +8.3 \text{ mm} \, . \tag{13.58}$$

Adding dc and cy to the IC above leads to the final instrument correction (in mm)

$$IC = +2.6 \text{ mm} - 2.18 (D/1000) - 8.34 \cos (2\pi D/U)$$

$$-0.98 \sin (2\pi D/U) \, , \tag{13.59}$$

where the arguments of the trigonometric functions are taken in radians and the distance D in metres.

13.4 Calibration of Modulation Frequency

The calibration of the master oscillators of EDM instruments is typically carried out in addition to baseline measurements. Frequency calibrations are usually not carried out by the EDM instrument user. Frequency calibration services are typically provided by EDM instrument service organizations, national standard laboratories, registered testing laboratories as well as some universities. All organizations can provide spot measurements of frequency at ambient temperatures in order to determine the ageing of the oscillator's crystal [see parameters A_{10} and A_{14} in Eq. (12.13)] and, if required, the warm-up effect [see parameters A_{15} and A_{16} in Eq. (12.13)]. It should be noted that frequency checks are straightforward as long as the distance meter is equipped with a frequency output socket. As most instruments do not have such sockets as standard features, frequency testing laboratories outside instrument-specific service organizations require information from the instrument manufacturer as far as the preferred mode of frequency measurement is concerned.

The determination of the frequency versus temperature characteristic [see Fig. 12.2 and parameters A_{10} to A_{13} of Eq. (12.13)] requires an additional piece of equipment in the form of a refrigerator/incubator which permits stabilization at any temperature within the EDM instrument's specified temperature range. Local agents of EDM instrument manufacturers usually do not own such devices and are therefore unable to determine the said temperature characteristic. Such facilities are available at EDM instrument factories and at some university departments.

It is important to realize that the parameter A_{10} of the instrument correction (IC) in Eq. (12.13) is caused in part only by the frequency offset of the main oscillator. Any parameter A_{10} determined by frequency measuring techniques should therefore be verified on baselines of known length. Traditionally, an exception has been made for long range distance meters employing HeNe lasers in combination with continuous modulation through electro-optical crystals. As there are few EDM calibration baselines of, say, 30 km length with known distances, the additive constant is determined on shorter baselines and the scale correction (A_{10}) is derived from frequency measurements alone.

13.4.1 Frequency Measuring Techniques

A laboratory for the frequency calibration of distance meters requires a frequency counter for frequencies to usually about 50 MHz and with a display of at least 8 significant digits. (For the frequency calibration of the Kern Mekometer ME 3000, a 500 MHz counter is necessary.) The internal or external time base of the frequency counter should have an accuracy specification of about 1 part in 10^8 between 0 and 50 °C. The frequency counters should be calibrated themselves periodically against the (national) standard of time either by direct comparison with a counter of superior specifications or indirectly with the aid of time transfer techniques based on radio broadcasts, television transmissions or GPS time signals receivers.

The modulation frequency of distance meters may be tapped as follows for connection to the counter:

- cable to test points on printed circuit board
- permanent frequency output socket on EDM instrument
- induction loop either inside or outside instrument
- opto-coupler (EDM instruments with continuous emission)
- opto-coupler with phase locked loop (EDM instruments with chopped emission).

The first option is sometimes available without opening of the instrument (e.g. Topcon GTS-2). In most cases, the relevant test points must be ascertained from the manufacturer and the instrument opened. As the instrument requires sealing of the case on completion of the measurement, the method should normally not be used. Permanent frequency output sockets are available on some instruments, either as a standard feature or as an optional extra. It has been shown experimentally that the attachment of cables can change the modulation frequency by 0.2 ppm. It has also been established that frequency measurements on open and closed instruments may differ significantly. Differences of 0.7 ppm and 0.4 ppm were reported by Sobotta et al. (1980) and Rüeger (1982). It follows that cable connections and opening of instruments should be avoided where possible. Induction loops have been successfully used but may also require opening of the instrument due to the heavy shielding of modern instruments.

Opto-couplers are routinely used by some manufacturers for frequency testing and adjustment of EDM instruments in order to avoid secondary effects of cables and open cases as well as to improve the efficiency of servicing. Opto-couplers consist of a complete EDM receiver section with lens, photodiode and necessary drive and amplification circuits [see Rüeger (1982) for a full description]. As long as a distance meter features an operating mode with an uninterrupted emission of the modulated signal for extended periods of time, the opto-coupler is simply set up in front of the transmitting telescope. After alignment (collimation) of the two telescopes (of EDM instrument and opto-coupler), the opto-coupler supplies the modulation signal to the counter. The method is not applicable in this simple form whenever the distance meter transmits a chopped rather than a continuous signal. Phase locked loops have been built to bridge the missing gaps in the transmitted signal and to supply a continuous signal to the frequency counter (Sobotta et al. 1980). Some problems are experienced with instruments which switch the phase of the modulation signal or turn themselves off after a certain period with no distance measurements.

13.4.2 Calibration of Ageing and Warm-up Effects

As the modulation frequency of a distance meter is dependent on the temperature of its quartz crystal, periodic calibration of the frequency must be carried out under well-defined conditions. The calibration must occur at a specific ambient temperature, after a specific number of minutes of continuous operation as well as after (and during) operation in specific measuring mode (e.g. tracking mode) as all three parameters affect the oscillator's temperature. Failure to do so will result in the ageing results being biased by secondary effects. To establish the ageing rate [or parameter A_{14} in Eq. (12.13)], the frequency offset needs to be established repeatedly, for example at yearly or half-yearly intervals. If the frequency offset is the sole contributor to the scale parameter in the instrument correction IC of Eq. (12.13), it may be used to define A_{10} directly.

The effect of the actual (or measured) frequency (f_{act}) differing from the nominal modulation frequency (f_{nom}) can be compensated by the following corrections

$$S_{corr} = \frac{f_{nom}}{f_{act}} \, s_{meas} \tag{13.60}$$

or

$$S_{corr} = s_{meas} + \left(\frac{f_{nom} - f_{act}}{f_{act}} \right) s_{meas} \, , \tag{13.61}$$

where s_{corr} is the distance corrected for scale errors. It is evident that the scale parameter A_{10} of the instrument correction becomes then

$$A_{10} = \frac{f_{nom} - f_{act}}{f_{act}} \tag{13.62}$$

and is usually given in parts per million.

Example:

f_{act} = 14 985 412.0 Hz (HP 3800 S/N 1141A00110 on 6th October 1976, at 20 °C, after 60 min continuous operation)

f_{nom} = 14 985 453.0 Hz

Using Eqs. (13.60) and (13.61) yields

S_{corr} = 1.000 002 736 s_{meas}

and

$S_{corr} = s_{meas} + 2.736 \times 10^{-6} s_{meas}$

A_{10} = +2.7 ppm .

The warm-up effect on frequency may be determined by acclimatisation of the switched-off distance meter (e.g. over night, or for 24 h) at a desired temperature and by continuous monitoring of the modulation frequency for 1 to 2 h after switch-on. Again, the mode of the distance measurements carried out in parallel will affect the warm-up characteristic. It is advisable to determine the effect with the distance meter in the aim mode as well as in the tracking mode, for example. The coefficients A_{15} and A_{16} of the instrument correction in Eq. (12.13) may then be resolved by least-squares estimation. In most practical cases the warm-up effect is minimized by either always measuring with a "cold" (immediately after switch on) or a "warm" (fully warmed-up) distance meter in connection with frequency calibrations (A_{10} and A_{14}) under similar conditions.

13.4.3 Frequency Versus Temperature Characteristic

The frequency versus temperature characteristic (parameters A_{10} to A_{13}) in Eq. (12.13) can be established if a refrigerator/incubator unit is available. Again it has to be decided if the calibration of the curve is to be carried out with a cold or a warm instrument. The latter is often easier as the refrigerator/incubator unit needs not to be opened to switch the distance meter on and off. Sufficient time has to be given to the distance meter to settle (acclimatise) at a new ambient temperature after a temperature change is induced in the refrigerator/incubator.

Typically, 2 days are required to establish the characteristic between, say, $-20\,^\circ$C and $+50\,^\circ$C. The intervals between the setting of new temperatures depends on the response time of the distance meter as well as the response time of the refrigerator/incubator, typical values being 1.5 to 2 h for $10\,^\circ$C changes. Results of such tests are depicted in Fig. 12.2. The parameters A_{10} to A_{13} of the instrument correction may be determined by least-squares estimation. More details on such tests may be found in Rüeger (1978, 1982) and Rüeger and Pascoe (1989), for example.

13.5 Accuracy Specifications of EDM Instruments

Most manufacturers state the accuracy of their instruments in the following form

$$s = \pm(A + Bd) \ , \tag{13.63}$$

where A in mm, B in ppm, d = distance (in km).

The variance follows as

$$s^2 = (A + Bd)^2 \ . \tag{13.64}$$

Most manufacturers consider the Eq. (13.63) as a standard deviation at 66% confidence level, but use it as a tolerance in their predelivery testing procedures.

The term A includes:

1. phase resolution of EDM instrument
2. maximum amplitude of short periodic error (or average effect)
3. maximum effect of non-linear distance-dependent errors (or average effect)
4. accuracy of a built-in (preset) additive constant
5. compatibility of reflectors.

Considering short range EDM the term B is either:

1. the range of the typical frequency drift of the main oscillator within the specified temperature range (e.g. $-20\,^\circ$C $\rightarrow$ $+40\,^\circ$C) or
2. the maximum error which may be caused by the limited step interval of the "ppm dial".

Please note that there is a very careless mixing of random and quasi-random errors (e.g. short periodic errors) with purely systematic errors (frequency drift with temperature). Because of this, the Eqs. (13.63) and (13.64) should not be used as a priori variances in least-squares adjustments. For example, under homogeneous temperature conditions in the field, the distance proportional term is likely to be much smaller [see Eq. (13.1) instead].

As mentioned above, most manufacturers tend to specify accuracy rather than precision. Some manufacturers, however, specify precision rather than accuracy. In these cases, the oscillator specifications and linear as well as non-linear terms of the instrument correction are not taken into account.

The meaning of the accuracy specification in Eq. (13.63) may be further illustrated by the following test. In the case of the manual of a distance meter not prescribing the calibration of additive constant and scale correction by the user, the accuracy specification could be verified by the instrument user as follows: Firstly, 14 known lengths are selected which cover the total specified distance range of the distance meter. Secondly, each of these distances is measured once only across the entire specified temperature range of the distance meter, e.g. at $-20\,°C$, $-10\,°C$, $0\,°C$, $+10\,°C$, $+20\,°C$, $+30\,°C$, $+40\,°C$. This leads to a total of 98 observations. In a plot of the errors (measured distance minus true distance) versus distance, 68 errors should fall within the error band given by Eq. (13.63) and 95 errors should fall within the doubled error band of $\pm(2\,A+2\,B\,d)$. Naturally, the instrument-specific reflector would have to be used in such a test and the user's instructions would have to be followed closely. Such a test would clearly establish if an instrument fulfils the manufacturer's accuracy specification or not. The test is unfortunately of little practical use as it is restricted to localities with a yearly temperature cycle between $-20\,°C$ and $+40\,°C$ and by the fact that it would take one full year to verify the accuracy specification.

In future, manufacturers may adopt the German Standard DIN 18723, Part 6, on Accuracy Tests of Geodetic Instruments: Electro-Optical Short Range Distance Meters (DIN 1985) or a corresponding ISO standard. In such a case, precision rather than accuracy would be specified, as the German draft standard derives the one-term DIN standard deviation according to Eq. (13.16) from measurements in all combinations on a seven-station EDM calibration baseline of the Heerbrugg design and of unknown length. The DIN standard deviations are unaffected by scale errors, as the latter map into the unknown baseline distances. The additive constant is determined in the same process. The author hopes that the manufacturers of EDM instruments will specify the oscillator specifications separately after adoption of the DIN (or similar) precision specifications in order to convey some information on the accuracy of instruments.

Appendix A. First Velocity Correction
for Precise Electro-Optical Distance Measurement

In most cases of electro-optical EDM, the first velocity correction according to Eqs. (6.11) to (6.13) is sufficiently accurate. The said equations are based on the refractive index equations given by Eqs. (5.12) and (5.15).

For the reduction of high precision measurements, it is suggested to employ the first velocity correction given by Eq. (6.5)

$$d = \left(\frac{n_{REF}}{n_L} \right) d'$$
(6.5)

and to use the equations given below for the computation of the group refractive index for light and infrared rays (n_L).

The phase refractive index (n) is given by Peck and Reeder (1972) in a two-term Sellmeier form as

$$(n-1)\times 10^8 = \frac{5\,791\,817}{238.0185 - \sigma^2} + \frac{167\,909}{57.362 - \sigma^2} \, ,$$
(A.1)

where $\quad \sigma$ = vacuum wave number

$$= \frac{1}{\lambda}$$

λ = carrier wavelength in vacuum (in micrometres).

This formula is satisfactory for the representation of all experimental data from the farthest infrared IR down to the ultraviolet (UV) (or from 1960 nm to 230 nm). The root mean square deviation of the experimental data about Eq. (A.1) amounts to ± 0.0018 ppm between 230 and 690 nm and to ± 0.0017 ppm between 720 and 1690 nm (Peck and Reeder 1972). Equation (A.1) fits the data in the IR region much better than the Edlen (1966) formula.

Equation (A.1) applies to standard dry air, at 15 °C, with 0.033% CO_2 content. In order to derive the group refractive index at ambient conditions, the effect of the CO_2 content is removed by division by 1.000 162 (Owens 1967):

$$(n-1)\times 10^8 = \frac{5\,790\,878.9}{238.0185 - \sigma^2} + \frac{167\,881.8}{57.362 - \sigma^2} \, .$$
(A.2)

Following Owens' (1967) example further, the phase refractive index at ambient conditions becomes

$$(n-1)\times 10^8 = \left(\frac{1\,646\,386.0}{238.0185 - \sigma^2} + \frac{47\,729.9}{57.362 - \sigma^2} \right) D_S$$

$$+ (6487.31 + 58.058\,\sigma^2 - 0.71150\,\sigma^4 + 0.08851\,\sigma^6)\,D_w \, ,$$
(A.3)

where D_S = density factor of dry air
D_w = density factor of water vapour

and (Owens 1967)

$$D_S = \frac{P_S}{T} \left[1 + P_S \left(57.90 \times 10^{-8} - \frac{9.3250 \times 10^{-4}}{T} + \frac{0.25844}{T^2} \right) \right] \tag{A.4}$$

$$D_w = \frac{P_w}{T} \left\{ 1 + P_w [1 + (3.7 \times 10^{-4}) P_w] \right.$$

$$\left. \times \left[-2.37321 \times 10^{-3} + \frac{2.23366}{T} - \frac{710.792}{T^2} + \frac{7.75141 \times 10^4}{T^3} \right] \right\} , \tag{A.5}$$

where P = total atmospheric pressure
$P_S = P - P_w$
P_w = partial water vapour pressure
P_S = partial pressure of dry air containing 0.03% CO_2 (standard air)
T = absolute temperature (K) = 273.15 + t
t = temperature (°C).

The coefficient of D_w in Eq. (A.3) has been taken from Owens (1967).

The group refractive index n_L of light and infrared rays has been defined previously as [see Eq. (5.7)]

$$n_L = n + \sigma \left(\frac{dn}{d\sigma} \right) . \tag{A.6}$$

Considering that for

$$n = \frac{B}{b - \sigma^2} + \frac{C}{c - \sigma^2} \tag{A.7}$$

the second term in Eq. (A.6) becomes

$$\sigma \left(\frac{dn}{d\sigma} \right) = \frac{2B\sigma^2}{(b - \sigma^2)^2} + \frac{2C\sigma^2}{(c - \sigma^2)^2} \tag{A.8}$$

and

$$n + \sigma \left(\frac{dn}{d\sigma} \right) = B \left(\frac{b + \sigma^2}{(b - \sigma^2)^2} \right) + C \left(\frac{c + \sigma^2}{(c - \sigma^2)^2} \right) , \tag{A.9}$$

it can be shown that application of Eq. (A.6) to Eq. (A.3) yields the group refractive index of light and infrared rays (n_L)

$$(n_L - 1) \times 10^8 = \left[1\,646\,386.0 \left(\frac{238.0185 + \sigma^2}{(238.0185 - \sigma^2)^2} \right) + 47\,729.9 \left(\frac{57.362 + \sigma^2}{(57.362 - \sigma^2)^2} \right) \right] D_S$$

$$+ [6487.31 + 174.174\,\sigma^2 - 3.55750\,\sigma^4 + 0.61957\,\sigma^6] D_w , \tag{A.10}$$

where D_S and D_w are given by Eqs. (A.4) and (A.5).

The first term (dry air) of Eq. (A.10) is valid between 230 nm and 1690 nm (Peck and Reeder 1972), 240 K and 330 K ($-33\,°C$ to $+57\,°C$) and $0-4000$ mb (Owens 1967). The second term of Eq. (A.10) (water vapour) is valid between 361 and 644 nm, 250 K and 320 K ($-23\,°C$ and $+47\,°C$) and for water vapour pressures between 0 and 100 mb or $0-100\%$ relative humidity (Owens 1967). The above (simplified) equation provides n_L with an accuracy of two parts in 10^8, assuming the validity of the Lorenz-Lorentz equation (Owens 1967).

Owens (1967) published a general formula which is accurate to one part in 10^9 and includes a carbon dioxide term. This equation is, however, based on the Edlen (1966) refractive index equation which models the infrared region less satisfactorily than the simpler Peck and Reeder equation used in this context.

Although it is suggested that Eq. (A.10) be used, Owens (1967) "simplified equation" is listed for comparison:

$$(n_L - 1)\times 10^8 = \left[2371.34 + 683\,939.7\,\frac{(130 + \sigma^2)}{(130 - \sigma^2)^2} + 4547.3\,\frac{(38.9 + \sigma^2)}{(38.9 - \sigma^2)^2} \right] D_S$$

$$+ [6487.31 + 174.174\,\sigma^2 - 3.55750\,\sigma^4 + 0.61957\,\sigma^6]\,D_w \ . \quad (A.11)$$

The first bracket term is based on Edlen's (1966) formula. Equations (A.10) and (A.11) differ by less than 0.0015 ppm between 633 nm and 900 nm. Equation (A.10) is preferred as it contains less terms and provides a better fit of experimental data in the infrared region.

Appendix B. Saturation Water Vapour Pressure at Different Temperatures

(After List 1958)

Table 1. Saturation vapour pressure over water

Temperature (°C) Unit	0.0 mb	0.1 mb	0.2 mb	0.3 mb	0.4 mb	0.5 mb	0.6 mb	0.7 mb	0.8 mb	0.9 mb
−10	2.8627	2.8402	2.8178	2.7956	2.7735	2.7516	2.7298	2.7082	2.6868	2.6655
−9	3.0971	3.0729	3.0489	3.0250	3.0013	2.9778	2.9544	2.9313	2.9082	2.8854
−8	3.3484	3.3225	3.2967	3.2711	3.2457	3.2205	3.1955	3.1706	3.1459	3.1214
−7	3.6177	3.5899	3.5623	3.5349	3.5077	3.4807	3.4539	3.4272	3.4008	3.3745
−6	3.9061	3.8764	3.8468	3.8175	3.7883	3.7594	3.7307	3.7021	3.6738	3.6456
−5	4.2148	4.1830	4.1514	4.1200	4.0888	4.0579	4.0271	3.9966	3.9662	3.9361
−4	4.5451	4.5111	4.4773	4.4437	4.4103	4.3772	4.3443	4.3116	4.2791	4.2468
−3	4.8981	4.8617	4.8256	4.7897	4.7541	4.7187	4.6835	4.6486	4.6138	4.5794
−2	5.2753	5.2364	5.1979	5.1595	5.1214	5.0836	5.0460	5.0087	4.9716	4.9347
−1	5.6780	5.6365	5.5953	5.5544	5.5138	5.4734	5.4333	5.3934	5.3538	5.3144
0	6.1078	6.1523	6.1971	6.2422	6.2876	6.3333	6.3793	6.4256	6.4721	6.5190
1	6.5662	6.6137	6.6614	6.7095	6.7579	6.8066	6.8556	6.9049	6.9545	7.0044
2	7.0547	7.1053	7.1562	7.2074	7.2590	7.3109	7.3631	7.4157	7.4685	7.5218
3	7.5753	7.6291	7.6833	7.7379	7.7928	7.8480	7.9036	7.9595	8.0158	8.0724
4	8.1294	8.1868	8.2445	8.3026	8.3610	8.4198	8.4789	8.5384	8.5983	8.6586
5	8.7192	8.7802	8.8416	8.9033	8.9655	9.0280	9.0909	9.1542	9.2179	9.2820
6	9.3465	9.4114	9.4766	9.5423	9.6083	9.6748	9.7416	9.8089	9.8765	9.9449
7	10.013	10.082	10.151	10.221	10.291	10.362	10.433	10.505	10.577	10.649
8	10.722	10.795	10.869	10.943	11.017	11.092	11.168	11.243	11.320	11.397
9	11.474	11.552	11.630	11.708	11.787	11.867	11.947	12.027	12.108	12.190

Table 1 (continued)

Temperature (°C) Unit	0.0 mb	0.1 mb	0.2 mb	0.3 mb	0.4 mb	0.5 mb	0.6 mb	0.7 mb	0.8 mb	0.9 mb
10	12.272	12.355	12.438	12.521	12.606	12.690	12.775	12.860	12.946	13.032
11	13.119	13.207	13.295	13.383	13.472	13.562	13.652	13.742	13.833	13.925
12	14.017	14.110	14.203	14.297	14.391	14.486	14.581	14.678	14.774	14.871
13	14.969	15.067	15.166	15.266	15.365	15.466	15.567	15.669	15.771	15.874
14	15.977	16.081	16.186	16.291	16.397	16.503	16.610	16.718	16.826	16.935
15	17.044	17.154	17.264	17.376	17.487	17.600	17.713	17.827	17.942	18.057
16	18.173	18.290	18.407	18.524	18.643	18.762	18.882	19.002	19.123	19.245
17	19.367	19.490	19.614	19.739	19.864	19.990	20.117	20.244	20.372	20.501
18	20.630	20.760	20.891	21.023	21.155	21.288	21.422	21.556	21.691	21.827
19	21.964	22.101	22.240	22.379	22.518	22.659	22.800	22.942	23.085	23.229
20	23.373	23.518	23.664	23.811	23.959	24.107	24.256	24.406	24.557	24.709
21	24.861	25.014	25.168	25.323	25.479	25.635	25.792	25.950	26.109	26.269
22	26.430	26.592	26.754	26.918	27.082	27.247	27.413	27.580	27.748	27.916
23	28.086	28.256	28.428	28.600	28.773	28.947	29.122	29.298	29.475	29.652
24	29.831	30.011	30.191	30.373	30.555	30.739	30.923	31.109	31.295	31.483
25	31.671	31.860	32.050	32.242	32.434	32.627	32.821	33.016	33.212	33.410
26	33.608	33.807	34.008	34.209	34.411	34.615	34.820	35.025	35.232	35.440
27	35.649	35.859	36.070	36.282	36.495	36.709	36.924	37.140	37.358	37.576
28	37.796	38.017	38.239	38.462	38.686	38.911	39.137	39.365	39.594	39.824
29	40.055	40.287	40.521	40.755	40.991	41.228	41.466	41.705	41.945	42.187
30	42.430	42.674	42.919	43.166	43.414	43.663	43.913	44.165	44.418	44.672
31	44.927	45.184	45.442	45.701	46.961	46.223	46.486	46.750	47.016	47.283
32	47.551	47.820	48.091	48.364	48.637	48.912	49.188	49.466	49.745	50.025
33	50.307	50.590	50.874	51.160	51.447	51.736	52.026	52.317	52.610	52.904
34	53.200	53.497	53.796	54.096	54.397	54.700	55.004	55.310	55.617	55.926
35	56.236	56.548	56.861	57.176	57.492	57.810	58.129	58.450	58.773	59.097
36	59.422	59.749	60.077	60.407	60.739	61.072	61.407	61.743	62.081	62.421
37	62.762	63.105	63.450	63.796	64.144	64.493	64.844	65.196	65.550	65.906
38	66.264	66.623	66.985	67.347	67.712	68.078	68.446	68.815	69.186	69.559
39	69.934	70.310	70.688	71.068	71.450	71.833	72.218	72.605	72.994	73.385

40	73.777	74.171	74.568	74.966	75.365	76.767	76.170	76.575	76.982	77.391
41	77.802	78.215	78.630	79.046	79.465	79.885	80.307	80.731	81.157	81.585
42	82.015	82.447	82.881	83.316	83.754	84.194	84.636	85.079	85.525	85.973
43	86.423	86.875	87.329	87.785	88.243	88.703	89.165	89.629	90.095	90.564
44	91.034	91.507	91.981	92.458	92.937	93.418	93.901	94.386	94.874	95.363
45	95.855	96.349	96.845	97.343	97.844	98.347	98.852	99.359	99.869	100.38
46	100.89	101.41	101.93	102.45	102.97	103.50	104.03	104.56	105.09	105.62
47	106.16	106.70	107.24	107.78	108.33	108.88	109.43	109.98	110.54	111.10
48	111.66	112.22	112.79	113.36	113.93	114.50	115.07	115.65	116.23	116.18
49	117.40	117.99	118.58	119.17	119.77	120.37	120.97	121.57	122.18	122.79
50	123.40	124.01	124.63	125.25	125.87	126.49	127.12	127.75	128.38	129.01
51	129.65	130.29	130.93	131.58	132.23	132.88	133.53	134.19	134.84	135.51
52	136.17	136.84	137.51	138.18	138.86	139.54	140.22	140.91	141.60	142.29
53	142.98	143.68	144.38	145.08	145.78	146.49	147.20	147.91	148.63	149.35
54	150.07	150.80	151.53	152.26	152.99	153.73	154.47	155.21	155.96	156.71
55	157.46	158.22	158.97	159.74	160.50	161.27	162.04	162.82	163.59	164.38
56	165.16	165.95	166.74	167.53	168.33	169.13	169.93	170.74	171.55	172.36
57	173.18	174.00	174.82	175.65	176.48	177.31	178.15	178.99	179.83	180.68
58	181.53	182.38	183.24	184.10	184.96	185.83	186.70	187.58	188.45	189.34
59	190.22	191.11	192.00	192.89	193.79	194.69	195.60	196.51	197.42	198.34
60	199.26	200.18	201.11	202.05	202.98	203.92	204.86	205.81	206.76	207.71

Table 2. Saturation vapour pressure over ice

Temperature (°C) Unit	0.0 mb	0.1 mb	0.2 mb	0.3 mb	0.4 mb	0.5 mb	0.6 mb	0.7 mb	0.8 mb	0.9 mb
−50	0.03935	0.03887	0.03839	0.03792	0.03745	0.03699	0.03653	0.03608	0.03564	0.03520
−49	0.04449	0.04395	0.04341	0.04289	0.04236	0.04185	0.04134	0.04083	0.04033	0.03984
−48	0.05026	0.04965	0.04905	0.04846	0.04788	0.04730	0.04673	0.04616	0.04560	0.04504
−47	0.05671	0.05603	0.05536	0.05470	0.05405	0.05340	0.05276	0.05212	0.05150	0.05087
−46	0.06393	0.06317	0.06242	0.06168	0.06095	0.06022	0.05950	0.05879	0.05809	0.05740
−45	0.07198	0.07113	0.07030	0.06947	0.06865	0.06784	0.06704	0.06625	0.06547	0.06469
−44	0.08097	0.08003	0.07909	0.07817	0.07725	0.07635	0.07546	0.07457	0.07370	0.07283
−43	0.09098	0.08993	0.08889	0.08786	0.08684	0.08584	0.08484	0.08386	0.08289	0.08192
−42	0.1021	0.1010	0.09981	0.09866	0.09753	0.09641	0.09530	0.09420	0.09312	0.09204
−41	0.1145	0.1132	0.1119	0.1107	0.1094	0.1082	0.1070	0.1057	0.1045	0.1033
−40	0.1283	0.1268	0.1254	0.1240	0.1226	0.1212	0.1198	0.1185	0.1171	0.1158
−39	0.1436	0.1420	0.1404	0.1389	0.1373	0.1358	0.1343	0.1328	0.1313	0.1298
−38	0.1606	0.1588	0.1571	0.1553	0.1536	0.1519	0.1502	0.1485	0.1469	0.1452
−37	0.1794	0.1774	0.1755	0.1736	0.1717	0.1698	0.1679	0.1661	0.1642	0.1624
−36	0.2002	0.1980	0.1959	0.1938	0.1917	0.1896	0.1875	0.1855	0.1834	0.1814
−35	0.2233	0.2209	0.2185	0.2161	0.2138	0.2115	0.2092	0.2069	0.2047	0.2024
−34	0.2488	0.2461	0.2435	0.2409	0.2383	0.2357	0.2332	0.2307	0.2282	0.2257
−33	0.2769	0.2740	0.2711	0.2682	0.2653	0.2625	0.2597	0.2569	0.2542	0.2515
−32	0.3079	0.3047	0.3014	0.2983	0.2951	0.2920	0.2889	0.2859	0.2828	0.2799
−31	0.3421	0.3385	0.3350	0.3315	0.3280	0.3246	0.3212	0.3178	0.3145	0.3112
−30	0.3798	0.3759	0.3720	0.3681	0.3643	0.3605	0.3567	0.3530	0.3494	0.3457
−29	0.4213	0.4170	0.4127	0.4084	0.4042	0.4000	0.3959	0.3918	0.3877	0.3838
−28	0.4669	0.4621	0.4574	0.4527	0.4481	0.4435	0.4390	0.4345	0.4300	0.4256
−27	0.5170	0.5118	0.5066	0.5014	0.4964	0.4913	0.4863	0.4814	0.4765	0.4717
−26	0.5720	0.5663	0.5606	0.5549	0.5493	0.5438	0.5383	0.5329	0.5276	0.5222
−25	0.6323	0.6260	0.6198	0.6136	0.6075	0.6015	0.5955	0.5895	0.5836	0.5778
−24	0.6985	0.6916	0.6848	0.6780	0.6713	0.6646	0.6580	0.6515	0.6450	0.6386
−23	0.7709	0.7634	0.7559	0.7485	0.7412	0.7339	0.7267	0.7195	0.7125	0.7055
−22	0.8502	0.8419	0.8338	0.8257	0.8176	0.8097	0.8018	0.7940	0.7862	0.7785
−21	0.9370	0.9280	0.9190	0.9101	0.9013	0.8926	0.8840	0.8754	0.8669	0.8585

−20	1.032	1.022	1.012	1.002	0.9928	0.9833	0.9739	0.9645	0.9553	0.9461
−19	1.135	1.124	1.114	1.103	1.092	1.082	1.072	1.062	1.052	1.042
−18	1.248	1.236	1.225	1.213	1.201	1.190	1.179	1.168	1.157	1.146
−17	1.371	1.358	1.345	1.333	1.320	1.308	1.296	1.284	1.272	1.260
−16	1.506	1.492	1.478	1.464	1.451	1.437	1.424	1.410	1.397	1.384
−15	1.652	1.637	1.622	1.607	1.592	1.577	1.562	1.548	1.534	1.520
−14	1.811	1.795	1.778	1.762	1.746	1.730	1.714	1.698	1.683	1.667
−13	1.984	1.966	1.948	1.930	1.913	1.895	1.878	1.861	1.844	1.827
−12	2.172	2.153	2.133	2.114	2.095	2.076	2.057	2.039	2.020	2.002
−11	2.376	2.355	2.334	2.313	2.292	2.271	2.251	2.231	2.211	2.191
−10	2.597	2.574	2.551	2.529	2.506	2.484	2.462	2.440	2.419	2.397
−9	2.837	2.812	2.787	2.763	2.739	2.715	2.691	2.667	2.644	2.620
−8	3.097	3.070	3.043	3.017	2.991	2.965	2.939	2.913	2.888	2.862
−7	3.379	3.350	3.321	3.292	3.264	3.236	3.208	3.180	3.152	3.124
−6	3.685	3.653	3.622	3.591	3.560	3.529	3.499	3.468	3.438	3.409
−5	4.015	3.981	3.947	3.913	3.879	3.846	3.813	3.781	3.748	3.717
−4	4.372	4.335	4.298	4.262	4.226	4.190	4.154	4.119	4.084	4.049
−3	4.757	4.717	4.678	4.638	4.600	4.561	4.523	4.485	4.447	4.409
−2	5.173	5.130	5.087	5.045	5.003	4.961	4.920	4.878	4.838	4.797
−1	5.623	5.577	5.530	5.485	5.439	5.394	5.349	5.305	5.260	5.217

Appendix C. Parameters of the ICAO (International Civil Aviation Organization) Standard Atmosphere

Z = geometric altitude, T, t = atmospheric temperature, P = atmospheric pressure (ICAO 1964)

Z, m	T, K	t, °C	P, mb	P, mm Hg
− 500	291.400	18.250	1074.78	806.151
− 450	291.075	17.925	1068.49	801.436
− 400	290.750	17.600	1062.24	796.743
− 350	290.425	17.275	1056.01	792.073
− 300	290.100	16.950	1049.81	787.425
− 250	289.775	16.625	1043.65	782.799
− 200	289.450	16.300	1037.51	778.195
− 150	289.125	15.975	1031.40	773.614
− 100	288.800	15.650	1025.32	769.054
− 50	288.475	15.325	1019.27	765.516
0	288.150	15.000	1013.25	760.000
50	287.825	14.675	1007.26	755.505
100	287.500	14.350	1001.29	751.032
150	287.175	14.025	995.360	746.581
200	286.850	13.700	989.454	742.151
250	286.525	13.375	983.576	737.743
300	286.200	13.050	977.727	733.356
350	285.875	12.725	971.906	728.990
400	285.550	12.400	966.114	724.645
450	285.225	12.075	960.349	720.321
500	284.900	11.750	954.612	716.018
550	284.575	11.425	948.904	711.736
600	284.250	11.100	943.223	707.475
650	283.925	10.775	937.570	703.235
700	283.600	10.450	931.944	699.015
750	283.276	10.126	926.346	694.816
800	282.951	9.801	920.775	690.638
850	282.626	9.476	915.231	686.480
900	282.301	9.151	909.714	682.342
950	281.976	8.826	904.225	678.225
1000	281.651	8.501	898.762	674.127
1050	281.326	8.176	893.327	670.050
1100	281.001	7.851	887.918	665.993
1150	280.676	7.526	882.535	661.956
1200	280.351	7.201	877.180	657.939
1250	280.027	6.877	871.850	653.941
1300	279.702	6.552	866.547	649.964
1350	279.377	6.227	861.270	646.006
1400	279.052	5.092	856.020	642.068
1450	278.727	5.577	850.795	638.149

Z, m	T, K	t, °C	P, mb	P, mm Hg
1500	278.402	5.252	845.596	634.249
1550	278.077	4.927	840.423	630.369
1600	277.753	4.603	835.276	626.509
1650	277.428	4.278	830.155	622.667
1700	277.103	3.953	825.059	618.845
1750	276.778	3.628	819.988	615.042
1800	276.453	3.303	814.943	611.258
1850	276.128	2.978	809.923	607.492
1900	275.804	2.654	804.928	603.746
1950	275.479	2.329	799.958	600.018
2000	275.154	2.004	795.014	596.309
2050	274.829	1.679	790.094	592.619
2100	274.505	1.355	785.199	588.947
2150	274.180	1.030	780.328	585.294
2200	273.855	0.705	775.482	581.659
2250	273.530	0.380	770.661	578.043
2300	273.205	0.055	765.863	574.445
2350	272.881	-0.269	761.091	570.865
2400	272.556	-0.594	756.342	567.303
2450	272.231	-0.919	751.618	563.760
2500	271.906	-1.244	746.917	560.234
2550	271.582	-1.568	742.240	556.726
2600	271.257	-1.893	737.588	553.236
2650	270.932	-2.218	732.958	549.764
2700	270.607	-2.543	728.353	546.310
2750	270.283	-2.867	723.771	542.873
2800	269.958	-3.192	719.213	539.454
2850	269.633	-3.517	714.677	536.052
2900	269.309	-3.841	710.166	532.668
2950	268.984	-4.166	705.677	529.301

Appendix D. Data of a Selection of Electro-Optical Distance Meters as Required for the Derivation of the First Velocity Correction and for Calibration Purposes

All data have been supplied by the manufacturers or their Australian agents. The main modulation frequency and the unit length may be used to compute the reference refractive index according to Eq. (6.3). The carrier wavelength is required for the computation of the group refractive index at standard conditions according to Eq. (5.12). The parameters C and D refer to Eq. (6.11) repeated below and may be computed from Eqs. (6.12) and (6.13). It should be noted that slightly different results may be obtained when doing so as manufacturers may base their computations (see C and D in tables) on Edlen's rather than Barrell and Sears' equation.

For high precision distance meters, the refractive index should be calculated according to Appendix A.

The parameters C and D may be substituted in the following first velocity equation:

$$K' = \left(C - \frac{Dp}{(273.15+t)} + \frac{11.27\,e}{(273.15+t)} \right) 10^{-6} d' \ , \tag{6.11}$$

where p in mb, t in °C, e in mb.

Note: This first velocity correction applies only if the distance meter is set to zero ppm (or equivalent) during distance measurement.

Manu-facturer	Model	Carrier wave length (nm)	Main modulation frequency (Hz)	Unit length (m)	Reference refractive index	Terms of first velocity correction		Remarks
						C	D	
AGA	(see Geodimeter)							
Fennel	(see IBEO)							
Führer	(see IBEO)							
Geo-dimeter Sweden	Geodimeter 6A	550	29970000	5	1.0003092	309.2	82.15	
	Geodimeter 6BL	632.8	29970000	5	1.0003086	308.6	81.00	
	Geodimeter 8	632.8	29970000	5	1.0003086	308.6	81.00	
	Geodimeter 710	632.8	29970000	5	1.0003086	308.6	81.00	
	Geodimeter 10	910	14985530	10	1.000275	275	79.6	
	Geodimeter 12A	910	14985530	10	1.000275	275	79.6	
	Geodimeter 14	910	14985530	10	1.000275	275	79.6	
	Geodimeter 120	910	14985530	10	1.000275	275	79.6	
	Geodimeter 14A	910	14985530	10	1.000275	275	79.55	
	Geodimeter 110/110A	910	14985530	10	1.000275	275	79.55	
	Geodimeter 112	910	14985530	10	1.000275	275	79.55	
	Geodimeter 116	910	14985530	10	1.000275	275	79.55	
	Geodimeter 122	910	14985530	10	1.000275	275	79.55	
	Geodimeter 136/140	910	14985530	10	1.000275	275	79.55	
	Geodimeter 142	910(880)	14985543	10	1.000275	275	79.55	
	Geodimeter 16	910	14985530	10	1.000275	275	79.55	
	Geodimeter 114	880	14984651	10	1.000275	275	79.55	
	Geodimeter 210	910	14984629	10	1.000275	275	79.55	
	Geodimeter 216	910	14984629	10	1.000275	275	79.55	
	Geodimeter 220	910	14984629	10	1.000275	275	79.55	
	Geodimeter 420	910	14984629	10	1.000275	275	79.55	i The carrier wavelength can vary between 850 nm and 910 nm due to the use of different types of diodes from various sources
	Geodimeter 440	910	14984629	10	1.000275	275	79.55	
	Geodimeter 6000	880	14984651	10	1.000275	275	79.55	
	Geodimeter 510	(i)	(j)	10	1.000275	275	79.55	
	Geodimeter 520	(i)	(j)	10	1.000275	275	79.55	
	Geodimeter 540	(i)	(j)	10	1.000275	275	79.55	j The manufacturer selects the modulation
	Geodimeter 4400	(i)	(j)	10	1.000275	275	79.55	
	Geodimeter 408	(i)	(j)	10	1.000275	275	79.55	

Manufacturer	Model	Carrier wave length (nm)	Main modulation frequency (Hz)	Unit length (m)	Reference refractive index	Terms of first velocity correction		Remarks
						C	D	
Geo-dimeter Sweden	Geodimeter 412	(i)	(j)	10	1.000275	275	79.55	frequency (~15 MHz) ac-
	Geodimeter 422LR	(i)	(j)	10	1.000275	275	79.55	cording to the actual carrier
	Geodimeter 424	(i)	(j)	10	1.000275	275	79.55	wavelength (as well as the
	Geodimeter 444	(i)	(j)	10	1.000275	275	79.55	10 m unit length) so that the
	Geodimeter 460	(i)	(j)	10	1.000275	275	79.55	same C and D terms can be
	Geodimeter 610	(i)	(j)	10	1.000275	275	79.55	used for all instruments
	Geodimeter 620	(i)	(j)	10	1.000275	275	79.55	
	Geodimeter 640	(i)	(j)	10	1.000275	275	79.55	
Geotronics (see Geodimeter)								
Hewlett-Packard USA	HP 3800B	910	14985453	10	1.0002783	278.3	79.15	
	HP 3805A	910	14987103[b]	10	1.0002783	278.3	79.15	[b] Frequency set to +110ppm
	HP 3808A	835	14987090[b]	10	1.0002792	279.2	79.46	
	HP 3820A	835	14987090[b]	10	1.0002792	279.2	79.46	
	HP 3801A	910	14987103[b]	10	1.0002783	278.3	79.15	
	HP 3859A	840	14987090[b]	10	1.0002783	278.3	79.15	
	HP 3810B	840	14987090[b]	10	1.0002783	278.3	79.15	
IBEO Germany	FEN 2000	905	(14983482)	–	1.00027345	273.45	79.08	Pulse distance meter
	FEN 4000	905	(14983482)	–	1.00027345	273.45	79.08	Pulse distance meter
	FEN 10000	905	(14983482)	–	1.00027345	273.45	79.08	Pulse distance meter
	Pulsar Survey	905						Pulse distance meter
	Pulsar 50/100	905						Pulse distance meter
	Pulsar 500/1000	905						Pulse distance meter
	Pulsar Minifix	905						Pulse distance meter
Kern Switzer-land	DM 500	875	14985400	10	1.0002816	281.6	79.26	
	DM 501	900	14985400	10	1.0002819	281.9	79.19	
	DM 502	860	14985400	10	1.000282	282	79.2	
	DM 102	860	14985400	10	1.000282	282	79.2	
	DM 503	860	14985400	10	1.000282	282	79.2	

Manufacturer	Instrument							Notes
Kern	DM 104	860	14985400	10	1.0002822	282.2	79.4	
Switzer-land	DM 150	860	14985400	10	1.0002822	282.2	79.4	
	DM 504	860	14985400	10	1.0002822	282.2	79.4	
	DM 550	860	14985400	10	1.000282	282.2	79.4	
	Mekometer ME5000	632.6	460 − 510MHz	~0.30	1.000284515	284.515	~80.91	See Appendix A
Keuffel	Ranger V	632.8	14984980	10	1.000310	310	81.01	
and Esser	Rangemaster II	632.8	14984980	10	1.000310	310	81.01	
U.S.A.	Autoranger I	910	14983482[c]	10	1.0003104	310.4	79.19	[c] Frequency set to − 100 ppm
	Autoranger S	865	14983482[c]	10	1.0003104	310.4	79.40	
	Autoranger II	865	14983482[c]	10	1.0003104	310.4	79.40	
Leica	Wild DI 3	875	7492700	20	1.000282	282	79.2	
Switzerland	Wild DI 3S	885	7492700	20	1.000282	282	79.2	
	Wild TC 1	885	4871444	30.7692	1.000282	282	79.2	
	Wild DI4/4L/4S	885	4870255	30.7692	1.000282	282	79.2	
	Wild TC 1 L	885	4870255	30.7692	1.000282	282	79.2	
	Wild DI 20	835	4495620	33.3333	1.0002822	282.2	79.43	
	Wild DI 5 S	845	4870255	30.7692	1.000282	282	79.2	
	Wild DI1000	865	7492700	20	1.000282	282	79.2	
	Wild DI2000	850	14835546	10.10101	1.0002818	281.8	79.4	
	Wild DI3000	865	(15000000)	−	1.0002815	281.5	79.3	Pulse distance meter
	Wild TC2000	850	4870255	30.7692	1.000282	282	79.2	
	Wild TC1600(I)	850	14835546	10.10101	1.000282	282	79.4	
	Wild TC1600(II)	850	50000000	3	1.0002818	281.8	79.39	
	Wild TC1000/1010	850	50000000	3	1.0002818	281.8	79.39	
	Wild TC1610	850	50000000	3	1.0002818	281.8	79.39	
	Wild TC2002	850	50000000	3	1.0002818	281.8	79.39	
	Wild DI1001	850	50000000	3	1.0002818	281.8	79.39	
	Wild DI1600	850	50000000	3	1.0002818	281.8	79.39	
	Wild DI2002	850	50000000	3	1.0002818	281.8	79.39	

Manu-facturer	Model	Carrier wave length (nm)	Main modulation frequency (Hz)	Unit length (m)	Reference refractive index	Terms of first velocity correction		Remarks
						C	D	
Leica Switzer-land	Wild DI3000S	860	(15000000)	—	1.0002815	281.5	79.3	Pulse distance meter
	Wild DI3010S	860	(15000000)	—	1.0002815	281.5	79.3	Pulse distance meter
	Wild DIOR3002S	860	(15000000)	—	1.0002815	281.5	79.3	Pulse distance meter
	Wild 3012S	860	(15000000)	—	1.0002815	281.5	79.3	Pulse distance meter
	Wild TC500	850	50000000	3	1.0002818	281.8	79.39	
	TC1100	850	50000000	3	1.0002818	281.8	79.39	
	TC1700	850	50000000	3	1.0002818	281.8	79.39	
	TC1800	850	50000000	3	1.0002818	281.8	79.39	
	TCA/TCM1100	850	50000000	3	1.0002818	281.8	79.39	
	TCA/TCM1800	850	50000000	3	1.0002818	281.8	79.39	
	TC400/TC600	850	50000000	3	1.0002818	281.8	79.39	
Nikon Japan	ND-20	820	14972947	10	1.000280	280	79.5	
	ND-21	820	14972947	10	1.000280	280	79.5	
	ND-26	820	14972947	10	1.000280	280	79.5	
	NTD-2/2S	820	14972947	10	1.000280	280	79.5	
	NTD-3	820	14972947	10	1.000280	280	79.5	
	NTD-4	820	14972947	10	1.000280	280	79.5	
	ND-30	820	14972947	10	1.000280	280	79.5	
	ND-31	820	14972947	10	1.000280	280	79.5	
	DTM-1	820	14972947	10	1.000280	280	79.5	
	DTM-5	820	14972947	10	1.000280	280	79.5	
	DTM-20	820	14972947	10	1.000280	280	79.5	
	C-100	850	14985520	10	1.000275	275	79.5	
	D-50	850	14985520	10	1.000275	275	79.5	
	DTM-A5/10/20	850	74927604	2	1.000275	275	79.5	
	ND-20F	810	14972947	10	1.000280	280	79.5	
	ND-21F	810	14972947	10	1.000280	280	79.5	
	DTM A5LG	850	74927604	2	1.000275	275	79.5	
	DTM A10LG	850	74927604	2	1.000275	275	79.5	
	DTM A20LG	850	74927604	2	1.000275	275	79.5	

Nikon	DTM-720	850	74 927 604	2	1.000275	275^m	79.5^m	^{m} Temperature and pressure measured internally for automatic first velocity correction
Japan	DTM-730	850	74 927 604	2	1.000275	275^m	79.5^m	
	DTM-750	850	74 927 604	2	1.000275	275^m	79.5^m	
	DTM-300	850	14 985 520	10	1.000275	275	79.5	
Pentax	PM-81	850	14 985 450	10	1.0002785	278.5^a	79.2^a	^{a} Applicable if PM-81 is set to 278 ppm
Japan	PX-06D	850	14 985 450	10	1.0002785	278.5^d	79.2^d	^{d} Applicable if a temperature of +15 °C and a pressure of 760 mmHg is entered into the PX-06D
	MD-14	815	14 985 450	10	1.0002785	278.5	79.2	
	MD-20	815	14 985 450	10	1.0002785	278.5	79.2	
	PTS-10	815	14 985 450	10	1.0002798	279.8	79.56	
	PX-20D	815	14 985 450	10	1.0002798	279.8	79.56	
	PX-10D	815	14 985 450	10	1.0002798	279.8	79.56	
	PTS-III	815	14 985 450	10	1.00027975	279.75	79.55	
	PTS-II	815	14 985 450	10	1.00027975	279.75	79.55	
	PCS-1S	815	14 985 450	10	1.00027975	279.75	79.55	
	PCS-2S	815	14 985 450	10	1.00027975	279.75	79.55	
	PTS-V2/V2C	815	14 985 450	10	1.00027975	279.75^n	79.55^n	^{n} Temperature and pressure measured internally for automatic first velocity correction
	PTS-V3/V3C	815	14 985 450	10	1.00027975	279.75^n	79.55^n	
	PTS-V5	815	14 985 450	10	1.00027975	279.75^n	79.55^n	
Precision	Beetle 500S	910	14 984 979	10	1.0003100	310.0	79.15	
Interna-	Beetle 1000S	910	14 984 979	10	1.0003100	310.0	79.15	
tional	Beetle 1600S	910	14 984 979	10	1.0003100	310.0	79.15	
U.S.A.	Citation CI-410	905	14 984 980	10	1.000310	310.0^e	79.21^e	^{e} Applicable if PPM value in Citation is set to 310 ppm
	Citation CI-450	905	14 984 980	10	1.000310	310.0^e	79.21^e	
Sokkia	SDM-1C	910	14 985 450	10	1.0002786	278.6	79.15	
Japan	RED-1(=SDM-1D)	830	14 985 450	10	1.0002786	278.6	79.49	
	SDM 3D, 5D	860	14 985 453	10	1.0002789	278.96	79.33	
	RED-2, RED-3	860	14 985 453	10	1.0002789	278.96	79.33	
	RED mini (SDM-300)	860	14 985 453	10	1.0002789	278.96	79.33	
	SDM-3E/3ER	860	14 985 453	10	1.0002789	278.96	79.33	
	SET 5	860	14 985 453	10	1.0002789	278.96	79.33	
	RED 2A	860	14 985 453	10	1.00027896	278.96	79.33	
	RED 2L	860	14 985 453	10	1.00027896	278.96	79.33	
	RED mini 2	810	14 985 453	10	1.00027896	278.96	79.33	
	SDM 3F/3FR	860	14 985 453	10	1.00027896	278.96	79.33	

Manu-facturer	Model	Carrier wave length (nm)	Main modulation frequency (Hz)	Unit length (m)	Reference refractive index	Terms of first velocity correction		Remarks
						C	D	
Sokia	SET 2	860	14985453	10	1.00027896	278.96	79.33	
Japan	SET 3	860	14985453	10	1.00027896	278.96	79.33	
	SET 4	860	14985453	10	1.00027896	278.96	79.33	
	SET 5	860	7492721	20	1.00027896	278.96	79.33	
	SET5E,SET6E	860	7492721	20	1.00027896	278.96	79.33	
	NET2,NET2B	860	74927210	2	1.00027896	278.96	79.33	
	SET4A	860	14985442	10	1.00027896	278.96	79.33	
	SET2B/3B/4B	860	14985442	10	1.00027896	278.96	79.33	
	SET2C/3C/4C	860	14985442	10	1.00027896	278.96	79.33	
	RED2LV	860	14985442	10	1.00027896	278.96	79.33	
	SET3E,SET4E	860	29970884	5	1.00027896	278.96	79.33	
	Mini AR	860	14985442	10	1.00027896	278.96	79.33	
	SET2CII	860	29970884	5	1.00027896	278.96	79.33	
	SET3CII	860	29970884	5	1.00027896	278.96	79.33	
	SET4CII	860	29970884	5	1.00027896	278.96	79.33	
	SET2BII	860	29970884	5	1.00027896	278.96	79.33	
	SET3BII	860	29970884	5	1.00027896	278.96	79.33	
	SET4BII	860	29970884	5	1.00027896	278.96	79.33	
	SET2000	860	29970884	5	1.00027896	278.96	79.33	
	SET3000	860	29970884	5	1.00027896	278.96	79.33	
	SET4000	860	29970884	5	1.00027896	278.96	79.33	
Sokkisha	(see Sokkia)							
Tellumat	(see Tellurometer)							
Telluro-meter	MA 100	930	74927600	1(2)	1.000274	274	79.13	
	CD 6	930	14985520	10	1.000274	274	79.17	
U.K.	MA 200	780	10000000[f]	1.67 − 1.36	1.000296	296	79.75	[f] Frequency of master oscillator

Topcon	DM-C2	820	14985437	10	1.0002811	281.1	80.00
Japan	DM-S1/S2/S3	820	14985437	10	1.0002796	279.6	79.51
	DM-C3	820	14985437	10	1.0002796	279.6	79.51
	GTS-1/GTS 10	820	14985437	10	1.0002796	279.6	79.51
	GTS-2	820	14985437	10	1.0002796	279.6	79.51
	ET-1	820	14985437	10	1.0002796	279.6	79.51
	DM-A2/DM-A3	820	14985437	10	1.0002796	279.6	79.51
	GTS-2B/2S	820	14985437	10	1.0002796	279.6	79.51
	GTS-3B	820	14985437	10	1.0002796	279.6	79.51
	GTS-3	820	14985437	10	1.0002796	279.6	79.51
	ET-2	820	14985437	10	1.0002796	279.6	79.51
	DM-A5	820	14985437	10	1.0002796	279.6	79.51
	GTS-4/4B	820	29970874	5	1.00027966	279.66	79.51
	GTS-5/05	820	29970874	5	1.00027966	279.66	79.51
	GTS-6A/6E	820	29970874	5	1.00027966	279.66	79.51
	GTS-6/6B	820	29970874	5	1.00027966	279.66	79.51
	ITS-1/1B	820	29970874	5	1.00027966	279.66	79.51
	CTS-1 (CS-20)	820	14985437	10	1.00027966	279.66	79.51
	DM-S3L	820	44956311	3.33333	1.00027966	279.66	79.51
	DM-H1	820	44956311	3.33333	1.00027966	279.66	79.51
	GTS-3D	820	14985437	10	1.00027966	279.66	79.51
	AP-S1	820	14985437	10	1.00027966	279.66	79.51
	GTS-301/302/303	820	29970874	5	1.00027966	279.66	79.51
	CTS-2	820	14985437	10	1.00027966	279.66	79.51
	AP-L1	820	14985437	10	1.00027966	279.66	79.51
	GTS-4A	820	29970874	5	1.00027966	279.66	79.51
	GTS-304	820	14985437	10	1.00027966	279.66	79.51
	CTS-2B	820	14985437	10	1.00027966	279.66	79.51
	GTS-3C	820	14985437	10	1.00027966	279.66	79.51
	GTS-701/702/703	820	29970874	5	1.00027966	279.66	79.51
	GTS-201D/202	820	14985437	10	1.00027966	279.66	79.51

VEB Carl   (see Zeiss)
Zeiss Jena

Wild   (see Leica)

Manufacturer	Model	Carrier wave length (nm)	Main modulation frequency (Hz)	Unit length (m)	Reference refractive index	Terms of first velocity correction		Remarks
						C	D	
Zeiss	RETA	860	14985570	10	1.0002705	270.5	79.35	
Germany	RECOTA	860	14985570	10	1.0002705	270.5	79.35	
	ELDI 1	910	14985100	10	1.0003020	302.0[g]	79.15[g]	[g] Use environmental correction dial and first velocity corrections chart as supplied with instruments; some corrections for the oscillator drift with temperature are then also included
	ELDI 2	910	14985100	10	1.0003020	302.0[g]	79.15[g]	
	ELDI 3	910	14985100	10	1.0003020	302.0[g]	79.15[g]	
	SM 4	910	14985100	10	1.0003020	302.0[g]	79.15[g]	
	ELTA 2	910	14985800	10	1.0002551	255.1[g]	79.15[g]	
	ELTA 4	910	14985800	10	1.0002551	255.1[g]	79.15[g]	
	ELTA 3	910	14985800	10	1.0002555	255.5[g]	78.96[g]	
	ELTA 20	910	14985800	10	1.0002555	255.5[g]	78.96[g]	
	ELTA 46R	910	14985800	10	1.0002555	255.5[g]	78.96[g]	
	RSM 3	910	14985800	10	1.0002555	255.5[g]	78.96[g]	
	SM 41	910	14985800	10	1.0002555	255.5[g]	78.96[g]	[h] Use on-board first velocity correction which also corrects for oscillator errors, A/D converter errors and average humidity (60%). Pulse distance meter
	E-Elta 4	860	14985800	10	1.0002551	255.2[h]	79.35[h]	
	E-Elta 3	860	14985800	10	1.0002551	255.2[h]	79.35[h]	
	ELDI 4	860	14985800	10	1.0002551	255.2[h]	79.35[h]	
	E-Elta 6	860	14985800	10	1.0002551	255.2[h]	79.35[h]	
	ELDI 10	905	–	–	1.0002539	253.9	79.35	
	E-Elta2	860	14985800	10	1.0002551	255.1[k]	79.34[k]	
	E-Elta5	860	14985800	10	1.0002551	255.1[k]	79.34[k]	[k] Temperature and pressure measured by instrument, 60% rel. hum. assumed
	E-Elta3 (since 1991)	860	14985800	10	1.0002551	255.1[k]	79.34[k]	
	E-Elta4 (since 1991)	860	14985800	10	1.0002551	255.1[k]	79.34[k]	
	Rec Elta 2	860	14985800	10	1.0002551	255.1[k]	79.34[k]	
	Rec Elta 3	860	14985800	10	1.0002551	255.1[k]	79.34[k]	
	Rec Elta 4	860	14985800	10	1.0002551	255.1[k]	79.34[k]	
	Rec Elta RL	900	–	–	1.0002539	253.9[k]	79.35[k]	Pulse Distance Meter
	Elta 50	860	14985800	10	1.0002551	255.1	79.34	
	Rec Elta 13 C	860	14985800	10	1.0002551	255.1	79.34	
	Rec Elta 14 C	860	14985800	10	1.0002551	255.1	79.34	

Appendix E. Technical Data of a Selection of Short Range Distance Meters

Some relevant technical data of a selection of short range distance meters is given in the following tables. The data have been supplied by the manufacturers or their Australian agents or have been gathered from information in promotional brochures.

The range of the distance meters is given for one prism and three prisms. Both distances refer to so-called average atmospheric conditions, the definition of which varies slightly between manufacturers.

The standard deviations are specified accuracies for single distance measurements using the most precise measuring modes. The definition has been given in Eq. (13.63) of Section 13.5. Values in brackets refer to alternative measuring modes.

Most manufacturers supply batteries of different capacities. The smallest are listed in the tables.

The beam divergence refers to the angle subtended by the 50% power points of the power profile across the beam.

The angular accuracy is relevant for tacheometers, only. For tacheometers with either electronic or optical circles, accuracies of zenith angles and horizontal directions are listed. For distance meters with on-board zenith angle sensors, the resolution of the zenith angle is listed, only.

Manu-facturer	Model	Range with		Radia-tion source	Stand. dev.		Weight		Power cons.	Battery		Beam div.	Angular accuracy	
		1 prism m	3 prisms m		mm	+ppm	Instr. kg	Battery kg	W	Volt V	Cap Ah	min	Hor.	Vert.
AGA	see Geodimeter													
Alpha Electronics U.S.A.	Alpha I	700	1500	Diode	5	5	2.5	0.6	9.6	12	?	?	–	–
	Alpha II	1200	2000	Diode	5	5	2.5	0.6	9.6	12	?	?	–	–
	Alpha III	2500	3000	Diode	5	5	2.5	0.6	9.6	12	?	?	–	–
	OMNI 1	?	2500	Diode	5	5	7.5	?	?	12	?	?	±10″	±10″
Benchmark Inc. U.S.A.	Surveyor I-X	?	1000	Diode	5	5	2.3	0.5	?	?	?	?	–	–
	Surveyor II-X	?	2000	Diode	5	5	2.3	0.5	?	?	?	?	–	–
	Surveyor III-X	?	3000	Diode	5	5	2.3	0.5	?	?	?	?	–	–
COM-RAD U.K.	Geomensor 204DME	5000	8000	XeFlash tube	0.3	0.3	26	?	24	12	?	3	–	–
Cubic Precision U.S.A.	DM 80	1000	1600	Diode	5	5 – 10	1.5	0.2	?	7.2	?	?	–	–
	DM 81	1000	1600	Diode	5	5	1.5	0.2	?	7.2	?	?	–	–
	Autoranger IIx	2000	3000	Diode	5	5	3.7	0.5	?	12	?	?	–	–
Geodimeter Sweden	Geodimeter 6A		< 5000	Tungsten lamp	5	1	16.3	–	30	12	–	7.0V 1.0H	–	–
	Geodimeter 710	1700	3500	Laser	5	1	(14.2)	10.9	100	12	60	0.7 – 35	±2″	±3″
	Geodimeter 10	700	1400	Diode	5	10	2.8	4	17	6	15	8.5	–	–
	Geodimeter 12A	1000	2000	Diode	5	5	2.8	4	17	6	15	8.5	–	–
	Geodimeter 14	4000	6000	Diode	5 (5 – 30)	10 (3)	2.8	4	17	6	15	8.5	–	–
	Geodimeter 120	1000	2000	Diode	5	7	2.6	1	?	12	2	8.6	–	±1′
	Geodimeter 14A	6000	8000	Diode	5 – 15	3	2.5	1	20	6	4	2.5	–	–
	Geodimeter 110	1400	2100	Diode	5	5	2.5	1	10	12	2	2.5	–	–
	Geodimeter 112	2000	3100	Diode	5	3	2.5	1	10	12	2	2.5	–	–
	Geodimeter 116	1000	1400	Diode	5	5	2.6	1	10	12	2	2.5	–	(±1′)
	Geodimeter 110A	800	1800	Diode	5	5	2.5	1	8.4	12	2	2.5	–	–
	Geodimeter 140	2500	3600	Diode	5	3	7.5	?	12	?	?	8	±2″	±2″

Manufacturer	Instrument													
Geodimeter	Geodimeter 16	5000	7000	Diode	5–15	3	2.5	2	?	6	4	7	–	–
Sweden	Geodimeter 136	1000	1400	Diode	5	5	8.5	?	?	?	?	8	±3″	±3″
	Geodimeter 122	2500	3600	Diode	5	3	2.8	1	?	12	2	8	–	±20″
	Geodimeter 114	8000	10000	Diode	5	1	2.7	1	8.5	12	2	0.7	–	–
	Geodimeter 210	2300	4000	Diode	5	3	1.1	0.25	<4	12	0.4	2.5	–	–
	Geodimeter 216	1000	1700	Diode	5	5	1.3	0.25	<4	12	0.4	2.5	–	±3″
	Geodimeter 220	2300	4000	Diode	5	3	1.3	0.25	<4	12	0.4	2.5	–	±20
	Geodimeter 420	1000	1600	Diode	5	3	6.9	1	6	12	1	2.5	±3″	±3″
	Geodimeter 440	2300	3500	Diode	3	3	6.9	1	6	12	1	2.5	±1″	±1″
	Geodimeter 142	2500	3600	Diode	2	3	9.5	1	12	12	2	2.5	±1″	±1″
	Geodimeter 422	2300	?	Diode	5	3	?	?	?	?	?	?	±2″	±2″
	Geodimeter 408	1000	1600	Diode	5	5	7.9	–	6	12	1	8.6	±5″	±5″
	Geodimeter 412	1600	2200	Diode	5	5	7.9	–	6	12	1	8.6	±3″	±3″
	Geodimeter 422LR	3300	4500	Diode	5	5	7.9	–	6	12	1	8.6	±2″	±2″
	Geodimeter 424	2300	3200	Diode	5	5	7.9	–	6	12	1	8.6	±2″	±2″
	Geodimeter 444	3300	4500	Diode	5	5	7.9	–	6	12	1	8.6	±1″	±1″
	Geodimeter 460	2300	3200	Diode	5	5	9.5	–	7–8	12	1	8.6	±2″	±2″
	Geodimeter 510	1200	1800	Diode	3	2	7.1	0.3	3.6–6	12	1.0	5.2	±3″	±3″
	Geodimeter 520	1800	2500	Diode	3	2	7.1	0.3	3.6–6	12	1.0	5.2	±2″	±2″
	Geodimeter 540	2500	3500	Diode	2	2	7.1	0.3	3.6–6	12	1.0	5.2	±1″	±1″
	Geodimeter 4400	2300	3200	Diode	3	3	9.5	–	7–8	12	8.6	–	±2″	±2″
	RPU 4000 (One-Man System)	1600	–	Diode	6	–	2.7	0.5	3	10	1.3	–	±2″	±2″
	Geodimeter 640}	{ 2500	3500}	{ Diode	2	2	7.2	0.3	4.8–10.8	12	1.2	5.5	±1″	±1″
	Geodimeter 620}	{ 1800	2500}	{ Diode	3	2	7.2	0.3	4.8–10.8	12	1.2	5.5	±2″	±2″
	Geodimeter 610}	{ 1200	1800}	{ Diode	3	2	7.2	0.3	4.8–10.8	12	1.2	5.5	±3″	±3″
Geotronics	See Geodimeter													
Hewlett-Packard U.S.A.	HP 3805A	500	1600	Diode	7	10	10.4	1.0	14	12	2	3	–	–
	HP 3810A	500	1600	Diode	5	10	11.9	1.0	18	12	2	3	±20″	±30″
	HP 3820A	1000	3000	Laser diode	5	5	10.7	–	(1.5)	3.6	?	3	±2″	±4″
	HP 3800B	1500	2500	Diode	5	7	7.7	6.0	12	12	8	3.5	–	–
	HP 3808A	3000	6000	Laser diode	5	1	9.2	1.0	16	12	2	2	–	–

Manu-facturer	Model	Range with		Radiation source	Stand. dev.		Weight		Power cons.	Battery		Beam div.	Angular accuracy	
		1 prism m	3 prisms m		mm	+ppm	Instr. kg	Battery kg	W	Volt V	Cap Ah	min	Hor.	Vert.
Hewlett-Packard U.S.A.	HP 3810B	2000	5000	Laser diode	5	1	12.0	1.0	16	12	2	3	±5″	±20″
	HP 3850A	2000	8000	Laser diode	5	1	12.7	?	14	12	?	3	–	–
IRI Inc. U.S.A.	Stinger II	1200	2500	Diode	5	5	1.13	0.5	?	7.2	?	?	–	–
	MTS 10	600	1500	Diode	5	5	5.5	1.1	?	7.2	2	?	±10″	±10″
Kern Switzerland	DM 500	300	500	Diode	5	5	1.6	2.8	?	5	10	4	–	–
	DM 501	1000	1600	Diode	5	5	1.6	2.8	?	5	10	4	–	–
	Mekometer ME3000	1500	2500	Xenon lamp	0.2	1	18.7	8.8	18	12	10	?	–	–
	DM 502	1200	2000	Diode	5	5	1.6	2.8	3.5	5	7	4	–	–
	DM 102	1000	1700	Diode	5	5	1.7	2.5	3.5	5	7	4	–	–
	DM 503	2500	3500	Diode	3	2	1.6	2.5	4.5	5	7	4	–	–
	DM 104	1000	2000	Diode	5	5	1.7	2.5	5	5	7	4	–	–
	DM 150	1000	2000	Diode	5	5	1.8	2.5	6.5	5	7	4	–	±10″
	DM 504	2500	3500	Diode	3	2	1.6	2.5	5	5	7	4	–	–
	DM 550	2500	3500	Diode	5	2	1.6	2.5	6.5	5	7	4	–	±10″
	Mekometer ME5000	4000	8000	HeNe Laser	0.2	0.2	11.0	3.6	24	12	7	?	–	–
Keuffel and Esser U.S.A.	Autoranger	700	1600	Diode	5	6	2.4	0.9	10.5	12	1.8	8	–	–
	Autoranger-A	300	500	Diode	2 – 3	0	3.0	1.0	10.5	12	1.8	8.5	–	–
	Autoranger S	1600	3200	Diode	5	6	2.4	0.9	10.5	12	2.5	3.5	–	–
	Autoranger II	1600	3200	Diode	5	6	3.1	–	10.5	12	1.5	3.5	–	–
	Autoranger III	–	3600	Diode	5	2	3.1	0.7	14	12	1.0	?	–	–
Leica Switzerland	Wild DI3	400	600	Diode	5	5	1.8	3.0	14	12	7	5	–	–
	Wild DI3S	1000	1600	Diode	5	5	2.0	2.3	17	12	1.8	5	–	–
	Wild TC1 (Tachymat)	700 (1000)	1000 (1600)	Diode	5 (5 – 15	5 5)	9.8	3.0	16	12	7.0	6	±2″	±3″

Leica	Wild DI4 (Distomat)	1000	1600	Diode	5	5	2.2	2.3	5	12	2	4	–	–
Switzerland	Wild DI4L	2500	3500	Diode	5	5	1.9	0.4	3.75	12	0.5	4	–	–
	Wild DI20	6000	7000	Diode	5	1	3.7	0.4	5	12	0.5	2.5	–	–
	Wild TC1L	2500	3500	Diode	5	5	9.8	0.4	$8-19$	12	0.5	4	$\pm 2''$	$\pm 3''$
	Wild DI4S	300	500	Diode	2	5	1.1	0.4	5	12	0.5	4	–	–
	Wild DI5	2500	3500	Diode	3	2	1.9	0.9	5	12	2	?	–	–
	Wild DI5S	2500	3500	Diode	3	2	1.9	–	5	12	0.4	2.5	–	–
	Wild DI1000	800	1200	Diode	5	5	1.1	–	2.4	12	0.2	8.3	–	–
	Wild DI2000	2000	2800	Diode	1	1	1.2	–	1	12	0.1	2.5	–	–
	Wild TC2000	2000	2800	Diode	3	2	22.9	–	9.5	12	0.8	4.7	$\pm 0.5''$	$\pm 0.5''$
	Wild TC1600(I)	2000	2800	Diode	3	2	5.5	0.2	5	12	0.4	4.7	$\pm 1.8''$	$\pm 1.8''$
	Wild TC1000	2000	2800	Diode	3	2	5.5	0.2	$0.7-4.8$	12	0.6	2.5	$\pm 3''$	$\pm 3''$
	Wild TC1610	2500	3500	Diode	2	2	5.6	0.2	$0.7-4.8$	12	0.6	2.5	$\pm 1.5''$	$\pm 1.5''$
	Wild TC1600(II)	2000	2800	Diode	3	2	5.5	0.2	$0.7-4.8$	12	0.6	2.5	$\pm 1.5''$	$\pm 1.5''$
	TC1010	2000	2800	Diode	3	2	5.5	0.2	$0.7-4.8$	12	0.45	2.5	$\pm 3''$	$\pm 3''$
	Wild TC2002	2000	2800	Diode	1	1	8.4	0.8	$1.6-5.4$	12	2.0	2.5	$\pm 0.5''$	$\pm 0.5''$
	Wild DI1001	800	1100	Diode	5	5	1.0	0.9	4	12	2.0	1.4	–	–
	Wild DI1600	2500	3500	Diode	3	2	1.1	0.9	4	12	2.0	2.5	–	–
	Wild DI2002	2500	3500	Diode	1	1	1.1	0.9	4	12	2.0	2.5	–	–
	Wild TC500	700	1100	Diode	5	5	4.2	0.2	$0.7-3.6$	12	0.6	3.5	$\pm 6''$	$\pm 6''$
	TC1100	2500	3500	Diode	2	2	6.1	0.3	$1.2-5.0$	12	1.1	2.5	$\pm 3''$	$\pm 3''$
	TC1700	2500	3500	Diode	2	2	6.4	0.3	$1.2-5.0$	12	1.1	2.5	$\pm 1.5''$	$\pm 1.5''$
	TC1800	2500	3500	Diode	2	2	6.4	0.3	$1.2-5.0$	12	1.1	2.5	$\pm 1''$	$\pm 1''$
	TCM1100	2500	3500	Diode	2	2	6.2	0.3	$1.2-6$	12	1.1	2.5	$\pm 3''$	$\pm 3''$
	TCM1800	2500	3500	Diode	2	2	6.5	0.3	$1.2-6$	12	1.1	2.5	$\pm 1''$	$\pm 1''$
	TC400	700	1100	Diode	5	5	4.2	0.2	$0.7-3.6$	12	0.6	3.5	$\pm 10''$	$\pm 10''$
	TC600	1100	1600	Diode	3	3	4.2	0.2	$0.7-3.6$	12	0.6	3.5	$\pm 5''$	$\pm 5''$
	TCA1100	2500	3500	Diode	2	2	6.8	0.3	$1.2-9$	12	1.1	2.5	$\pm 3''$	$\pm 3''$
	TCA1800	2500	3500	Diode	2	2	7.1	0.3	$1.2-9$	12	1.1	2.5	$\pm 1''$	$\pm 1''$
Nikon	ND-160	1000	1600	Diode	5	5	2.3	0.7	7.6	8.4	1.8	?	–	–
Japan	ND-250	1600	2500	Diode	5	5	2.3	0.7	7.6	8.4	1.8	?	–	–
	ND-20	700	1000	Diode	5	5	2.2	(0.4)	3	8.4	0.5	?	–	–
	ND-21	1000	1600	Diode	5	5	2.2	(0.4)	3	8.4	0.5	?	–	–
	ND-26	2000	3000	Diode	5	5	2.3	(0.4)	3	8.4	0.5	?	–	–
	NTD 2/2S	1600	2300	Diode	5	5	6.6	(0.4)	3	8.4	0.5	?	$\pm 10/5''$	$\pm 10/5''$
	NTD 3	1600	2300	Diode	5	5	6.6	(0.4)	3	8.4	0.5	?	$\pm 5''$	$\pm 5''$
	NTD 4	1600	2300	Diode	5	5	6.6	(0.4)	3	8.4	0.5	?	$\pm 3''$	$\pm 3''$

Manu-facturer	Model	Range with		Radia-tion source	Stand. dev.		Weight		Power cons. W	Battery		Beam div. min	Angular accuracy	
		1 prism m	3 prisms m		mm	+ppm	Instr. kg	Battery kg		Volt V	Cap Ah		Hor.	Vert.
Nikon	DTM-1	1600	2300	Diode	5	5	8.3	1.1	<5.5	8.4	2	2	±2″	±3″
Japan	DTM-5	1600	2300	Diode	5	5	6.4	0.9	<5	8.4	1.2	?	±5″	±5″
	DTM-20	800	1200	Diode	5	5	?	0.9	<5	8.4	1.2	?	±10″	±10″
	ND-30	1600	2500	Diode	5	5	2.4	0.8	7.6	8.4	1.2	?	–	–
	ND-31	1900	3000	Diode	5	5	2.4	0.8	7.6	8.4	1.2	?	–	–
	C-100	500	800	Diode	5	5	6	0.75	2.5	7.2	1.7	6.4	±6″	±6″
	D-50	300	500	Diode	5	10	5.8	0.75	2.5	7.2	1.7	6.4	20″	20″
	DTM-A5	2400	3100	Diode	2	2–3	6.3	0.7	3.3	7.2	1.7	5.8	±2″	±2″
	DTM-A10	2200	2900	Diode	3	3	6.3	0.7	3.3	7.2	1.7	5.8	±3″	±3″
	DTM-A20	1600	2300	Diode	3	3	6.3	0.7	3.3	7.2	1.7	5.8	±5″	±5″
	ND-20F	1000	1600	Diode	5	5	2.2	0.2	5	8.4	0.5	?	–	–
	ND-21F	700	1000	Diode	5	5	2.2	0.2	5	8.4	0.5	?	–	–
	DTM A5LG	2700	3600	Diode	2	2–3	6.3	0.7	3.3	7.2	1.7	5.8	±2″	±2″
	DTM A10LG	2500	3300	Diode	3	3	6.3	0.7	3.3	7.2	1.7	5.8	±3″	±3″
	DTM A20LG	2000	2800	Diode	3	3	6.3	0.7	3.3	7.2	1.7	5.8	±4″	±4″
	DTM-750	2700	3600	Diode	2	2–3	6.9	0.7	5.2	7.2	1.7	5.8	±2″	±2″
	DTM-730	2500	3300	Diode	3	3	6.9	0.7	5.2	7.2	1.7	5.8	±3″	±3″
	DTM-720	2000	2800	Diode	3	3	6.9	0.7	5.2	7.2	1.7	5.8	±4″	±4″
	DTM-300	800	1100	Diode	5	3	5.0	0.3	2.2	7.2	1.7	6.4	±5″	±5″
Pentax	PM-81	1400	2000	Diode	5	5	3	–	5	8.4	0.6	2.2	–	–
Japan	PX-06D	1400	2000	Diode	5	5	6.3	0.8	4.2	8.4	1.1	2.5	±6″	±6″
	MD-14	1000	1400	Diode	5	5	2.2	0.18	4	8.4	0.45	2.5	–	–
	MD-20	1400	2000	Diode	5	5	2.2	0.18	4	8.4	0.45	2.5	–	–
	PTS-10	1400	2000	Diode	5	5	7.7	0.6	4	8.4	1.5	2.5	<5″	<5″
	PX-20D	1000	1600	Diode	5	5	6.3	0.8	4.2	8.4	1.2	?	±20″	±20″
	PX-l0D	1400	2000	Diode	5	5	6.3	0.8	?	8.4	1.2	?	±10″	±10″
	PTS-II	1600	2400	Diode	5	3	6.9	0.7	?	8.4	?	?	±5–7″	±5–7″
	PTS-III 05	1800	2600	Diode	5	3	6.2	0.7	3.8	8.4	1.5	6.5	±3″	±3″
	PTS-III 10	1200	1800	Diode	5	3	6.2	0.7	3.8	8.4	1.5	6.5	±5″	±5″
	PTS-II 05	1600	2400	Diode	5	3	6.2	0.7	3.8	8.4	1.5	6.5	±5″	±5″

Pentax	PTS-II 10	1000	1600	Diode	5	3	6.2	0.7	3.8	8.4	1.5	6.5	±7″	±7″
Japan	PCS-1S	700	1000	Diode	5	3	4.7	0.3	2.9	7.2	1.4	7	±5″	±5″
	PCS-2S	700	1000	Diode	5	3	4.7	0.3	2.9	7.2	1.4	7	±5″	±5″
	PTS-V2/V2c	2400	3100	Diode	2	2	6.2	0.16	2.4	6	1.1	5	±2″	±2″
	PTS-V3/V3c	2200	2900	Diode	3	2	6.2	0.16	2.4	6	1.1	5	±3″	±3″
	PTS-V5	1900	2500	Diode	3	3	6.2	0.16	2.4	6	1.1	5	±5″	±5″
Precision	Beetle 500S	300	500	Diode	10	−	2.7	2.3	17	12	4	8	−	−
International	Beetle 1000S	500	1000	Diode	10	−	2.7	2.3	17	12	4	8	−	−
U.S.A.	Beetle 1600S	700	1600	Diode	10	−	2.7	2.3	17	12	4	8	−	−
	Super Beetle 1400S	700	1400	Diode	4.5	5	2.5	1.8	18−24	12	4	8.5	−	−
	Super Beetle 2000S	1000	2000	Diode	4.5	5	2.5	1.8	18−24	12	4	8.5	−	−
	Citation CI-450	1600	2300	Diode	5	5	2.8	2.3/3.0	9	12	2(7)	?	−	−
	Citation Cl-410	1000	1500	Diode	5	5	2.6	2.3/3.0	9	2	2(7)	?	−	−
Sokkisha	SDM-1C	1000	1600	Diode	5	5	2.8	1.5	15	5	3.5	4	−	−
Japan	RED-1 (=SDM-1D)	1300	2000	Diode	5	5	3.5	−	<12	12	1.0	3	−	−
	SDM 5A	500	800	Diode	5	5	2.1	0.4	6	12	1.2	6	−	−
	SDM 3D	1000	1600	Diode	5	5	8.5	−	8	8.4	1.5	2.5	±10″	±10″
	RED-2	1800	2600	Diode	5	5	2.0	0.3	?	6	?	?	−	−
	SDM-300 (RED mini)	300	500	Diode	5	5	>0.9	−	6.6	6	1.6	?	−	−
	RED-3	1800	2600	Diode	5	5	2	0.4	5.5	6	1.8	7	±10″	±10″
	SDM-3E	800	1400	Diode	5	5	7.2	0.4	5.0	6	1.8	7	±10″	±10″
	SDM-3ER	800	1400	Diode	5	5	7.2	0.4	5.5	6	1.8	7	±10″	±10″
	SET 5	800	1400	Diode	5	5	7.5	0.4	6.0	6	1.8	7	±5″	±5
	RED 2A	2300	3200	Diode	5	3	2	0.3	5.4	6	1.3	2	−	−
	RED 2L	4500	6400	Diode	5	3	2	0.3	5.4	6	1.3	2	−	−
	RED mini 2	500	1200	Diode	5	5	1	0.2	3.6	6	1	3.4	−	−
	SDM 3F/3FR	1300	2100	Diode	5	3	7.3	0.2	4.8/6	6	1	3.4	±5″	±5″
	SET 2	2300	3100	Diode	3	2	7.6	0.2	4.2	6	?	3.4	±2″	±2″
	SET 3	2200	3000	Diode	5	3	7.6	0.2	4.2	6	?	3.4	±3″	±3″
	SET 4	1300	2100	Diode	5	3	7.6	0.2	4.2	6	?	3.4	±5″	±5″
	SET5	800	1000	Diode	5	5	6.4	−	?	6	?	?	±7″	±7″
	SET5E	800	1000	Diode	5	3	5.6	0.24	?	6	?	?	±5″	±5″
	NET2	2−100 to re-		Diode	1	2	7.3	−	?	6	?	?	±2″	±2″
	NET2B	2−100 flective sheeting		Diode	0.8	1	7.3	0.24	?	6	?	?	±2″	±2″

Manu-facturer	Model	Range with		Radia-tion source	Stand. dev.		Weight		Power cons.	Battery		Beam div.	Angular accuracy	
		1 prism m	3 prisms m		mm	+ppm	Instr. kg	Battery kg	W	Volt V	Cap Ah	min	Hor.	Vert.
Sokkia	SET2B/2C	2400	3100	Diode	3	2	7.4	–	?	6	?	?	±2″	±2″
Japan	SET3B/3C	2200	2900	Diode	5	3	7.4	–	?	6	?	?	±3″	±3″
	SET4B/4C	2400	3100	Diode	5	3	7.4	–	?	6	?	?	±5″	±5″
	SET2BII/2CII	2400	3100	Diode	3	2	7.5	0.24	?	6	?	?	±2″	±2″
	SET3BII/3CII	2200	2900	Diode	3	3	7.5	0.24	?	6	?	?	±3″	±3″
	SET4BII/4CII	1200	1700	Diode	5	3	7.4	0.24	?	6	?	?	±5″	±5″
	SET3E	1600	2100	Diode	3	2	5.5	0.24	?	6	?	?	±3″	±3″
	SET4E	1600	2100	Diode	3	3	5.5	0.24	?	6	?	?	±5″	±5″
	SET5A	1000	1200	Diode	5	3	5.6	0.24	?	6	?	?	±5″	±5″
	SET6	700	1000	Diode	5	5	4.8	0.24	?	6	?	?	±10″	±10″
	SET6E	500	800	Diode	5	5	4.8	0.24	?	6	?	?	±7″	±7″
	RED2LV	6000	8000	Diode	5	5	2.2	0.35	?	6	?	?	–	–
	Mini AR	800	1200	Diode	5	5	1.8	0.2	?	6	?	?	–	–
	SET2000	2400	3100	Diode	2	2	5.7	–	?	6	?	?	±2″	±2″
	SET3000	2200	2900	Diode	2	2	5.7	–	?	6	?	?	±3″	±3″
	SET4000	1600	2100	Diode	2	2	5.7	–	?	6	?	?	±5″	±5″
Sokkisha	(see Sokkia)													
Tellumat	(see Tellurometer)													
Tellurometer	MA 100	–	1000	Diode	1.5	1	17	–	18	12	?	15	–	–
U.K.	CD6	800	1500	Diode	5	5	3.7	–	15	12	?	20	–	–
	MA 200	1000	–	Diode	0.5	0.5	9.0	–	6	12	?	6	–	–
Topcon	DM-C2	1400	2000	Diode	5	5	2.8	0.7	7	8.4	1.8	7	–	–
Japan	GTS-1/GTS-10	700	1000	Diode	5	10	5	0.7	7	8.4	1.8	7	±10″	±10′
	DM-S1	1000	1400	Diode	5	5	2.9	–	7	8.4	1.8	7	–	–
	DM-C3	1600	2500	Diode	5	5	2.9	–	7	8.4	1.8	7	–	–
	GTS-2	1100	1700	Diode	5	5	6	0.8	9.2	8.4	1.1	?	±10″	±10″
	DM-S2	1600	2000	Diode	5	5	1.5	0.4	6.3	8.4	1	7	–	–
	DM-S3	2200	3900	Diode	5	5	1.5	0.4	6.3	8.4	1	7	–	–

Topcon	ET-1	1400	2000	Diode	5	5	7.5	0.8	6.3	8.4	1	7		±2″	±3″
Japan	DM-A2/DM-A3	700	1000	Diode	5	5	1.3	1	0.7	8.4	2	7		–	–
	GTS-2B/2S	1400	2000	Diode	5	5	6	1	1	8.4	2	7		±6″	±6″
	GTS-3B	2000	2800	Diode	5	3	5.2	1	1	8.4	2	?		±2″	±2″
	GTS-3	2000	2800	Diode	5	5	5.2	1	?	8.4	2	?		±3″	±3″
	ET-2	2200	3000	Diode	5	3	7.5	1	5	8.4	1.8	?		±1″	±1″
	DM-A5	750	1100	Diode	5	3	0.7	0.2	?	8.4	1	?		–	–
	GTS-2R	1800	2500	Diode	5	3	5.2	0.8	5	8.4	1.2	?		±6″	±6″
	GTS-4	2400	3100	Diode	2	2	5.8	1.1	6	8.4	1.8	2.8		±2″	±2″
	GTS-4B	2000	2700	Diode	2	2	5.8	1.1	6	8.4	1.8	2.8		±5″	±5″
	GTS-6/6A	2400	3100	Diode	2	2	5.8	1.1	6	8.4	1.8	2.8		±2″	±2″
	GTS-6B/6E	2000	2700	Diode	2	2	5.8	1.1	6	8.4	1.8	2.8		±5″	±5″
	ITS-1	2400	3100	Diode	2	2	5.8	1.1	6	8.4	1.8	2.8		±2″	±2″
	ITS-1B	2000	2700	Diode	2	2	5.8	1.1	6	8.4	1.8	2.8		±5″	±5″
	CTS-1 (CS-20)	500	800	Diode	5	5	4	0.3	3.3	7.2	1.2	3.2		±20″	±20″
	DM-S3L	4300	5400	Diode	5	3	2.3	0.5	5.5	8.4	1.1	3.4		–	–
	DM-H1	800		Diode	1	2	2.2	0.5	5	8.4	1.1	3.4		–	–
	AP-S1	1400		Diode	5	5	15.5	–	50	–	–	3.9		±5″	±5″
	GTS-301	2400	3100	Diode	2	2	5.2	0.9	3.3	7.2	2.8	2.8		±2″	±2″
	GTS-302	2200	2900	Diode	2	2	5.2	0.9	3.3	7.2	2.8	2.8		±3″	±3″
	GTS-303	1200	2000	Diode	2	2	5.2	0.9	3.3	7.2	2.8	2.8		±5″	±5″
	AP-L1	700		Diode	3	2	9.2	–	27.4	12	8	4		±3″	±3″
	CTS-2	600	900	Diode	3	5	4	0.3	2.9	7.2	1.2	3.2		±10″	±10″
	GTS-4A	2400	3100	Diode	2	2	5.8	1	6	8.4	1.8	2.8		±1″	±1″
	GTS-304	1000	1800	Diode	3	5	5.2	0.9	3.3	7.2	2.8	2.8		±5″	±5″
	CTS-2B	350		Diode	3	5	4	0.3	2.9	7.2	1.2	3.2		±10″	±10″
	GTS-3B	1600	2200	Diode	3	7	5.2	1	6	8.4	1.8	3.9		±5″	±5″
	GTS-701	2400	3100	Diode	2	2	5.8	1	4.1	7.2	2.8	2.8		±2″	±2″
	GTS-702	2200	2900	Diode	2	2	5.8	1	4.1	7.2	2.8	2.8		±3″	±3″
	GTS-703	1200	2000	Diode	2	2	5.8	1	4.1	7.2	2.8	2.8		±5″	±5″
	GTS-201D	900	1200	Diode	3	2	4.5	0.3	2.5	7.2	1.4	4.5		±5″	±5″
	GTS-202	600	900	Diode	3 – 5	5	4.5	0.3	2.5	7.2	1.4	4.5		±6″	±6″
VEB Carl Zeiss Jena	(see Zeiss)														
Wild	(see Leica)														

Manufacturer	Model	Range with		Radiation source	Stand. dev.		Weight		Power cons. W	Battery		Beam div. min	Angular accuracy	
		1 prism m	3 prisms m		mm	+ppm	Instr. kg	Battery kg		Volt V	Cap Ah		Hor.	Vert.
Zeiss Germany	EOT 2000	1000	1500	Diode	10	–	10.5	?	?	12	?	?	±1″	±1″
	RETA	1000	1500	Diode	5	2	10.2	0.8	4–8	12	1.8	2.5	±3″	±3″
	RECOTA	1000	1500	Diode	5	2	11.7	0.8	4–8	12	1.8	2.5	±1.6″	±1.6″
	ELDI 1	2000	2500	Diode	5–10	2	8	0.4	5	9	1.8	2	–	–
	ELDI 2	700	1000	Diode	5	2	4.2	0.4	5	9	1.8	4	–	–
	ELDI 3	400	700	Diode	5–10	2	3.8	0.4	5	9	1.8	4	–	–
	SM 4	700	1000	Diode	5–10	2	7.8	0.4	5	9	1.8	4	±3″	± 3″
	Reg ELTA 14	500	800	Diode	5–10	–	25	2.5	14	12	6	2	±3″	±5″
	ELTA 4	700	1200	Diode	5	2	6.5	–	5	9	1.8	1.2	±3″	±3″
		1000	1600		10	2								
	ELTA 2	1200	1600	Diode	5	2	12	–	7	9	1.8	1	±0.6″	±0.6″
			2000		10	2								
	ELTA 3	?	1200	Diode	5	2	13.5	–	?	9	1.8	?	±2″	±2″
	ELTA 20	?	1600	Diode	5	2	13.5	–	?	9	1.8	?	±1″	±1″
	RSM 3	1000	1600	Diode	5	2	7.5	–	?	9	1.8	?	±2″	±3″
	SM 41	?	1000	Diode	10	2	6.5	–	?	9	1.8	?	±3″	±3″
	ELTA 46	?	1600	Diode	5	2	6.5	–	5.6	9	1.8	2.5	±3″	±3″
	E-Elta4	1000	1500	Diode	3	2	4.8	0.36	2–3.5	4.8	1.8	3.5	±3″	±3″
	E-Elta3	1600	2000	Diode	3	2	5.3	0.36	2–3.5	4.8	1.8	3.5	±2″	±2″
	ELDI 4	1000	1500	Diode	5	3	0.8	0.36	3	4.8	1.8	5	–	–
	E-Elta6	1000	1500	Diode	5	3	4.7	0.36	3	4.8	1.8	5	±5″	±5″
	E-Elta5	1000	1500	Diode	5	3	4.8	0.36	3	4.8	1.8	5	±5″	±5″
	E-Elta2	1800	2500	Diode	2	2	5.0	0.36	3	4.8	1.8	3.5	±0.6″	±0.6″
	E-Elta3 (since 1991)	1600	2000	Diode	3	3	4.8	0.36	1.1	4.8	1.8	3.5	±1.5″	±1.5″
	E-Elta4 (since 1991)	1600	2000	Diode	3	3	4.8	0.36	1.1	4.8	1.8	3.5	±3″	±3″
	Rec Elta 2	1800	2500	Diode	2	2	5.9	0.36	1.1	4.8	1.8	3.5	±0.6″	±0.6″
	Rec Elta 3	1600	2000	Diode	3	3	5.9	0.36	1.1	4.8	1.8	3.5	±2″	±2″
	Rec Elta 4	1600	2000	Diode	3	3	5.9	0.36	1.1	4.8	1.8	3.5	±3″	±3″
	Rec Elta 5	1000	1500	Diode	5	3	5.9	0.36	1.1	4.8	1.8	5	±5″	±5″
	Elta 50	800	1200	Diode	5	3	3.5	0.25	0.7	6.0	1.1	5	±5″	±6″
	Rec Elta 13 C	2000	2400	Diode	3	2	5.2	0.36	1.4	4.8	2.4	3.5	±1.5″	±1.5″
	Rec Elta 14 C	1600	2000	Diode	3	2	5.2	0.36	1.4	4.8	2.4	3.5	±2.5″	±2.5″

Appendix F. Technical Data of a Selection of Pulse Distance Meters

Some relevant technical data of a selection of pulse distance meters is given in the following tables. The data have been supplied by the manufacturers or their Australian agents or have been gathered from information in promotional brochures.

The range of the distance meters is given for passive reflecting surfaces, an acrylic reflector and three glass prisms. The distances refer to so-called average atmospheric conditions. The range to passive reflecting surfaces depends heavily on the structure of such and on the ambient light.

The standard deviations are specified accuracies for single distance measurements in the most precise measuring modes. The definition has been given in Eq. (13.63) of Section 13.5.

Most manufacturers supply batteries of different capacities. The smallest are listed in the tables.

The beam divergence refers to the angle subtended by the 50% power points of the power profile across the beam.

The length of the emitted IR pulses is given, as is the pulse repetition rate. The latter becomes relevant when tracking moving reflectors.

Manufacturer	Model	Range with Passive Reflector m	Range with Plastic m	Range with Triple prism m	Wave length of laser diode nm	Standard dev. mm	Standard dev. +ppm	Weight Instr. kg	Weight Battery kg	Power cons. W	Battery Volt V	Battery Cap. Ah	Beam div. min	Pulse Length ns	Rep. rate Hz
Cubic Precision U.S.A.	Red-Dot	30	800	2500	?	5	5	?	?	?	?	?	?	?	?
Eumig Austria	LP8010–01µP	80	—	—	890	10	—	9.5	—	85	12	—	22	30	2500
	LP8015–03	15	—	—	900	30	—	7.2	—	60	12	—	22	30	800
	LP80130–10	100	300	—	900	100	1000	6.9	—	30	12	—	12	30	120
	LP80200–10	120	300	—	890	100	1000	6.9	—	30	12	—	10	30	120
	LP80300–10	150	300	—	890	100	1000	6.9	—	25	12	—	22	35	120
	LP80400–20	250	500	—	900	200	1000	6.9	—	30	12	—	22	35	120
	LP80130/ 1000–20	80	—	1000	900	200	1000	6.9	—	30	12	—	12	30	120
	LP80200/ 2000–20	120	—	2000	890	200	1000	6.9	—	30	12	—	9	30	120
Fennel	(see IBEO)														
Führer	(see IBEO)														
Ferrotron Germany	RTF 03	2–20	—	—	904	2–10	—	80	—	?	?	?	?	?	?
IBEO Germany	Fen 2000	—	150	2000	905	5	5	3.5	—	10	12	1.1	6	5	var.
	Fen 10000	—	—	3000	905	100	—	3.1	—	10	12	1.1	6	5	var.
	Fen 4000	—	—	3000	905	5	5	3.5	—	10	12	1.1	6	5	var.
	Pulsar	—	—	8000	905	5	5	1.7	0.38	?	6	0.2	1	?	?
	Pulsar Survey	—	500	8000	905	5	5	1.7	—	7.8	6	1.1	3.4	10	40
	Pulsar 1000	1000	1500	>12000	905	<30	5	1.8	—	7.8	6	1.1	1	10	40
	Pulsar Minifix	—	1000	12000	905	10	5	1.7	—	7.8	6	1.1	3.4	10	40
	Pulsar 50	70	—	8000	905	3	5	1.7	—	?	6	—	1	?	?
	Pulsar 100	140	—	5000	905	5	5	1.7	—	?	6	—	2.4	?	?
	Pulsar 500	500	—	12000	905	10	5	1.85	—	?	6	—	2	?	?

Manufacturer	Model														
Inner-space U.S.A.	Model 604	–	–	3000	?	300	150	8.1	–	?	12	–	8.6	?	?
Keuffel & Esser U.S.A.	Pulseranger	100	–	>3000	905	30	30	3.6	–	?	12	–	?	?	?
Krupp Atlas	Polarfix	–	–	5000	905	100	200	19	–	45	12	–	0.7/14	20	435
Germany	Polartrack	–	–	10000	905	150	5	20	–	2.7	12	28	–	13	1260
Laser Atlanta U.S.A.	ProSurvey 1000	610	–	1370	800–900	75–375	–	1.9	–	4.8	12	2.3	18	?	?
Laser Technology USA	Criterion 100/200	457	–	12200	?	94	–	2.8	–	3.8	9.6	3.2	10	?	?
	Criterion 300/400	457	–	12200	?	94	–	2.8	–	3.8	9.6	3.2	10	?	?
	Hydro II	–	–	12200	?	152	–	???	–	?	12	–	17	?	?
Leica Switzerland	Wild DI3000	–	>300	11000	865	5	1	1.7	–	4	12	0.33	2.6	12	2000
	Wild DIOR3002	200	–	5000	865	5	–	1.7	–	4	12	0.33	40	?	?
	Wild DI3000S	–	–	11000	860	3	1	1.7	0.9	4	12	2	2.4	11	?
	Wild DI3010S	–	–	11000	860	3	1	1.7	0.9	4	12	2	2.4	11	?
	Wild DIOR3002S	300	1300	7000	860	3–5	1	1.7	0.9	4	12	2	7.2	11	?
	Wild DIOR3012S	300	1300	7000	860	3–5	1	1.7	0.9	4	12	2	8.9	11	?
Millnar Holland	Esprit	100	150	5200	904	250	500	2.5	5.5	12	12	6	14	25	250
Riegl Austria	LD 90–210	10	–	–	?	10	–	3.5	–	12	12	–	10	?	40
	LD 90–230	30	–	–	?	20	–	3.5	–	12	12	–	10/1.5	?	10
	LD 90–250	50	1000	10000	?	10	–	3.5	–	12	12	–	10	?	40

Manu- facturer	Model	Range with			Wave length of laser diode nm	Standard dev.		Weight		Power cons. W	Battery		Beam div. min	Pulse	
		Passive Reflector m	Plastic m	Triple prism m		mm+ppm		Instr. kg	Battery kg		Volt V	Cap. Ah		Length ns	Rep. rate Hz
Riegl Austria	LD 90−2100	100	1000	10000	?	50	−	3.5	−	12	12	−	10/ 1.5	?	20
	LD 90−2300	300	−	−	?	100	−	3.5	−	12	12	−	10	?	20
	LD 90−2-HT	10	−	−	?	10	−	20	−	12	12	−	10	?	40
	LR 90−201	100	500	5000	?	10	5	2.1	0.3	7	11−28	−	?	?	?
	LR 90−202	100	800	6500	?	10	5	2.1	0.3	7	11−28	−	7	?	?
	LR 90−205	300	1000	10000	?	100	10	2.1	0.3	7	11−28	−	7/1.8	?	?
Vyner U.K.	RF 2K	50	300	2500	904	500	100	3.0	3.0	?	12	−	7.0	25	400
	RF 4K	100	500	4500	904	500	100	3.0	3.0	?	12	−	7.0	25	300
Wild	(see Leica)														
Zeiss Germany	ELDI 10	−	500	8500	905	5	3	1.5	0.36	4	4.8	1.8	2.8	5000	?
	Rec Elta RL	200	500	6000	900	5	3	5.2	0.36	1.2	4.8	1.8	8.0	12	1000

Appendix G. Technical Data of a Selection of Long Range Distance Meters

Some relevant technical data of a selection of long range distance meters is given in the following table. Separate tables are given for electro-optical and microwave instruments. The data have been supplied by the manufacturers or their Australian agents or have been gathered from information in promotional brochures.

The standard deviations are specified accuracies for single distance measurements in the most precise measuring modes. The definition has been given in Eq. (13.63) of Section 13.5.

Long range distance meters usually have a minimum as well as a maximum range. Both values are listed. It should be noted that the maximum range refers to excellent conditions during day-time, and, in the case of electro-optical instruments, to a large number of prisms (possibly as many as 40).

The column "phase measurement" makes a distinction between analogue and digital phase measurement. The former is carried out manually.

The beam divergence refers to the angle subtended by the 50% power points of the power profile across the beam. In the case of microwave distance meters, different antennae with different divergence may be available.

Manufacturer	Model	Radiation source	Carrier frequency or wavelength	Unit length m	Stand. dev.		Range km	Weight kg	Power consumption W	Phase measurement	Beam divergence
					mm	+ ppm					
Microwave Instruments:			GHz								
Microfix South Africa	100 C	Gunn-diode	16.25	?	15	3	0.02 − 60	3.4	6	analogue	3.56°/ 6°/45°
Siemens-Albis Switzerland	SIAL MD 60	Klystron	10.3	1	10	3	0.2 − 150	15	38	digital	6°
Tellurometer/ Tellumat U.K.	MRA 5	Gunn-diode	10 − 10.5	5	10	3	0.1 − 50	6/12	42	digital	6°
	CA 1000	Gunn-diode	10.1 − 10.4	3	22	5	0.05 − >30	1.6	4.5	analogue	20°
	CMW 6	?	16.0 − 16.5	?	10	3	0.1 − 50	7	<20	digital	4.5°
	CMW 20	?	34.5 − 34.9	0.9	5	3	0.02 − 25	4	12	digital	3°
	MRA 6	?	?	?	10	3	0.10 − 50	4	?	?	?
	MRA 7	Gunn-diode	16.25/17.5	1.8737	15	3	0.02 − 50	4	10	?	6°
Electro-optical Instruments:			nm								
AGA Geotronics Sweden	Geodimeter 8	5 mW laser	632.8	5	5	1	60	23	75	analogue	20″
	Geodimeter 600	1 mW laser	632.8	5	5	1	0.015 − 40	15	26	analogue	10″
	Geodimeter 78	1.8 mW laser	632.8	10	10	1	10	8.2	60 − 72	digital	40″
	Geodimeter 6000	Diode	880	10	5	1	0.0002 − 21	2.7	8.5	digital	114″
Keuffel & Esser U.S.A.	Rangemaster II	5 mW laser	632.8	10	5	1	60	18	60	digital	20″
	Ranger V	3 mW laser	632.8	10	10	2	25	16	57	digital	?
	Ranger V-A	HeNe laser	632.8	10	5	2	up to 25 km	20	60	digital	?
	Rangemaster III	HeNe laser	632.8	10	5	1	up to 60 km	20	60	digital	?
Nikon Japan	NLD-3	He-Ne laser	632.8	?	5	1	up to 60 km	19	47 − 61	?	4″ − 62″
Spectra Physics U.S.A.	Geodolite 3 G	5 mW laser	632.8	3	1	1	70	49	400	digital	?
Terra Technology Corp. U.S.A.	Terrameter LDM2	HeNe laser + HeCd laser	632.8 441.6	0.05	0.1	0.1	1.0 − 20	42.2	575	?	40″

Appendix H. Critical Dimensions of a Selection of Reflectors

(All data supplied by manufacturers)

Manufacturer	Reflector type	a mm	b mm
AGA	Old (-30 mm)	41	35
	New ("zero const")	40.5	62
Hewlett Packard	Single prism (HP 11410E)	50.3	47.8
Kern	DM 500$-$504 prism	81.0	34.5
	ME 5000 prism	40.00*	26.00*
Nikon	Single prism set	50	75
Pentax	MPU-54, MPU-7A	39.5	49.5
Sokkisha	PR-1, PR-3, PR-9	57.6	57.0
	APO (-40 mm)	50.0*	35.45*
	APO S (-30 mm)	50.0	45.45
	APO S2 (0 mm)	50.0	75.45
Topcon	Prism 2 on tilting prism holder 3 (0 mm)	48	72.8
	(-30 mm)	48	42.8
Wild	GDR 11/GDR 31	60	22
	GPR 1/GPR 1P	39.6*	26.2*
Zeiss	TR 2, TR 7, TR 19	40*	27*
(Oberkochen)	KTR 1, KTR 3, KTR 7	41*	27*

Asterisks indicate prism systems which minimize pointing errors on distances and angles [see Eq. (10.27)]. Some nominal absolute reflector constants are given in brackets. The parameters a and b are defined as follows: a = height of prism corner above front plane of prism (refer to Fig. 10.4, Sect. 10.2.4), b = distance between horizontal and vertical axis of reflector and front plane of prism.

The manufacturing tolerances of parameters a and b are usually better than 0.5 mm. However, the tolerance for the Kern ME 5000 prism is given as ± 0.05 mm.

References

Aeschlimann H, Stocker R (1974) Instrumental errors in electro-optical distance measurements. Paper 505.4, Proc, 14th FIG Congress, Washington DC

AGA (1969) Geodimeter model 6A operational manual. Publication 571/071, AGA, Lindigö, 53 pp

AGA Products Australia (1975) Newsletter, No 13

Allan AL (1967) The accurate measurement of a short line by Geodimeter model 6. Surv Rev 19(145):98−102

Angus-Leppan PV (1974) Refraction over snow and ice surfaces. Unisurv G-21, School of Surveying, University of New South Wales, Sydney, pp 73−97

Angus-Leppan PV (1979) Network adjustment by the ratio method and its meteorological basis. Aust J Geod Photogram Surv 31:15−26

Angus-Leppan PV, Brunner FK (1980) Atmospheric temperature models for short-range EDM. Can Surv 34:153−165

Anon (1984) Documents concerning the new definition of the metre. Metrologia 19:163−177

Anthes RA, Panofsky HA, Cahir JJ, Rango A (1975) The atmosphere. Merrill, Columbus, Ohio, USA, 339 pp

Arecchi FT, Schulz-Dubois EO (eds) (1972) Laser handbook, vol 1 and 2. North-Holland, Amsterdam

Arnold DA (1979) Method of calculating retroreflector-array transfer functions. Special Report 382, Smithonian Astrophysical Observatory, Cambridge, Mass, USA, 165 pp

Baarda W (1962) A generalization of the concept, strength of figure. Report submitted to the members of I.A.G. Special Study Group 1.14. In: Appendix, Netherlands Geodetic Commission, vol 2, no 4, 1967, pp 55−74

Baarda W (1967) Statistical concepts in geodesy, vol 2, no 4, 1967. Netherlands Geodetic Commission, Publications on Geodesy, New Series, pp 5−54

Barak M (ed) (1980) Electrochemical power sources: primary and secondary batteries. IEE Energy Series 1, Institution of Electrical Engineers, London, 498 p

Barrell H, Sears JE (1939) The refraction dispersion of air for the visible spectrum. Philos Trans R Soc Lond A Math Phys Sci 238:1−64

Benwell GL, Murnane AB, Sprent A (1985) Results of EDM Calibrations. Aust Surv 32:422−432

Best AC, Knighting E, Pedlow RH, Stormonth K (1952) Temperature and humidity gradients in the first 100 m over south-east England. Geophys Mem 89:60

Billard B (1988) Assessment of various configurations for locating the mean sea surface reference in laser hydrography. Mar Geod 12:1−20

Bittel H (1935) Über den Brechungsindex von Gasgemischen. Ann Phys 5(23):61−89

Bjerhammar A (1971) The second generation of electro-optical distance measuring instruments. Proc, 13th Int Congress of Surveyors (FIG), paper 502.1, Wiesbaden, FRG, 11 pp

Bloom GS, Golomb SW (1976) Numbered complete graphs, unusual rulers and assorted applications. In: Alavi Y, Lick DR (eds) Theory and applications of graphs. Proceedings, Michigan 1976, Lecture Notes in Mathematics, no 642. Springer, Berlin Heidelberg New York, pp 53−65

Bolsakov VD, Deumlich F, Golubev AN, Vasilev VP (1985) Elektronische Streckenmessung. Verlag Bauwesen, Berlin, G.D.R., 255 pp

Bomford G (1975) Geodesy, 3rd edn. Oxford University Press, reprint, London

Born M, Wolf E (1970) Principles of optics. Electro-magnetic theory of propagation, interference and diffraction of light, 4th edn. Pergamon, London

Borutta G, Maass KD (1986) Überwachungsmessungen zur Erfassung kurz- und langzeitiger Pfeilerbewegungen von Festpunkten eines Lagenetzes. Vermessungstechnik 34:164–166

Bradsell RH (1976) Georan I, a compact two-colour EDM instrument − Part II: Electronic principles. Surv Rev 23:219–233

Brettenbauer K (1975) Phänomenologische Deutung der Rinnerschen Ortskonstanten. Vermessung − Photogrammetrie − Kulturtechnik (VPK) (Fachblatt 3/4) 73:231–233

Brown T, Irving S, Arneson G (1981) 100% Thru-put improvement in PC shop automatic crane control systems with non-contact position feedback and optimized proportional control. Technical paper, 24th Annual Meeting of the Institute for Interconnecting and Packaging Electronic Circuits (IPC), April 1981, Washington DC, ITC-TP-371, 21 pp

Brunner FK (1973) The effect of deviation of vertical in trigonometric levelling with steep sights. Österr Z Vermessungswesen Photogram 61:126–134 (in German)

Brunner FK (1975a) Coefficients of refraction on a mountain slope. Unisurv G-22, School of Surveying, University of New South Wales, Sydney, pp 81–96

Brunner FK (1975b) Trigonometric levelling with measured slope distances. Proceedings, 18th Australian Survey Congress, Institution of Surveyors, Australia, Perth, pp 79–94

Brunner FK (1984) Modelling of atmospheric effects on terrestrial geodetic measurements. In: Brunner FK (ed) Geodetic refraction. Springer, Berlin Heidelberg New York, pp 143–162

Brunner FK, Angus-Leppan PV (1976) On the significance of meteorological parameters for terrestrial refraction. Unisurv G-25, School of Surveying, University of New South Wales, Sydney, pp 95–108

Brunner FK, Fraser CS (1977) Application of the atmospheric turbulent transfer model (TTM) for the reduction of microwave EDM. Unisurv G-27, School of Surveying, University of New South Wales, Sydney, pp 3–26

Brunner FK, Fraser CS (1978) An atmospheric turbulent transfer model for EDM reduction. In: Richardus P (ed) Proceedings, international symposium on EDM and the influence of atmopheric refraction, 23–28 May 1978, Wageningen, Netherlands Geodetic Commission, pp 304–315

Brunner FK, Groenhout KI (1979) Microwave distance measurements during rainfall. Aust Surv 29:569–580

Buck AL (1981) New equations for computing vapour pressure and enhancement factor. Appl Meteorol 20:1527–1532

Burnside CD (1991) Electromagnetic distance measurement, 3rd ed. BSP Prof. Books, Oxford, 278 pp

Burrus CA (1972) Radiance of small-area high-current-density electroluminescent diodes. Proceedings of the IEEE, February 1972, pp 231–232

Byers TJ (1984) Making photovoltaics work for you: 20 selected solar projects. Micro text publications, Prentice-Hall, Englewood Cliffs, NJ, USA, 173 pp

Carslaw HS (1950) Introduction to the theory of Fourier's series and integrals, 3rd edn. Dover, NY, 368 pp

Caspary WF (1987) Concepts of network and deformation analysis. School of Surveying, University of New South Wales, Sydney, Australia, 185 pp

Chaborski H (1978) Laserentfernungsmesser. Swiss Patent no 628995, submitted 25 May 1978, 5 pp

Chang RF, Currie DG, Alley CO (1971) Far-field pattern for corner reflectors with complex reflection coefficients. J Opt Soc Am 61:431–438

Ciddor PE (1987) A 70-metre laser interferometer for the calibration of survey tapes and EDM equipment. Aust Surv 33:493–502

Cooper MAR (1987) Control surveys in civil engineering. Collins, London, 381 pp

Couchmann HD (1974) A method of evaluating cyclic errors in EDM equipment. Aust Surv 26:113–115

Covell PC (1979) Periodic and non-periodic errors of short-range distance meters. Aust J Geod Photogram Surv 30:79–101

Covell PC, Rüeger JM (1982) Multiplicity of cyclic errors in electro-optical distance meters. Surv Rev 26(203):209–224

Dabrowski W, Maier V (1977) Determination of horizontal distance and height difference from combined applications of EDM instruments and theodolites. Allgem Vermessungs-Nachr (AVN) 84:111–117 (in German)

Daino B, Piazzolla S, Spano P (1976) Spatial dependence of time delay in light-emitting diodes. J Appl Phys 47:2773–2775

Dalcher A (1975) Simplified theory of the deformation of a reflector due to unequal temperature. Vermessungs – Programmetrie – Kulturtechnik (VPK) (Fachblatt 3/4) 73:237–239 (in German)

Deichl K (1984) Der Brechungsindex für Licht und Mikrowellen. Allgem Vermessungs-Nachr (AVN) 91:85–100

DIN Deutsches Institut für Normung (1985) Deutsche Norm DIN 18723 Teil 6, Genauigkeitsuntersuchungen an geodätischen Instrumenten: Elektrooptische Distanzmesser für den Nahbereich. Entwurf, Oktober 1985, 5 pp

Dodson AH, Deeth CP, Ashkenazi V (1981) The use of the laser interferometer for scale calibration of the Nottingham baseline. Manuscripta Geod 6:63–74

Edge RCA (1960) Report of IAG special study group no 19 on electronic distance measurement using ground instruments. Presented at the XIIth General Assembly of the International Union of Geodesy and Geophysics (IUGG), Helsinki, 1960

Edlen B (1953) The dispersion of standard air. J Opt Soc Am 43:339–344

Edlen B (1966) The refractive index of air. Metrologia 2:71–80

Ehbets H, Stöckli B, LeHelloco G (1983a) Le Distomat Wild DI20. Vermessung – Programmetrie – Kulturtechnik (VPK) 81:125–126

Ehbets H, Stöckli B, LeHelloco G (1983b) The Distomat Wild DI20. Wild Heerbrugg, CH-9435 Heerbrugg, Switzerland, 6 pp

Elmiger A (1977) Three-dimensional adjustment of terrestrial geodetic networks. Proceedings, XVth International Congress of Surveyors, FIG, Stockholm 5:251–257 (in German)

Elmiger A, Siegerist C (1977) Practical tests and experiences with the Mekometer. Proceedings, 7th international course for high precision engineering surveys. THD Schriftenreihe Wissenschaft and Technik, Darmstadt, 1:203–219

Emenike EN (1982a) A simplified approach to accurate EDM instrument calibration. Surv Rev 26:273–278, 27:140

Emenike EN (1982b) On the calibration of EDM instruments. Aust Surv 31:175–185

Engler ML (1983) Terrain profiling using an inertial survey system integrated with an airborne laser altimeter. Technical papers, 1983 ACSM-ASP Fall convention, Salt Lake City, Utah, 19–23 September 1983, pp 595–606

Erickson KE (1983) Light modulator employing electro-optic crystals. US Patent no 4379620, 12 April 1983

Essen L, Froome KD (1951) The refractive indices and dielectric constants of air and its principal constituents at 24000 Mc/s. Proc Phys Soc (Lond) B-64:862–875

Fischer W (1971) Protokoll der 117. Sitzung der Schweizerischen Geodätischen Kommission vom 18. Juni 1971, p 45

Fraser CS (1979) Identification of optimum meteorological conditions for EDM via potential refractivity. Can Surv 33:27–38

Fraser CS (1984) The turbulent transfer model applied to Geodolite measurements. Can Surv 38:79–90

Frerking ME (1978) Crystal oscillator design and temperature compensation. Van Nostrand-Reinhold, New York, 240 pp

Fritschen LJ, Gay LW (1979) Environmental instrumentation. Springer, Berlin Heidelberg New York, 216 pp

Froome KD (1971) Mekometer III: EDM with sub-millimetre resolution. Surv Rev 161:98–118

Gardner M (1983) Wheels, life and other mathematical amusements. Freeman, New York, pp 152–165

General Electric (1975) Nickel-cadmium battery. Application engineering handbook. General Electric Company, Publ GET 3148A, 2nd edn. Gainesville, Florida, 240 pp

Geotronics AB (1987) Geodimeter 140 SMS Slope Monitoring System – system description. Publ no 571140305, Geotronics AB, Sweden, 10 pp

Gervaise J (1984) Results of the geodetic measurements carried out at CERN with the Terrameter. In: Technical papers, meeting of commission 6 of FIG, 10–11 March 1983, Washington DC, pp 23–32

Geotronics AB (1985) Data sheet – Geodimeter 220. Publ no 571140207, July 1985

Giger K (1983) Vorrichtung zur Laufzeitmessung von Impulssignalen. German patent, DE 33 22 145, submitted 20. 6. 1984, 19 pp

Glicksman R (1975) Semiconductor laser diodes. Application note AN-4993, Radio Corporation of America (RCA), Solid State Division, Electro-optics and devices, Lancaster, PA 17604, USA, pp 11

Grimm K, Frank P, Giger K (1986a) Distanzmessung nach dem Laufzeitmessverfahren mit geodätischer Genauigkeit. Vermessung − Photogrammetrie − Kulturtechnik (VPK) 84: 627−631

Grimm K, Frank P, Giger K (1986b) Time-pulse distance measurement with geodetic accuracy. Wild Heerbrugg Ltd, Heerbrugg, Switzerland, May 1986, 14 pp

Gründig L, Teskey WF (1984) Improvement of network precision and reliability by the use of scaled distances. In: Technical papers, meeting of FIG commission 6, 10−11 March 1984, Washington DC, pp 104−113

Halmos F (1977) Zur Genauigkeitssteigerung der Distanzmessung. In: Festschrift Löschner, Veröffentlichung Nr 23, Geodätisches Institut, Rheinisch-Westfälische Technische Hochschule Aachen, FRG, pp 363−370

Halmos F (1980) A particular case of the calibration of distance meters and of the increase of the accuracy of distances. In: Parm T (ed) Proc, Symposium on high precision geodetic length measurements, 19−22 June 1978, Helsinki, Finland, Report of the Finnish Geodetic Institute, No 80:1, pp 215−232

Halmos F, Kadar I (1972) Die Bestimmung der Additionskonstanten elektronischer Entfernungsmesser. Vermessungstechnik (VT) 20:343−346

Halmos F, Kadar I (1976) Weitere Untersuchungen zur Bestimmung der Eichkonstanten von elektro-optischen Streckenmessgeräten kurzer Reichweite. Vermessungstechnik (VT) 24:309−312

Hamada T, Ohtomo F (1981) Electro-optical range finder using three modulation frequencies. European Patent Application, 5 March 1981, Publ no 0035 755-A3, 32 pp

Hayashi I (1984) Heterostructure lasers. IEEE transactions on electron devices, ED-31: 1630−1641

Heck B, Heil E (1981) Zur Bestimmung der Nullpunktskorrektur elektronischer Distanzmesser durch Streckenmessung in allen Kombinationen. Allgem Vermessungs-Nachr (AVN) 88:286−298

Heister H (1988) Zur Fehlausrichtung von Tripelprismen. Z Vermessungswesen (ZfV) 113: 249−258

Hewlett-Packard (1973) Hewlett-Packard prisms. Insight, No 2:1−2. Hewlett-Packard Civil Engineering Products, Loveland, Colorado, USA

Hines RH (1976) A geodetic and survey infrared distance measurement instrument. Proc of SPIE (Soc. of Photo-Optical Instrumentation Engineers), vol 95. Modern utilization of infrared technology II − Civilian and Military, 26−27 August 1976, San Diego, California, pp 204−211

Hoar GJ (1981) Satellite surveying: theory − applications. Publication Mx-TM-3346-81. Magnavox Advanced Products and Systems Company, Torrance, California, 90503, 140 pp

Hölscher HD (1980) Extension of the 432 m Väisälä Base at Pienaarsrivier to 864 m and 1728 m. In: Parm T (ed) Proc, Symposium on high precision geodetic length measurements, 19−22 June 1878, Helsinki, Report no 80:1 of the Finnish Geodetic Institute, Helsinki, pp 206−214

Höpcke W (1964) Über die Bahnkrümmung elektromagnetischer Wellen und ihren Einfluss auf die Streckenmessungen. Z Vermessungswesen 89:183−200

Hoskins GW (1982) Navy navigation satellite system status. Proc, Third international geodetic symposium on satellite doppler positioning, 8−12 February 1982, Las Cruces, NM, USA, pp 825−841

Hübner E (1978) Theoretische und experimentelle Untersuchungen zur Ermittlung des vertikalen Tagesgangs der bodennahen Refraktion zur Ableitung von Einsatzkriterien für Laserleitstrahlsysteme. Geodätische und geophysikalische Veröffentlichungen, Reihe III, Heft 41, Dresden, pp 188−206

Huggett GR (1979) Two-colour Terrameter. Symposium on recent crustal movements, XVII General Assembly, IUGG, Canberra, December 1979, 16 pp

Huggett GR (1981) Two-colour Terrameter. Tectonophysics 71:29−39
Huggett GR, Slater LE (1975) Precision electro-magnetic distance measuring instrument for determining secular strain and fault movement. Tectonophysics 29:1−4
Humphries JG (1989) Scale accuracy of the CR 204 Geomensor. Paper presented at the 1989 Conference of Southern African Surveyors, 17 pp
ICAO (1964) Manual of the ICAO standard atmosphere extended to 32 kilometres (10500 feet), 2nd edn. International Civil Aviation Organization, doc 7488/2
Iribarne JV, Cho HR (1980) Atmospheric physics. Reidel, Boston
IUGG (International Union of Geodesy and Geophysics) (1975) Resolutions, XVIth General Assembly, 1975, Grenoble, France. Bull Géod 118:365
Jäger R (1985) Zur Anwendung von Streckenverhältnisbeobachtungen in Überwachungsnetzen und auf Eichstrecken. Allgem Vermessungs-Nachr (AVN) 92:53−65
Johnson D (1981) Reflective prisms for surveying use. POB, Wayne Mich, USA, 6:26−33
Jones HE (1971) Systematic errors in Tellurometer and Geodimeter measurements. Can Surv 25:406−423
Kääriäinen J, Konttinen R, Lu Q, Du ZY (1986) The Chang Yang standard baseline. Publications of the Finnish Geodetic Institute, No 105, Helsinki, 36 pp
Kääriäinen J, Konttinen R, Németh Z (1988) The Gödöllö standard baseline. Publications of the Finnish Geodetic Institute, No 109, Helsinki, 66 pp
Kahmen H (1978) Electronic methods of measurement in geodesy. Fundamentals and applications, 2nd ed. Wichmann, Karlsruhe, 406 pp (in German)
Kahmen H, Faig W (1988) Surveying. De Gruyter, Berlin New York, 578 pp
Kahmen H, Steudel J (1988) Das automatisch zielende Meßsystem Georobot II. In: Schädelbach K, Ebner H (eds) Ingenieurvermessung 88, Proc of Xth Int Kurs für Ingenieurvermessung, 12−17 September 1988, Munich. Dümmler, Bonn, pp A7/1−A7/14
Kahmen H, Zetsche H (1974) Comparative investigations into electro-optical short range distance meters. Z Vermessungswesen (ZfV) 99:68−81 (in German)
Kasser M (1982) Le Georan II, Nouvel appareil de mesure de distances à deux couleurs. In: Proc, OPTO 82, 16−18 November 1982, Paris, France, ESI, pp 176−178
Kasser M (1985) Recent advances in high precision trigonometric motorized levelling (NIPREMO) in IGN France. In: Proc, Third International Symposium on the North American Vertical Datum (NAVD Symposium '85), 21−26 April 1985, Rockville, Maryland, USA, NOAA, Rockville, pp 57−64
Kendal B (1979) Manual of avionics − an introduction to the electronics of civil aviation. Granada, London, 276 pp
Kennie TJM (1983) Some tests of retro-reflective materials for electro-optical distance measurements. Surv Rev 27:3−12, 51−61
Kern (1974) Test line for electro-optical distance measuring instruments. Kern Bull 20:6−9
Kern (1987) Landslip under surveillance. Kern Bull 40:11−13
Keuffel, Esser (1976) Electronic distance measuring equipment. Catalog 10b, Morristown NJ
Kneissl M (1956) Höhenmessung − Tachymetrie, Band III. In: Jordan, Eggert, Kneissl (eds) Handbuch der Vermessungskunde, 10th edn. Metzlersche Verlagsbuchhandlung, Stuttgart, FRG, 776 pp
Kobold F (1958) Geodätische Methoden zur Bestimmung von Geländebewegungen und von Deformationen an Bauwerken. Schweizerische Bauzeitung 76:163−167, 182−187
Kobold F (1961) Measurement of displacement and deformation by geodetic methods. Journal of the Surveying and Mapping Division, Proc of the American Society of Civil Engineers 87(SU2):37−67
Kohlrausch F (1955) Praktische Physik, Band 1, 19th edn, 4th imp. In: Henning F (ed) Teubner, Leipzig, GDR
Kondo (1985) Article on ET-1. J (Jpn) Assoc Precise Surv Appl Technol (APA) 28(1):73−78
Konttinen R (1988) Baseline multiplication using the Kern Mekometer ME 3000. Report no 88:1 of the Finnish Geodetic Institute, Helsinki, 10 pp
Kopeika NS et al (1983) Wavelength tuning of GaAs LED's through surface effects. IEEE Transactions on electron devices, ED-30:334−347
Kressel H, Ettenberg M (1982) Semiconductor lasers. In: Weber MJ (ed) CRC handbook of laser science and technology, vol 1 (Lasers and masers). CRC, Boca Raton, Florida, 552 pp

Krupp Atlas Elektronik (1983) Atlas Polarfix — Das automatische Laser Polarortungssystem. Publ no BL 9153 T 029, July 1983, Bremen, FRG

Kukkamäki TJ (1968) Ohio standard baseline. Report no 107, Dept of Geodetic Science. The Ohio State University, Columbus, Ohio, 43210, USA, 59 pp

Kukkamäki TJ (1978) Väisälä interference comparator. Publication no 87 of the Finnish Geodetic Institute, Helsinki, 1978, 49 pp

Küpfer HP (1968) Ground reflections and measurement technique in microwave distance measurement. Schweiz Z Vermessung, Photogrammetrie Kulturtechnik 66:290−309, 329−348 (in German)

Küpfer HP, Hossmann M (1971) The concept of the new microwave distance meter Distomat DI 60. A contribution to the solution of the reflection problem. Allgem Vermessungs-Nachr (AVN) 78:50−64 (in German)

Landolt-Börnstein (1962) Zahlenwerte und Funktionen aus Physik, Chemie, Astronomie, Geophysik und Technik, vol 2. In: Hellwege KH, Hellwege M (eds) Eigenschaften der Materie in ihren Aggregatzuständen, Part 8: Optische Konstanten. Springer, Berlin Heidelberg New York

Langbein JO, Linker MF, McGarr AF, Slater LE (1987) Precision of two-colour geodimeter measurements: results from 15 months of observations. J Geophys Res 92(B11): 11644−11656

Lau KY (1988) High speed semiconductor laser technology. Solid State Technology 31:91−95

Laurila SM (1976) Electronic surveying and navigation. Wiley, New York, 545 pp

Leitz H (1977) Zur Genauigkeit und Reichweite von elektro-optischen Distanzmessern am Beispiel des Zeiss Eldi 2. Z Vermessungswesen (ZfV) 102:152−156

Le Roux ZP (1976) Reduction of eccentric distance measurement. S Afr Surv J 15:62−63

Levine J (1984) Multiple wavelength electromagnetic distance measurements. In: Brunner FK (ed) Geodetic refraction. Springer, Berlin Heidelberg New York, 45−51 pp

Lindner K, Ritter B (1985) Geodätische Arbeiten auf den Filchner-Ronne und Ekström-Schelfeisen 1979 bis 1982. Polarforsch 55:1−26

List RJ (1958) Smithsonian meteorological tables, 6th revised edn. Publication 4014. Smithsonian Miscellaneous Collection 114, Washington

Liu LS, Klinger JH (1979) Interferometry. In: Hopkins GW (ed) Proceedings of SPIE, vol 192. The International Society for Optical Engineering, pp 17−26

McLean RE (1984) Retrodirective prism constants: an investigation. New Zealand Surv 30:636−643

Mamon G et al (1978) Pulsed GaAs laser terrain profiler. Appl Opt 17:868−877

Mantell CL (1983) Batteries and energy systems, 2nd edn. McGraw-Hill Book, NY, 319 pp

Maurer W, Schirmer W, Schwarz W (1988) Wirkungsweise und Einsatzmöglichkeiten des Mekometer ME 5000. In: Schädelbach K, Ebner H (eds) Ingenieurvermessung 88 — Beiträge zum X Internationalen Kurs für Ingenieurvermessung, 12−17 September 1988, München. Dümmlers, Bonn, pp A 10/1−A 10/14

Meier D, Loser R (1986) Das Mekometer ME 5000 — Ein neuer Präzisionsdistanzmesser. Allgem Vermessungs-Nachr (AVN) 93:182−190

Meier D, Loser R (1988) Versuche mit einem Zwei-Farben-Distanzmesser. In: Schädelbach K, Ebner H (eds) Ingenieurvermessung 88 — Beiträge zum X Internationalen Kurs für Ingenieurvermessung, 12−17 September 1988, München. Dümmlers, Bonn, pp A 11/1−A 11/12

Meier D, Rüeger JM (1984) Accuracy of Mekometer scale. Can Surv 38:27−33

Melles Griot (1975) Optics guide. Melles Griot, 1770 Kettering Street, Irvine, Cal 92714, USA

Menzies GH (1972) Eccentric distance measurement. S Afr Surv J 13:3−7

Michelson AA (1927) Measurement of the velocity of light between Mount Wilson and Mount San Antonio. Astrophys J 82:1−22

Moritz H (1967) Application of the conformal theory of refraction. Österr Z Vermessungswesen. Sonderband 25:323−334 (in German)

Münch KH (1973) The infrared distance meter Kern DM 1000. Allgem Vermessungs-Nachr (AVN) 80:201−207 (in German)

Münch KH (1974) The electronic distance meter Kern DM 500: part of a modern survey system. Allgem Vermessungs-Nachr (AVN) 81:48−54 (in German)

Murray FW (1967) On the computation of saturation vapour pressure. J Appl Meteorol 6:203

Myres JAL (1983) Laser airborne depth sounder (LADS). Proc, Conference of Commonwealth Surveyors, 1983, Paper no C4, pp 1–21

NASA (National Aeronautics and Space Administration) (1983) Transportable laser ranging system (TLRS-2) commences tracking. Crustal Dynamics Project Bull, vol 1, no 3, April 1983, pp 1–2

National Physical Laboratory (1960) The refractive index of air for radio waves and microwaves. National Physical Laboratory, London

National Standards Laboratory (1972) The Australian standard for the measurement of physical quantities. CSIRO, Melbourne

Norton TG (1986) Analysis of EDM calibration measurements: program calory. Aust Surv 33:44–54

NSW Dept of Lands (1976) Manual of the New South Wales integrated survey grid. NSW Department of Lands, Sydney, 166 pp

Ohtomo F (1983) Optical distance-measuring method and apparatus therefore. European patent application, publ no 0091665, date of filing: 7 April 1983. Tokyo Kogaku Kikai Kabashiki Kaisha, 60 pp

Olson GH, Ban VS (1987) InGaAsP: the next generation in photonics materials. Solid State Technol 30:99–105

Owens JC (1967) Optical refractive index of air: dependence on pressure, temperature and composition. Appl Opt 6:51–59

Pauli W (1968) Messung einer kurzen Basis mit dem EOS durch Streckenmessung in allen Kombinationen. Vermessungstechnik (VT) 16:242–246

Pauli W (1969) Vorteile eines kippbaren Reflektors bei der elektro-optischen Streckenmessung. Vermessungstechnik 17:412–415

Pauli W (1973) Advantages of a tiltable reflector for electro-optical telemetry. Surv News, Zeiss, Jena, GDR 26:9–14

Pauli W (1977a) Calibration of and baselines for electro-optical distance meters. Vermessungstechnik (VT) 25:265–267 (in German)

Pauli W (1977b) Ausgleichende Sinuslinie für die Eichkorrekturkurve elektro-optischer Streckenmessgeräte. Vermessungstechnik (VT) 25:168–169

Pauli W et al (1979) Establishment of a baseline of VEB Carl-Zeiss Jena for the calibration of electro-optical distance meters. Vermessungstechnik (VT) 27:135–137 (in German)

Peck ER (1948) Theory of the corner cube interferometer. J Opt Soc Am 38:1015–1024

Peck ER, Reeder K (1972) Dispersion of air. J Opt Soc Am 62:958–962

Pelli T, Meier-Bukowiecki Y (1986) Luftbelastung. Schweiz Ing Architekt 104:1211–1216

Plessey Australia Pty Ltd (1975) Nickel cadmium cells. Plessey Components, product data PD 2098

Querzola B (1979) High accuracy distance measurement by two-wavelength pulsed laser sources. Appl Opt 18:3035–3047

Radio Corporation of America (1974) Electro-optics handbook, EOH-11. RCA Commercial Engineering, Haarison, NJ, 255 pp

Raphael D (1983) Surveillance surveys conducted by the design engineering and environment department of the state electricity commission of Victoria. In: Rivett L (ed) Proc, Symposium on the Surveillance of Engineering Structures, Dept of Surveying, University of Melbourne, 14–15 November 1983, 15 pp

Reissmann G (1982) Elektronische Streckenmessung in allen Kombinationen und Bestimmung der Additionskonstanten. Vermessungstechnik (VT) 30:311–313

Richter H (1970) Visibility and range of electro-optical distance meters. Vermessungstechnik (VT) 18:31–35 (in German)

Rinner K, Benz F (1966) Die Entfernungsmessung mit elektro-magnetischen Wellen und ihre geodätische Anwendung, vol 6. In: Jordan, Eggert, Kneissel (eds) Handbuch der Vermessungskunde, 10th edn, Metzlersche Verlagsbuchhandlung, Stuttgart, 1038 pp (in German)

Ritter B (1987) Einige Ergebnisse des motorisierten trigonometrischen Nivellements 1987 im Ritscher-Hochland/Antarktis. Geodätische Schriftenreihe der Technischen Universität Braunschweig, Nr 7, pp 49–59

Robertson KD (1971) Measurements of small movements in large structures. Surv Rev 20(160):74–84

Rüeger JM (1976) Problems in the joint determination of cyclic error and additive constant of electro-optical distance meters. Allgem Vermessungs-Nachr (AVN) 83:338–344 (in German)

Rüeger JM (1976a) The effect of phase inhomogeneities during the use of plane mirrors for electro-optical distance measurement. Z Vermessungswesen (ZfV) 101:391–392 (in German)

Rüeger JM (1976b) Remarks on the joint determination of zero error and cyclic error for EDM instrument calibration. Aust Surv 28:96–103

Rüeger JM (1976c) An aid for the design of baselines for electronic distance measuring instruments. Vermessung – Photogrammetrie – Kulturtechnik (Mitteilungsblatt) 74:249–251 (in German)

Rüeger JM (1977) Design and use of baselines for the calibration of EDM instruments. Proc, 20th Australian Survey Congress, Institution of Surveyors, Australia, Darwin, pp 175–189

Rüeger JM (1978) Computation of cyclic errors of EDM instruments using pocket calculators. Aust Surv 29:268–284

Rüeger JM (1978a) Misalignment of EDM reflectors and its effects. Aust Surv 29:28–36

Rüeger JM (1978b) Design of baselines of the Schwendener type for electro-optical distance meters. Vermessungswesen Raumordnung (VR) 40:315–324 (in German)

Rüeger JM (1978c) Variation with temperature of frequencies of short range EDM instruments. Unisurv G-28, School of Surveying, University of NSW, Sydney, pp 47–58

Rüeger JM (1978d) Frequency calibration of short range EDM instruments. Vermessung – Photogrammetrie – Kulturtechnik 76:125–130 (in German)

Rüeger JM (1980a) Legal requirements for the calibration of EDM instruments. Proc, 22nd Australian Survey Congress (Hobart), Institution of Surveyors, Australia, Sydney

Rüeger JM (1980b) Recent developments in electronic distance measurement. Aust Surv 30:170–177

Rüeger JM (1982) Quartz crystal oscillators and their effects on the scale stability and standardization of electronic distance meters. Unisurv Report S-22, School of Surveying, University of New South Wales, Sydney, Australia, 150 pp

Rüeger JM (1983) Deformation measurement and analysis as applied to EDM baselines. In: Rivett L (ed) Proc, Symposium on the surveillance of engineering structures, Dept of Surveying, University of Melbourne, 14–15 November 1983, 27 pp

Rüeger JM (1984) A test of atmospheric temperature models in EDM, using a Wild Distomat DI 20. Aust J Geod Photogram Surv 41:61–96

Rüeger JM (1985) Traceability of electronic distance measurements to national standards. Proc, 27th Australian Survey Congress, 23–30 March 1985, Alice Springs, Australia, pp 149–163

Rüeger JM (1986) Measurement and computation of short periodic errors of EDM instruments. Surv Mapping 46:231–241

Rüeger JM (1987) Zielsetzungen der Kalibrierung elektronischer Distanzmesser. Vermessungswesen-Raumordnung (VR) 49:433–444

Rüeger JM (1988) Introduction to electronic distance measurement. 2nd edn, 5th impr. School of Surveying, University of New South Wales, Sydney, Australia, 150 pp

Rüeger JM (1989) Permanent and inexpensive reflectors for monitoring surveys. In: Proc, Symposium on surveillance and monitoring surveys, Dept of Surveying University of Melbourne, 9–10 November 1989, pp 63–73

Rüeger JM, Brunner FK (1981) Practical results of EDM-height traversing. Aust Surv 30:363–372

Rüeger JM, Brunner FK (1982) EDM-height traversing versus geodetic levelling. Can Surv 36:69–88

Rüeger JM, Ciddor PE (1987) Short range performance of precision distance meters Kern Mekometer ME 3000 and Com-Rad Geomensor 204 DME. Aust Surv 33:480–492

Rüeger JM, Ciddor PE (1989) Close range performance of precision distance meters. Manuscripta Geod 14:43–48

Rüeger JM, Pascoe RW (1989) Perfomance of a Wild Distomat DI 3000 distance meter. In: Proc, 31st Australian Survey Congress, 1–7 April 1989, Hobart, Australia, pp 307–323

Rüeger JM, Siegerist C, Stähli W (1975) Investigations into electro-optical short range distance meters. Vermessung − Photogrammetrie − Kulturtechnik (Mitteilungsblatt) 73:73−76, 93−100, 126−133, 270 (in German)

Rüeger JM, Brunner FK, Becek K (1989) EDM monitoring using a local scale parameter model. In: Proc, Symposium on surveillance and monitoring surveys, Dept of Surveying, University of Melbourne, 9−10 November 1989, 12 pp

Saastamoinen J (1962) The effect of path curvature of light waves on the refractive index. Can Surv 16:90−100

Saastamoinen J (1963) Some experimental results on the NASM-4B Geodimeter. Can Surv 17:47−52

Saastamoinen J (1964) Curvature correction in electronic distance measurement. Bull Géod 73:265−269

Saastamoinen J (ed) (1969) Surveyors guide to electromagnetic distance measurement. Hilger, Bristol

Savage JC, Prescott WH (1973) Precision of Geodolite distance measurements for determining fault movements. J Geophys Res 78:6001−6008

Savage JC, Prescott WH, Lisowski M, King N (1979) Geodolite measurements of deformation near Hollister, California, 1971−1978. J Geophys Res 84(B13):7599−7615

Savage JC, Prescott WH, Lisowski M (1987) Deformation along the San Andreas Fault 1982−1986 as indicated by frequent Geodolite measurements. J Geophys Res 92(B6): 4785−4797

Schlichting R (1980) Vereinfachte Additionswertbestimmung eines EDM nach der Methode der Streckenmessung in allen Kombinationen. Vermessungsingenieur 31:158−162

Schneuwly B, Celio T (1988) Topomat − Ein Vermessungsroboter. In: Schädelbach K, Ebner H (eds) Ingenieurvermessung 88, Proc, Xth Int Kurs für Ingenieurvermessung, 12−17 September 1988, München, Dümmlers, Bonn, pp A8/1−A8/10

Schwarz W (1981) Kalibrierung elektro-optischer Distanzmesser. Vermessungswesen-Raumordnung (VR) 43:65−90

Schwarz W (1983) Die Additionskorrektion bei elektro-optischen Distanzmessern. Allgem Vermessungs-Nachr (AVN) 90:54−63

Schwarz W, Witte B (1986) Die Prüfung elektro-optischer Distanzmesser am Geodätischen Institut der RWTH Aachen. In: Festschrift E. Hektor, Veröffentlichung des Geodätischen Instituts der Rheinisch-Westfälischen Technischen Hochschule Aachen, Nr. 40, pp 15-1−15-45

Schwendener HR (1971) Elektronische Distanzmesser für kurze Strecken − Genauigkeitsfragen und Prüfverfahren. Schweiz Z Vermessung, Photogrammetrie Kulturtechnik 69:59−67

Schwendener HR (1972) Electronic distancers for short-ranges: accuracy and checking procedures. Surv Rev 21(164):273−281

Slater LE (1979) Can the ratio of single-wavelength EDM data improve the resolution of small changes in line length? A comparison with multiwavelength EDM data. J Geophys Res 84(B7):3659−3663

Sobotta C, Schwarz W, Witte B (1980) Frequenzprüfung elektro-optischer Entfernungsmesser mit Photodioden. Allgem Vermessungs-Nachr (AVN) 87:257−265

Solarex (1986) How to determine the solar system to suit your needs. Solarex Pty Ltd, POBox 204, Chester Hill NSW 2162, Australia, 4 pp

Spindler, Hoyer Co (1976) Fine optical construction elements, List 31. Publication B4.576, Göttingen, 58 pp

Sprent A (1980) EDM calibration in Tasmania. Aust Surv 30:213−227

Sprent A, Zwart P (1978) EDM calibration − a scenario. Aust Surv 29:157−169

Staiger R (1987) Die Anwendung von Golomb-Linealen zur Anlage von Prüfstrecken für elektronische Distanzmesser. Allgem Vermessungs-Nachr (AVN) 94:361−367

Stock HMP (1982) Lasers, interferometry and the measurement of length. In: Report on the regional workshop on metrology for developing countries, 30 August−10 September 1982, Sydney, Australia, vol 2. Proc, part 3: length standards and measurements. CSIRO, Division of Applied Physics, Sydney, Australia, pp 122−157

Swenson LS (1987) Michelson and measurement. Phys Today, May 1987, 40(5):24−30

Sze SM (1969) Physics of semiconductor devices. Wiley-Interscience, New York

Vincenty T (1969) The scale problem in adjustments of linear measurements. Surv Rev 20(154):164–170

Vincenty T (1975) A note on the reduction of measured distances to the ellipsoid. Surv Rev 23:40–42

Vincenty T (1979) Methods of adjusting relative lateration networks. Surv Rev 25(193):103–117

Walsh CJ (1987) Measurements of absolute distances to 25 m by multiwavelength CO_2 laser interferometry. Appl Opt 26:1680–1687

Webb EK (1984) Temperature and humidity structure in the lower atmosphere. In: Brunner FK (ed) Geodetic refraction. Springer, Berlin Heidelberg New York, pp 85–141

Webb PP, McIntyre RJ, Conradi J (1974) Properties of avalanche diodes. RCA Rev 35:234–277

Werner APH (1968) Lecture notes on history of surveying. Aust Surv, 24 June 1968

Whalen CT (1985) Trigonometric motorized levelling at the National Geodetic Survey. In: Proc, Third International Symposium on the North American Vertical Datum (NAVD Symposium '85), 21–26 April 1985, Rockville, Maryland, USA, NOAA, Rockville, pp 65–80

Wild Heerbrugg (1988) Facts about power sources for surveying instruments, Product Information, Geodesy 6/88, 8 pp

Wilkening G (1986) The measurement of the refractive index of the air. Paper presented at IMEKO symposium on laser applications in precision measurement. Laser Measurement Working Group, Budapest, Hungary, 19–21 November 1986, 8 pp

Wilkins GA (1980) Project MERIT – a review of the techniques to be used during the project MERIT to monitor the rotation of the earth. Royal Greenwich Observatory, England, 77 pp

Witte B, Schwarz W (1982) Calibration of electro-optical rangefinders – experience gained and general remarks relative to calibration. Surv Mapping 42:151–162

Zetsche H (1979) Electronic distance measurement (EDM). Wittwer, Stuttgart, 434 pp (in German)

Addendum:

Preston-Thomas H (1991) The International Temperature Scale of 1990 (ITS-90). Metrologia 27:3–10, 107

Subject Index